Sustainability Economics

Lin Zhang

Sustainability Economics

Springer

Lin Zhang
School of Energy and Environment
City University of Hong Kong
Kowloon, Hong Kong

ISBN 978-3-031-87828-2 ISBN 978-3-031-87829-9 (eBook)
https://doi.org/10.1007/978-3-031-87829-9

This Springer imprint is published by the registered company Springer Nature Switzerland AG
The registered company address is: Gewerbestrasse 11, 6330 Cham, Switzerland

Preface

In an era marked by unprecedented environmental challenges and socio-economic transformations, the imperative for a comprehensive understanding of sustainability economics has never been more critical. *Sustainability Economics* seeks to bridge the gap between traditional economic theories and the urgent sustainability issues that define our times.

The journey of this book begins with an exploration of the foundational concepts in economics and their relevance to sustainability. The book delves into the economic origins of sustainability challenges, providing a critical evaluation of how current economic systems contribute to environmental degradation and resource depletion. By understanding these roots, readers can better appreciate the complexities and interdependencies that characterize the global economy.

The second part of this book addresses energy-related sustainability issues. Energy is the lifeblood of modern economies, and its sustainable management is crucial for future prosperity. The book examines the dynamics of energy demand and supply, the implications of energy efficiency and rebound effects, and the complexities of energy markets. These chapters highlight the critical role of energy in the pursuit of sustainable development and the need for innovative solutions to balance economic growth with environmental stewardship.

Environmental sustainability is the focus of the third part. Here, the book investigates the origins of environmental problems, the principles of dynamic resource allocation, and the management of air and water resources. Land use management is also discussed, emphasizing the need for sustainable practices to preserve the natural environment. These chapters underscore the importance of integrating environmental considerations into economic decision-making processes.

In the fourth part, the book turns its attention to societal aspects of sustainability. Topics such as poverty and inequality, environmental justice, and the social dimensions of energy are explored. Urban studies and city sustainability, as well as the intricate interactions between humans and their environment, are also considered. These discussions highlight the interconnectedness of social and environmental issues and the need for holistic approaches to sustainability.

The final part of the book is dedicated to policy and practice. Various economic policy instruments, business practices, and green finance initiatives that can drive sustainable development are analyzed. Additionally, non-economic instruments and their potential to foster a more sustainable future are explored. These chapters provide practical insights and strategies for policymakers, businesses, and individuals committed to advancing sustainability.

Sustainability Economics stands out for its integrated approach and multidisciplinary perspectives. This book draws from diverse fields—economics, environmental science, social science, and policy studies. This multidisciplinary lens enriches discussions on sustainability, allowing readers to appreciate how economic decisions intersect with various aspects of sustainability challenges. By highlighting the interconnectedness of economic decisions and social well-being, readers emerge not only informed but also motivated to advocate for a more equitable and sustainable future.

One of the most distinctive features of this book is its practical guidance. Emphasizing real-world applications, it explores the latest advancements in energy efficiency, climate policy, and green finance, providing actionable insights that can be applied by businesses, policymakers, and communities. By addressing societal issues like poverty, inequality, and environmental justice, the book raises public awareness and inspires collective action toward a more sustainable future.

Sustainability Economics is designed to be an essential resource for students, researchers, policymakers, and anyone interested in understanding and addressing the economic dimensions of sustainability. By bringing together a wide range of perspectives and insights, the book aims to contribute to the ongoing dialogue on how to achieve a more sustainable and equitable world.

Readers are invited to embark on this journey toward a more sustainable and equitable world. Together, they can explore the economic pathways that lead to a future where prosperity and sustainability go hand in hand.

The completion of this book would not have been possible without the support and encouragement of many individuals. I would like to express my deepest gratitude to my colleagues and collaborators, for their invaluable insights and suggestions. Special thanks to my family and friends for their unwavering support throughout this journey. I am also grateful to my students for their diligent assistance, which have been instrumental in the development of this work. Lastly, I extend my appreciation to the editorial team at Springer for their professionalism and dedication in bringing this project to fruition.

Kowloon, Hong Kong
January 2025

Lin Zhang

Competing Interests The author has no competing interests to declare that are relevant to the content of this manuscript.

Contents

Part V Policy-and Practice-Related Sustainability Issues

Abbreviations

AI	Artificial Intelligence
ATC	Average Total Cost
BCA	Benefit-Cost Analysis
BTUs	British thermal units (= 1.055 kJ)
CAC	Command-and-Control
CBD	Convention on Biological Diversity
CCS	Carbon Capture and Storage
COP	Conference of the Parties
COPD	Chronic Obstructive Pulmonary Disease
CSR	Corporate Social Responsibility
EC	Environmental Cost
EIA	Energy Information Administration
EKC	The Environmental Kuznets Curve
EMS	Energy Management System
EPA	Environmental Protection Agency
ESG	Environmental, Social, and Governance
ETS	Emissions Trading System
EVs	Electric Vehicles
FAO	Food and Agriculture Organization
GDP	Gross Domestic Product
GHG	Greenhouse Gases
GIS	Geographic Information Systems
HIA	Health Impact Assessment
IEA	International Energy Agency
IOT	Input-Output Table
IoT	Interests-of-Things
IPCC	Intergovernmental Panel on Climate Change
LDCs	Least Developed Countries
LIA	Land Intensification Allowance
LNG	Liquefied Natural Gas
LPG	Liquefied Petroleum Gas

LSA	Land Suitability Analysis
MAC	Marginal Abatement Cost
MB	Marginal Benefit
MC	Marginal Cost
MDC	Marginal Damage Cost
MEC	Marginal External Cost
MSC	Marginal Social Cost
NASA	National Aeronautics and Space Administration
NDCs	Nationally Determined Contributions
NPV	Net Present Value
OECD	Organization for Economic Co-operation and Development
OPEC	Organization of the Petroleum Exporting Countries
OTC	Over the Counter
PV	Photovoltaics
R&D	Research and Development
SDGs	Sustainable Development Goals
SVC	Short-term Variable Cost
t	Ton (= 1000 kg)
TOU	Time-of-Use
TWh	Terawatt-Hours
UN	The United Nations
UNEP	The United Nations Environment Programme
UNFCCC	The United Nations Framework Convention on Climate Change
VOCs	Volatile Organic Compounds
WHO	World Health Organization
WTP	Willingness-to-Pay

Part I
The Basis

Chapter 1
Introduction

Sustainable development is development that meets the needs of the present without compromising the ability of future generations to meet their own needs.
"Our Common Future", Brundtland Report (1987)

1.1 The History of Sustainability

The concept of sustainability has evolved significantly over time, reflecting humanity's growing awareness of the need to balance economic development with environmental protection and social equity. This section explores the historical milestones and key developments that have shaped our understanding of sustainability.

1.1.1 Early Concepts and Evolution

The roots of sustainability can be traced back to ancient civilizations, where the idea of living in harmony with nature was integral to many cultures. Indigenous communities around the world practiced sustainable agriculture and resource management long before the term "sustainability" was coined. These early practices were based on a deep understanding of local ecosystems and a commitment to preserving resources for future generations.

For instance, the ancient Egyptians developed an intricate system of irrigation to manage the Nile River's annual flooding. This system allowed them to control water distribution, ensuring that crops received adequate water without depleting the river's resources. The Egyptians also practiced crop rotation and used natural fertilizers to maintain soil fertility, demonstrating an early understanding of sustainable agricultural practices.

L. Zhang, *Sustainability Economics*, https://doi.org/10.1007/978-3-031-87829-9_1

Similarly, the ancient Mesopotamians, who lived in the Fertile Crescent between the Tigris and Euphrates rivers, developed advanced irrigation techniques to support their agriculture. They constructed canals and dikes to manage water flow and prevent soil salinization, which could render the land infertile. These practices allowed them to sustain agricultural productivity over long periods, supporting large urban populations.

In the Americas, the Inca civilization of South America is another example of early sustainable practices. The Incas built terraces on the steep slopes of the Andes Mountains to create arable land and prevent soil erosion. These terraces also helped to conserve water by reducing runoff and allowing it to seep into the soil. The Incas' agricultural practices were closely tied to their understanding of the local environment and their efforts to maintain ecological balance.

Indigenous communities in North America also practiced sustainable resource management. The Native American tribes of the Great Plains, for example, relied on the buffalo for food, clothing, and tools. They hunted buffalo in a way that ensured the herds remained healthy and abundant, taking only what they needed and using every part of the animal. This practice reflected a deep respect for the natural world and an understanding of the need to preserve resources for future generations.

In Asia, the traditional rice farming systems of Southeast Asia provide another example of sustainable agriculture. Farmers in this region developed complex irrigation systems to manage water resources efficiently. They also practiced intercropping, growing different crops together to enhance soil fertility and reduce pest infestations. These practices allowed them to maintain high levels of productivity without depleting the land.

These early practices were often guided by a worldview that emphasized the interconnectedness of all living things and the importance of maintaining balance within the natural world. Many indigenous cultures had sophisticated methods for managing water resources, rotating crops, and maintaining soil fertility, ensuring that their agricultural practices did not deplete the land. For example, the traditional Hawaiian practice of "ahupua'a" involved managing land and water resources within a watershed to ensure that all parts of the ecosystem were healthy and productive. This holistic approach to resource management reflects a deep understanding of the interconnectedness of natural systems.

The ancient Greeks also contributed to early sustainability concepts. Philosophers like Plato and Aristotle wrote about the importance of managing natural resources wisely. Plato, in particular, lamented the deforestation of the hills around Athens, recognizing the long-term consequences of environmental degradation. His observations highlight an early awareness of the need to balance human activities with the preservation of natural resources.

In medieval Europe, the practice of sustainable agriculture continued with the development of the three-field system. This system involved rotating crops between three fields, with one field left fallow each year to restore its fertility. This practice helped to maintain soil health and increase agricultural productivity, supporting growing populations.

These historical examples demonstrate that the principles of sustainability are not new. They have been practiced by various cultures throughout history, reflecting a deep understanding of the need to live in harmony with nature. These early practices have laid the foundation for modern sustainability concepts, emphasizing the importance of preserving resources for future generations and maintaining ecological balance. As we face contemporary sustainability challenges, these historical lessons remind us of the value of traditional knowledge and the importance of integrating it with modern scientific and technological advancements.

1.1.2 Key Milestones in Sustainability

The modern concept of sustainability began to take shape in the mid-twentieth century, driven by growing concerns about environmental degradation and resource depletion. Several key milestones have marked the evolution of sustainability.

Table 1.1 presents the historical milestones in urban sustainability, illustrating the progressive understanding and commitment to sustainable practices in urban development.

In the 1960s and 1970s, there was a significant environmental awakening. The publication of Rachel Carson's "Silent Spring" in 1962 highlighted the dangers of pesticide use and sparked widespread environmental awareness. This book is often credited with launching the modern environmental movement, as it brought to light the harmful effects of chemicals on the environment and human health.

In 1970, the founding of Earth Day further amplified public engagement with environmental issues. This global event mobilized millions of people worldwide, advocating for the protection of the planet and signifying a major shift in public consciousness toward environmental sustainability. These early initiatives laid the foundation for the evolving integration of sustainability within urban planning and policy frameworks.

Table 1.1 Key milestones in sustainability

Year	Key milestones
1962	Publication of "Silent Spring" by Rachel Carson
1970	Establishment of Earth Day
1972	United Nations Conference on the Human Environment (Stockholm)
1987	Brundtland Report: "Our Common Future"
1992	United Nations Conference on Environment and Development (Earth Summit) - Agenda 21
2015	Adoption of the Sustainable Development Goals (SDGs)

The Stockholm Conference and the Emergence of Global Environmental Governance

In 1972, the United Nations held a Conference on the Human Environment in Stockholm, which was a pivotal moment in the history of global environmental governance. It was the first major international conference to address environmental issues on a global scale, highlighting the necessity of international cooperation to tackle environmental challenges. This conference laid the groundwork for future efforts to integrate sustainability principles into urban development practices (Baste & Watson, 2022).

One of the key outcomes of the Stockholm Conference was the creation of the United Nations Environment Programme (UNEP), which has since become a leading institution in promoting global environmental sustainability. The conference also underscored the interconnectedness of environmental, social, and economic issues, paving the way for the adoption of the sustainable development paradigm.

The Brundtland Report and the Formalization of Sustainable Development

The notion of "sustainable development" was further refined in the 1987 Brundtland Report, officially titled *Our Common Future*, published by the World Commission on Environment and Development. This report offered a comprehensive definition of sustainable development as development that "meets the needs of the present without compromising the ability of future generations to meet their own needs" (Brundtland, 1987). This definition has since become the cornerstone of sustainability discourse across academic and policy domains, particularly in urban studies and planning.

The Brundtland Report advocated for a holistic approach to development, balancing environmental, social, and economic considerations. It called for a shift in development strategies to ensure that economic growth does not exacerbate environmental degradation or perpetuate social inequalities. Additionally, the report emphasized the need for international cooperation to address global environmental challenges, such as climate change and biodiversity loss.

Agenda 21 and Local Agenda 21

The 1990s saw global commitments and frameworks being established. The 1992 Earth Summit in Rio de Janeiro resulted in the adoption of Agenda 21, a comprehensive plan of action for sustainable development. This document outlined strategies for achieving sustainable development at local, national, and global levels, covering a wide range of issues from poverty eradication to environmental protection. The establishment of the United Nations Framework Convention on Climate Change (UNFCCC) and the Convention on Biological Diversity (CBD) provided frameworks for addressing climate change and biodiversity loss. These conventions represented significant steps toward international cooperation on environmental issues, setting the stage for future agreements and initiatives.

Agenda 21 encouraged local authorities to develop tailored Local Agenda 21 plans that addressed the unique needs and challenges of their communities (Parnell, 2016). These localized plans aimed to integrate sustainability principles into urban planning and decision-making processes, promote active public participation, and foster partnerships between governments, private sector entities, and civil society organizations.

The Sustainable Development Goals (SDGs)

In the 2000s, sustainability began to be integrated into policy and practice. The Millennium Development Goals (MDGs), adopted in 2000, included several targets related to environmental sustainability. These goals aimed to address some of the most pressing global challenges, such as poverty, hunger, and disease, while also promoting environmental sustainability. The 2002 World Summit on Sustainable Development in Johannesburg emphasized the need for integrating economic, social, and environmental dimensions of sustainability. This summit built on the outcomes of the 1992 Earth Summit and reinforced the importance of sustainable development as a global priority.

In 2015, the United Nations formally adopted the 2030 Agenda for Sustainable Development, which includes 17 goals aimed at addressing global challenges such as poverty, inequality, climate change, and environmental degradation. These goals represent a universal call to action to end poverty, protect the planet, and ensure that all people enjoy peace and prosperity by 2030. The Paris Agreement, also adopted in 2015, represented a landmark commitment by countries to limit global warming and enhance climate resilience. This agreement set ambitious targets for reducing greenhouse gas emissions and adapting to the impacts of climate change, reflecting the urgent need for collective action to address the climate crisis (Fig. 1.1).

The history of sustainability is a testament to humanity's evolving understanding of the interconnectedness of economic, environmental, and social systems. From ancient practices to modern frameworks, the journey toward sustainability has been

Fig. 1.1 17 Sustainable Development Goals of the United Nations. *Note* The content of this book has not been approved by the United Nations and does not reflect the views of the United Nations or its officials or Member States. *Source* United Nations Sustainable Development Goals website https://www.un.org/sustainabledevelopment/

marked by significant milestones and a growing recognition of the need for a holistic approach to development. As we continue to face global challenges, the lessons from this history will be crucial in guiding future efforts to achieve a sustainable and equitable world. The ongoing evolution of sustainability concepts and practices underscores the importance of continuous learning, adaptation, and collaboration in our quest for a sustainable future.

1.2 The Principles of Sustainability

Sustainability is a multifaceted concept that aims to balance economic growth, environmental protection, and social equity. Within this broad framework, the principles of weak and strong sustainability offer distinct perspectives on how to achieve and maintain this balance. Understanding these principles is crucial for developing effective policies and practices that promote long-term sustainability.

1.2.1 Weak Sustainability

Weak sustainability is based on the idea that natural and human-made capital are largely interchangeable. According to this principle, as long as the total stock of capital (natural plus human-made) remains constant or increases, sustainability can be achieved. This perspective assumes that technological advancements and human ingenuity can compensate for the depletion of natural resources. For example, if a forest is cut down, the loss of natural capital can be offset by planting new trees or investing in other forms of capital, such as infrastructure or education.

In daily life, weak sustainability can be seen in practices like recycling and the use of renewable energy. For instance, when we recycle materials like paper, plastic, and metal, we are substituting natural resources with human-made processes to reduce the need for new raw materials. Similarly, the adoption of solar panels and wind turbines to generate electricity represents a shift from non-renewable fossil fuels to renewable energy sources, demonstrating how technological innovation can help maintain the overall capital stock.

Proponents of weak sustainability argue that economic growth and development can continue indefinitely, provided that the overall capital stock is maintained. They emphasize the role of innovation and technological progress in addressing environmental challenges and believe that market mechanisms can effectively allocate resources and drive sustainable outcomes. This approach is often associated with neoclassical economic theories, which prioritize efficiency and market-based solutions.

However, critics of weak sustainability point out that not all forms of natural capital can be easily replaced or substituted. Certain ecosystems and biodiversity, for instance, provide unique and irreplaceable services that are essential for human well-being. The loss of these natural assets can have profound and irreversible consequences, which cannot be fully mitigated by human-made capital. Therefore, relying solely on weak sustainability may lead to the degradation of critical natural resources and undermine long-term sustainability goals.

1.2.2 Strong Sustainability

In contrast, strong sustainability emphasizes the non-substitutability of natural capital. This principle asserts that natural capital must be preserved and maintained independently of human-made capital. Strong sustainability recognizes that certain ecological functions and services are vital for the survival and well-being of both current and future generations. These include clean air and water, fertile soil, stable climate, and biodiversity, which cannot be replicated or replaced by technological means.

In everyday life, strong sustainability can be observed in practices such as conservation efforts and sustainable agriculture. For example, protecting a wetland area from development ensures that the ecosystem services it provides, such as water filtration, flood control, and habitat for wildlife, are preserved. Similarly, sustainable farming practices, such as crop rotation, organic farming, and reduced use of chemical fertilizers and pesticides, help maintain soil health and biodiversity, ensuring that agricultural land remains productive for future generations.

Advocates of strong sustainability argue that preserving natural capital is essential for maintaining the resilience and stability of ecological systems. They stress the importance of setting strict limits on resource extraction and environmental degradation to ensure that natural capital is not depleted beyond its regenerative capacity. This approach often involves implementing regulatory measures, conservation strategies, and sustainable management practices to protect and restore natural ecosystems.

Strong sustainability also highlights the ethical and moral dimensions of sustainability. It emphasizes the responsibility of current generations to safeguard the environment for future generations and to ensure that all people have access to essential natural resources. This perspective aligns with the principles of intergenerational equity and environmental justice, which seek to address the disparities in resource distribution and environmental impacts.

1.2.3 Integrating Weak and Strong Sustainability

While weak and strong sustainability represent different approaches to achieving sustainability, they are not mutually exclusive. In practice, a balanced and integrated approach that incorporates elements of both principles may be necessary to address the complex and interconnected challenges of sustainability. For instance, promoting technological innovation and economic development can complement efforts to conserve natural capital and enhance ecosystem resilience.

Policymakers and practitioners can draw on the strengths of both weak and strong sustainability to develop comprehensive strategies that support sustainable development. This may involve investing in green technologies, promoting sustainable consumption and production patterns, and implementing policies that protect critical natural resources. By recognizing the value of both human-made and natural capital, it is possible to create a more sustainable and equitable future.

In daily life, this integrated approach can be seen in urban planning and community initiatives. For example, cities can enhance their natural capital and promote economic development by investing in green infrastructure, such as parks and urban forests. These initiatives not only improve the quality of life for residents but also support environmental sustainability. Additionally, community gardens and local food initiatives can contribute to both economic and environmental goals by offering fresh, locally grown produce and minimizing the carbon emissions linked to food transportation.

The principles of weak and strong sustainability offer valuable insights into the different ways of achieving and maintaining sustainability. Weak sustainability focuses on the interchangeability of natural and human capital, while strong sustainability emphasizes the preservation of natural capital as essential for long-term well-being. By integrating these principles, it is possible to develop holistic and effective strategies that promote sustainable development and address the pressing environmental, economic, and social challenges of our time. Understanding and applying these principles is crucial for creating a sustainable future that benefits both current and future generations. The ongoing evolution of sustainability concepts and practices underscores the importance of continuous learning, adaptation, and collaboration in our quest for a sustainable future.

1.3 The Challenges of Sustainability

Sustainability, as a concept and practice, aims to balance economic development, environmental protection, and social equity. However, achieving this balance is fraught with numerous challenges that span across various domains. These challenges are complex and interconnected, often requiring multifaceted solutions and collaborative efforts. This section discusses the primary challenges of sustainability.

1.3.1 Environmental Challenges

One of the most pressing challenges of sustainability is environmental degradation. Human activities, particularly since the Industrial Revolution, have significantly altered natural ecosystems. Deforestation, pollution, and the overexploitation of natural resources have led to the loss of biodiversity and the degradation of ecosystems. Forests, which act as carbon sinks and biodiversity hotspots, are being cleared at alarming rates for agriculture, urban development, and logging. This deforestation not only contributes to climate change by releasing stored carbon dioxide but also disrupts the habitats of countless species, leading to a loss of biodiversity.

Pollution is another critical environmental challenge. Industrial processes, agricultural activities, and urbanization have resulted in the contamination of air, water, and soil. Air pollution, caused by emissions mainly from factories and power generation, poses severe health risks to humans and animals. Water pollution, from agricultural runoff and industrial wastewater discharges, affects freshwater ecosystems and the availability of clean drinking water. Soil pollution, often due to the use of pesticides and industrial waste, reduces soil fertility and affects food security.

Climate change, primarily caused by greenhouse gases in the atmosphere, stands as one of the most pressing environmental challenges today. Activities such as burning fossil fuels, deforestation, and industrial processes have elevated levels of carbon dioxide and other greenhouse gases, resulting in global warming. The consequences of climate change are extensive and severe, including rising sea levels, more frequent and intense natural disasters, and alterations in weather patterns. These changes pose significant threats to ecosystems, human health, and economic stability, making climate change a crucial issue for sustainability.

1.3.2 Economic Challenges

Economic challenges are closely linked to environmental issues and are a significant barrier to achieving sustainability. One of the primary economic challenges is the reliance on non-renewable resources. Fossil fuels, such as coal, oil, and natural gas, have been the backbone of industrial economies. However, these resources are finite, and their extraction and use have severe environmental impacts. Transitioning to renewable energy sources, such as solar, wind, and hydroelectric power, is essential for sustainability but requires substantial investment and technological innovation.

Another economic challenge is the concept of economic growth itself. Traditional economic models prioritize continuous growth, often measured by Gross Domestic Product (GDP). However, this growth-centric approach can lead to the overexploitation of natural resources and environmental degradation. Sustainable development requires a rethinking of economic models to prioritize well-being, equity, and environmental health over mere economic expansion. This shift involves the adoption of sustainability-oriented measures such as the Genuine Progress Indicator (GPI), which considers social and environmental factors alongside economic performance.

Income inequality is another significant economic challenge. The benefits of economic growth are often unevenly distributed, leading to disparities in wealth and access to resources. This inequality can exacerbate environmental degradation, as marginalized communities may rely on unsustainable practices for their livelihoods.

Addressing income inequality is crucial for sustainability, as it ensures that all individuals have the resources and opportunities to participate in sustainable practices and benefit from sustainable development.

1.3.3 Social Challenges

Social challenges are integral to the sustainability discourse, as they directly impact human well-being and equity. One of the primary social challenges is poverty. Poverty limits access to essential resources such as clean water, nutritious food, and healthcare, which makes it difficult for individuals to engage in sustainable practices. Moreover, impoverished communities are often the most vulnerable to environmental degradation and climate change impacts. Addressing poverty is essential for sustainability, as it enables individuals to improve their quality of life while adopting sustainable practices.

Education is another critical social challenge. Education plays a vital role in raising awareness about sustainability and equipping individuals with the knowledge needed to implement sustainable practices. However, access to quality education is uneven, with many communities lacking the resources and infrastructure to provide comprehensive education. Promoting education for sustainable development is crucial for empowering individuals and communities to contribute to sustainability efforts.

Cultural attitudes and behaviors also pose challenges to sustainability. Consumerism, driven by the desire for material goods and convenience, often leads to unsustainable consumption patterns. This culture of consumption is fueled by advertising and societal norms that equate success and happiness with material wealth. Shifting cultural attitudes toward sustainability requires promoting values such as conservation, mindfulness, and community well-being. This shift can be facilitated through education, media, and policy initiatives that encourage sustainable lifestyles.

1.3.4 Technological Challenges

Technological innovation is a double-edged sword in the context of sustainability. On one hand, technological advancements have the potential to address many sustainability challenges by improving efficiency, reducing waste, and enabling the use of renewable resources. On the other hand, technology can also contribute to environmental degradation and social inequality if not managed responsibly.

One of the primary technological challenges is the development and deployment of clean technologies. While renewable energy sources such as solar, wind, and hydroelectric power offer green alternatives to fossil fuels, their adoption requires significant investment and infrastructure development. Additionally, the production and disposal of renewable energy technologies, such as solar panels and batteries, can have environmental impacts if not managed properly. Ensuring the sustainability of clean energy technologies involves addressing these lifecycle impacts and promoting circular economy principles.

Another technological challenge is the digital divide. Access to technology and the internet is unevenly distributed, with many communities lacking the infrastructure and resources to benefit from technological advancements. This digital divide can exacerbate social and economic inequalities, limiting opportunities for education, employment, and participation in sustainability initiatives. Bridging the digital divide is essential for ensuring that all individuals and communities can benefit from and contribute to technological solutions for sustainability.

1.3.5 Political and Institutional Challenges

Political and institutional challenges are significant barriers to achieving sustainability. Effective governance and strong institutions are essential for implementing and enforcing sustainability policies and practices. However, political will and institutional capacity often fall short of what is needed to address complex sustainability challenges.

One of the primary political challenges is the lack of international cooperation. Sustainability issues, such as climate change and biodiversity loss, are global in nature and require coordinated efforts across countries. However, geopolitical tensions, differing national interests, and economic disparities can hinder international cooperation. Strengthening global governance mechanisms and fostering collaboration among countries is crucial for addressing transboundary sustainability challenges.

Corruption and lack of transparency are also significant institutional challenges. Corruption can undermine environmental regulations, allowing unsustainable practices to continue unchecked. Lack of transparency in decision-making processes can erode public trust and hinder the implementation of sustainability policies. Strengthening institutions, promoting transparency, and combating corruption are essential for creating an enabling environment for sustainability.

Policy coherence is another critical challenge. Sustainability requires the integration of environmental, economic, and social considerations into all policy areas. However, policies in different sectors, such as energy, agriculture, and transportation, are often developed in isolation and may conflict with sustainability goals. Promoting policy coherence involves aligning policies across sectors and levels of government to ensure that they support rather than undermine sustainability objectives.

1.3.6 Ethical and Moral Challenges

Sustainability also poses ethical and moral challenges, as it involves making decisions that affect current and future generations, as well as other species and ecosystems. One of the primary ethical challenges is intergenerational equity. Sustainability requires balancing the needs of the present and future generations. This involves making decisions that do not deplete resources or degrade the environment, ensuring that future generations have the same opportunities for well-being and prosperity. However, achieving intergenerational equity can be challenging, as it requires long-term thinking and the willingness to make sacrifices in the present for the benefit of the future.

Environmental justice is another critical ethical challenge. Environmental degradation and climate change often disproportionately affect marginalized and vulnerable communities, exacerbating existing social and economic inequalities. Addressing environmental justice involves ensuring that all individuals and communities have equal access to environmental benefits and are not disproportionately burdened by environmental harms. This requires recognizing and addressing the social dimensions of sustainability and promoting inclusive and equitable policies and practices.

The intrinsic value of nature is also an important ethical consideration. Many sustainability challenges arise from the exploitation and degradation of natural resources for human benefit. Recognizing the intrinsic value of nature involves acknowledging that ecosystems and species have value beyond their utility to humans and deserve protection for their own sake. This perspective aligns with the principles of deep ecology and biocentrism, which advocate for the preservation of nature as an ethical imperative.

1.4 The Sustainability Journey

The journey toward sustainability is a complex and evolving narrative that encompasses historical developments, theoretical principles, and multifaceted challenges. This chapter has explored these dimensions, providing a comprehensive understanding of sustainability's past, its guiding principles, and the obstacles that must be overcome to achieve a sustainable future.

The history, principles, and challenges of sustainability underscore the complexity and interconnectedness of achieving a sustainable future. Addressing these challenges requires a holistic and integrated approach that draws on the strengths of both weak and strong sustainability. It involves promoting technological innovation and economic development while preserving natural capital and ensuring social equity. Policymakers, practitioners, and individuals must collaborate to develop comprehensive strategies that support sustainable development, recognizing the value of both human-made and natural capital.

As we continue to face global challenges, the lessons from the history of sustainability, the guiding principles of weak and strong sustainability, and the multifaceted challenges we encounter will be crucial in guiding future efforts. By understanding and addressing these dimensions, we can work toward a future that is equitable, prosperous, and sustainable for all. The ongoing evolution of sustainability concepts and practices highlights the importance of continuous learning, adaptation, and collaboration in our quest for a sustainable future.

References

Baste, I. A., & Watson, R. T. (2022). Tackling the climate, biodiversity and pollution emergencies by making peace with nature 50 years after the Stockholm Conference. *Global Environmental Change, 73*, 102466.

Brundtland, G. H. (1987). *Our common future*. World Commission on Environment and Development.

Convention on Biological Diversity (CBD). (1992). *Convention on biological diversity*. United Nations.

Parnell, S. (2016). Defining a global urban development agenda. *World Development, 78*, 529–540.

Chapter 2
Economics Theory and Evaluation Method

The ideas of economists and political philosophers, both when they are right and when they are wrong, are more powerful than is generally understood. Indeed, the world is ruled by little else.
John Maynard Keynes (1936)

2.1 Demand and Supply

A deep understanding for the notions of demand and supply is important for grasping how markets operate. These concepts form the backbone of economic theory and are crucial for analyzing how the processes by which prices and quantities of goods and services are established in a market. This section will introduce you to the basic principles of demand and supply, their determinants, and how these forces interact to achieve market equilibrium.

2.1.1 What Is Demand?

Demand refers to the quantity of a service or good that consumers are both willing and able to purchase at various price levels over a designated period. In economic theory, this relationship is graphically depicted by the demand curve, which generally slopes downward from left to right. This downward slope indicates that as the price of a good decreases, the quantity demanded tends to increase, and conversely, when the price rises, the quantity demanded diminishes. This fundamental principle is referred to as the law of demand.

Determinants of Demand

Numerous factors can influence demand in the following:

L. Zhang, *Sustainability Economics*, https://doi.org/10.1007/978-3-031-87829-9_2

- *Price of the Good*: The most direct factor affecting demand is the price of the good itself. As prices fall, consumers are generally more willing to buy more of the good.
- *Income of Consumers*: Higher levels of income typically increase the demand for services and goods, as consumers have more purchasing power. Suppose the average income of consumers rises due to economic growth. With higher incomes, people have more purchasing power and are more likely to buy smartphones, even if the price remains the same. This increase in demand shifts the demand curve to the right.
- *Prices of Related Goods*: The demand for a good may be influenced by the prices of related goods, such as substitutes and complements. For example, if the price of coffee rises, the demand for tea (a substitute) might increase.
  - *Substitutes:* If the price of tablets (a substitute for smartphones) increases, consumers might opt to buy smartphones instead. This would increase the demand for smartphones, shifting the demand curve to the right.
  - *Complements:* If the price of mobile data plans (a complement to smartphones) decreases, more people might buy smartphones to take advantage of the cheaper data plans. This would also increase the demand for smartphones, shifting the demand curve to the right.
- *Tastes and Preferences*: The changes in consumer preferences may significantly affect demand. For instance, a rise in health consciousness can increase the demand for organic foods. If a new trend or technological advancement makes smartphones more desirable (e.g., the introduction of 5G technology), more consumers will want to buy the latest smartphones. This change in consumer preferences increases demand, shifting the demand curve to the right.
- *Expectations*: If consumers, in the future, expect prices to increase, they may adjust their current demand to circumvent higher costs later. If consumers expect that the price of smartphones will rise in the future, they might decide to purchase smartphones now rather than later. This expectation increases current demand, resulting in a rightward shift of the demand curve.
- *Number of Buyers*: An increase in the number of consumers can lead to higher demand for a service or good. If the population grows or if a new market segment (e.g., teenagers) starts buying smartphones, the overall number of buyers increases. This increase in the number of buyers raises the demand for smartphones, resulting in a rightward shift in the demand curve.

Reading

Demand for Smartphones

Imagine a market for smartphones. The demand curve for smartphones presents the relationship between the price of smartphones and the quantity demanded by consumers. Generally, as the price of smartphones declines, the quantity demanded tends to increase, and vice versa. This inverse relationship is graphically represented by a downward-sloping demand curve (Fig. 2.1).

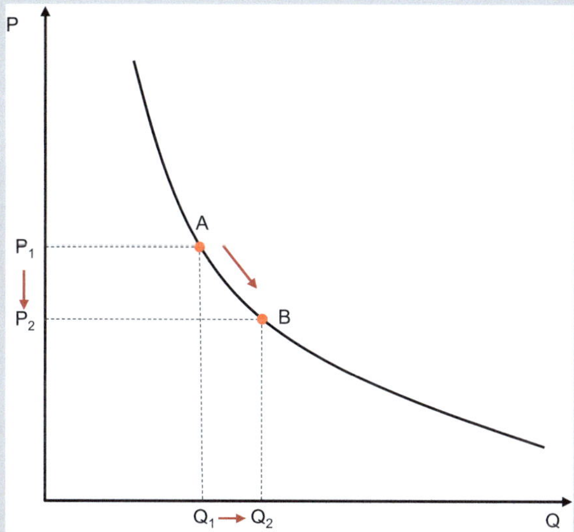

Fig. 2.1 Changing demand

2.1.2 *What Is Supply?*

Supply is defined as the quantity of a service or good that producers are both willing and able to offer for sale at different price levels over a specified timeframe. The relationship between the price of a good and the quantity supplied is typically illustrated by the supply curve, which usually slopes upward from left to right. This upward slope shows that as the price of a good rises, the quantity supplied also tends to increase; conversely, a decrease in price results in a lower quantity supplied. This fundamental principle is referred to as the law of supply.

Reading

Supply of Electric Cars

The supply curve for electric cars presents the relationship between the price of electric cars and the quantity supplied by manufacturers. Typically, as the price of electric cars increases, manufacturers are willing to produce and sell more cars, and vice versa. This direct relationship is illustrated by an upward-sloping supply curve (Fig. 2.2).

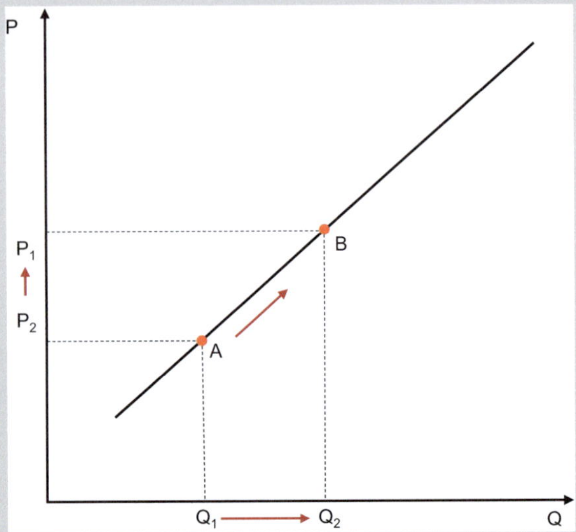

Fig. 2.2 Changing supply

Determinants of Supply

Several factors can influence supply, including:

- *Price of the Good*: The primary factor affecting supply is the price of the good itself. Higher prices offer an incentive to producers to supply more of the good. If the price of electric cars rises from $30,000 to $40,000, manufacturers are more likely to increase production because higher prices can lead to higher revenues. This increase in quantity supplied is a movement along the supply curve.
- *Input Prices*: The cost of inputs, including machinery, raw materials, and labor, can affect the supply of a good. Higher input prices can decrease supply as production becomes more expensive. Suppose the cost of lithium, a key component in electric car batteries, decreases. This reduction in input costs makes it cheaper to produce electric cars, allowing manufacturers to supply more at the same price. An increase in supply results in a rightward shift of the supply curve.
- *Technology*: Advancements in technology can enhance supply by increasing the efficiency of production processes. Advances in manufacturing technology, such as more efficient battery production methods, can reduce production costs and increase the supply of electric cars. This technological improvement shifts the supply curve to the right.

- *Expectations*: If producers, in the future, expect higher prices, they may decrease the current supply to maximize potential profits later. If manufacturers expect that the price of electric cars will increase in the future, they might limit their current output to sell a greater quantity at these elevated prices later. Such expectations can lead to a leftward shift in the supply curve, shifting the supply curve to the left.
- *Number of Sellers*: An increase in the number of producers can lead to a higher supply of a service or good. If more companies enter the market to produce electric cars, the overall supply will increase. This increase in the number of sellers shifts the supply curve to the right.
- *Government Policies*: Regulations, subsidies, and taxes may all impact supply. For example, a subsidy might increase supply by lowering production costs.
  - *Subsidies:* If the government offers subsidies to electric car manufacturers, production costs decrease, leading to an increase in supply. This shifts the supply curve to the right.
  - *Taxes and Regulations:* Conversely, if the government imposes higher taxes or stricter regulations on electric car production, the cost of production increases, leading to a decrease in supply. This shifts the supply curve to the left.

2.1.3 Market Equilibrium

Market equilibrium occurs where the quantity demanded equals the quantity supplied (Stackelberg, 2011). This point is named the equilibrium price and quantity, where the market remains balanced, exhibiting no inherent pressure for price changes unless shifts in demand or supply occur. Changes in the determinants of demand and supply can lead to shifts in the respective curves, thereby establishing a new equilibrium price and quantity.

- *Shift in Demand*: An increase in demand, such as that caused by rising consumer income, leads to a rightward shift of the demand curve. Consequently, this shift results in a higher equilibrium price and quantity. Conversely, a decrease in demand shifts the curve to the left, resulting in a lower equilibrium price and quantity.
- *Shift in Supply*: An increase in supply, often driven by technological advancements, causes the supply curve to shift to the right. This shift results in a lower equilibrium price and a higher equilibrium quantity. Conversely, a decrease in supply leads to a leftward shift of the curve, resulting in a higher equilibrium price and a lower equilibrium quantity.

2.1.4 Elasticity of Demand and Supply

Elasticity quantifies the responsiveness of quantity demanded or supplied to variations in price. The *Price Elasticity of Demand (PED)* specifically measures how the quantity demanded reacts to changes in price. When demand is elastic, even a minor price change results in a significant shift in the quantity demanded. Conversely, if demand is inelastic, the quantity demanded remains relatively unresponsive to price fluctuations. Understanding elasticity is crucial for analyzing consumer behavior and market dynamics within sustainability economics.

$$\in = \frac{\%\text{ change in quantity demanded}}{\%\text{ change in price}} = \frac{\Delta Q / Q \times 100}{\Delta P / P \times 100} = \frac{\Delta Q}{Q} \times \frac{P}{\Delta P} \tag{2.1}$$

where ΔQ is change in quantity demanded and Q denotes the initial quantity demanded; ΔP signifies the change in price and P is the initial price. For instance, consider a scenario where the price of a smartphone decreases from $800 to $700, leading to an increase in quantity demanded from 1000 units to 1200 units. Thus,

- Calculate the change in quantity demanded: $\Delta Q = 1200 - 1000 = 200$
- Calculate the change in price: $\Delta P = 700 - 800 = -100$
- Calculate the initial quantity demanded and price: $Q = 1000$, $P = 800$
- Substitute these values into the PED formula: $\in = \frac{200}{-1000} \times \frac{800}{-100} = -1.6$
- Interpretation: The price elasticity of demand in this example is -1.6. The negative sign indicates the inverse relationship between price and quantity demanded (as price decreases, quantity demanded increases). A PED of -1.6 means that for every 1% decrease in price, the quantity demanded increases by 1.6%.

Price Elasticity of Supply: This measures the responsiveness of quantity supplied to a change in price. If supply is elastic, a small change in price leads to a large change in quantity supplied. If supply is inelastic, quantity supplied is relatively unresponsive to price changes.

$$\eta = \frac{\%\text{ change in quantity supplied}}{\%\text{ change in price}} = \frac{\Delta Q_S / Q_S \times 100}{\Delta P / P \times 100} = \frac{\Delta Q_S}{Q_S} \times \frac{P}{\Delta P} \tag{2.2}$$

Suppose the price of electric cars increases from $30,000 to $35,000, and as a result, the quantity supplied increases from 500 to 700 units.

- Calculate the change in quantity supplied: $\Delta Q_S = 700 - 500 = 200$
- Calculate the change in price: $\Delta P = 35{,}000 - 30{,}000 = 5{,}000$
- Calculate the initial quantity supplied and price: $Q_S = 500$, $P = 30{,}000$
- Substitute these values into the PES formula: $\eta = \frac{200}{500} \times \frac{30{,}000}{5000} = 2.4$
- Interpretation: The price elasticity of supply in this example is 2.4. This means that for every 1% increase in price, the quantity supplied increases by 2.4%. A

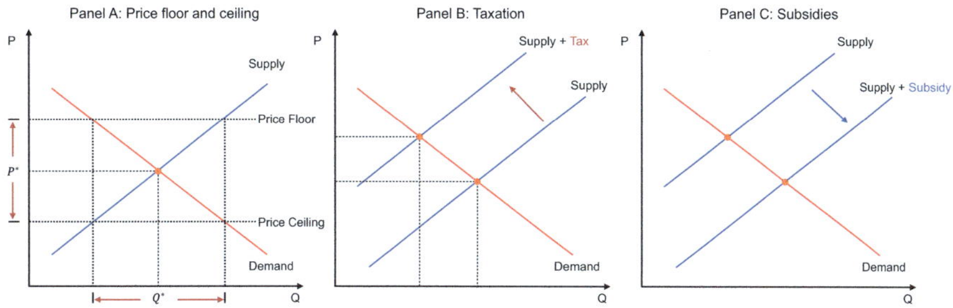

Fig. 2.3 Changing supply and demand curves

PES greater than 1 indicates that the supply is elastic, meaning suppliers are highly responsive to price changes.

2.1.5 *Applications of Demand and Supply*

A deep understanding for the demand and supply is significant for analyzing various economic issues and policies. Here are a few applications:

- *Price Controls*: Governments sometimes impose price ceilings (maximum prices) and price floors (minimum prices) to control prices. For example, rent controls are a type of price ceiling. These controls can lead to shortages or surpluses if the controlled price is not at the equilibrium level.
- *Taxation*: Taxes on goods and services can affect both demand and supply. For example, a tax on cigarettes can reduce demand by increasing the price consumers pay.
- *Subsidies*. Subsidies can increase supply by lowering production costs. For example, agricultural subsidies can result in an increase in the supply of crops (Fig. 2.3).

2.1.6 *Marginal Analysis*

The marginal effect refers to the change in the dependent variable Y, resulting from a one-unit change in the independent variable X at a specific value of X. Essentially, it represents the slope of the function at that particular point (Fig. 2.4).

$$\text{Marginal Effect} = \frac{\Delta Y}{\Delta X} \tag{2.3}$$

If ΔX is smaller, the exact slope can be found by taking the limit as $\Delta X \to 0$, and the limit is called the derivative of Y with respect to X: $\frac{\mathrm{d}Y}{\mathrm{d}X} = \lim_{\Delta X \to 0} \frac{\Delta Y}{\Delta X}$. Therefore,

Fig. 2.4 Marginal effect

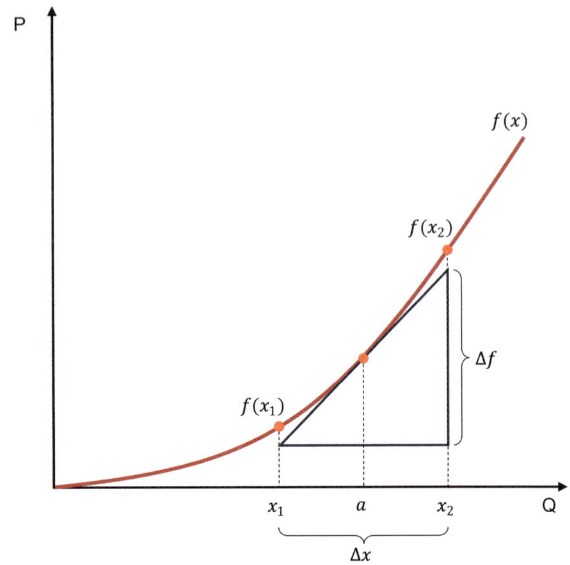

the marginal effect can be mathematically described using the concept of derivatives. For a function $Y = f(X)$, the marginal effect of X on Y at a particular value of X is given by the derivative of Y with respect to X:

$$\text{Marginal Effect} = \frac{\mathrm{d}Y}{\mathrm{d}X} \tag{2.4}$$

For a linear function, such as $Y = a + bX$, the marginal effect is constant and equal to the coefficient b:

$$\frac{\mathrm{d}Y}{\mathrm{d}X} = b \tag{2.5}$$

For a nonlinear function, such as $Y = a + bX + cX^2$, the marginal effect varies with X and is given by:

$$\frac{\mathrm{d}Y}{\mathrm{d}X} = b + 2cX \tag{2.6}$$

In this case, the marginal effect depends on the value of X, illustrating how the rate of change in Y varies as X changes.

Marginal analysis is a significant notion in economics that focuses on the rate of change of economic quantities. For instance, economists are not only interested in the absolute value of an economy's gross domestic product (GDP) at a given time but also in the rate at which GDP is growing or declining. This rate of change provides insights into the economic health and growth prospects of a country, helping policymakers and analysts make informed decisions. Similarly, manufacturers are concerned with more than just the total cost associated with a certain level of production. They are also interested in understanding how the total cost changes as the level of production varies. This involves examining the marginal cost, which is the additional cost incurred by producing one more unit of a commodity. By analyzing marginal costs, manufacturers can optimize production levels, minimize costs, and maximize profits. Marginal analysis, therefore, plays a vital role in both macroeconomic and microeconomic contexts, providing valuable information for decision-making and strategic planning.

Assume that the total cost of operating a business is denoted as $C(X)$, Marginal cost is the additional cost incurred when producing one more unit of a commodity. Mathematically, marginal cost is the derivative of the total cost function $C(X)$, with respect to the quantity of output X.

$$\text{Marginal cost} = \frac{\mathrm{d}C(X)}{\mathrm{d}X} \tag{2.7}$$

Marginal cost plays a crucial role in helping businesses assess the cost-effectiveness of increasing production. By comparing the marginal cost to the revenue generated from selling an additional unit, firms can determine whether expanding production is profitable. A clear understanding of marginal cost is essential for optimizing production levels and maximizing profitability.

Average cost, or unit cost, represents the total cost of production divided by the number of units produced. It serves as a measure of cost per unit of output, enabling businesses to evaluate their production efficiency. Average cost is calculated by summing fixed and variable costs and dividing this total by the overall quantity of output. Mathematically, it is expressed as:

$$\text{Average cost} = \frac{C(X)}{X} \tag{2.8}$$

The main difference between average cost and marginal cost lies in their focus. Average cost measures the cost per unit of output, taking into account the total cost and total quantity produced. Marginal cost, however, pays attention to the cost of producing one additional unit, providing a more dynamic view of cost changes with varying production levels. While average cost helps in setting pricing strategies and understanding overall production efficiency, marginal cost is essential for optimizing production decisions and maximizing profitability. By analyzing both average and marginal costs, businesses may make informed decisions that balance cost efficiency with production goals.

Total benefit and marginal benefit can be calculated similarly to total cost and marginal cost. Total benefit represents the overall gain or satisfaction derived from consuming a certain quantity of goods or services. Marginal benefit, on the other hand, is the additional benefit gained from consuming one more unit of a good or service. Mathematically, the marginal benefit is the derivative of the total benefit function with respect to the quantity consumed. By analyzing both total and marginal benefits, individuals and businesses can make informed decisions about consumption and production, optimizing their strategies to maximize overall satisfaction and efficiency.

2.2 Benefit–Cost Analysis

Understanding the concepts of benefit and cost is important for making informed decisions in both personal and professional contexts. These concepts are fundamental to economic theory and are widely used in various fields, including business, public policy, and environmental studies. This section will introduce you to the basic principles of benefit and cost, their evaluation methods, and their applications.

Benefits refer to the positive outcomes or gains from an action, decision, or investment. These can be tangible, such as financial profits, or intangible, such as improved health or increased satisfaction. *Costs*, on the other hand, refer to the negative outcomes or expenditures associated with an action, decision, or investment. These can include direct financial costs, time, effort, and opportunity costs. Benefit–cost analysis (BCA) is a systematic approach to evaluating the economic pros and cons of different actions or decisions (Mishan & Quah, 2020). It involves comparing the total expected benefits of an action to its total expected costs to determine whether the action is worthwhile.

2.2.1 *Net Present Value*

Present value (PV) is a core concept in finance that represents the current worth of a future sum of money or stream of cash flows, given a specific rate of return. The principle behind PV is the time value of money, which asserts that a dollar today is worth more than a dollar in the future due to its potential earning capacity. By discounting future cash flows to their present value, investors and financial analysts can make more informed decisions about the value of investments and projects. The formula for calculating the present value of a single future amount is:

$$PV = \frac{FV}{(1+r)^t} \tag{2.9}$$

where FV is the future value, r is the discount rate, and t is the number of periods. Present value is widely used in various financial applications, including bond pricing, capital budgeting, and retirement planning. For example, when valuing a bond, the present value of its future coupon payments and principal repayment is calculated to determine its fair price. One of the main advantages of using present value is its ability to offer a clear and comparable measure of value across different time periods. By converting future cash flows into their current value, decision-makers can assess the relative attractiveness of different investment opportunities. This helps in prioritizing projects and allocating resources more effectively.

This formula can be extended to calculate the current value of a series of future cash flows by summing the current values of each individual cash flow, or net present value (NPV). It represents the difference between the present value of cash inflows and the present value of cash outflows over a specified period. NPV is used to assess the value of an investment by considering the time value of money, which acknowledges that a dollar received today is worth more than a dollar received in the future due to its potential earning capacity.

To calculate NPV, future cash flows are discounted back to their present value using a discount rate, which typically is defined as the cost of capital or the required rate of return. The formula for NPV is:

$$\mathrm{NPV} = \sum_{t=0}^{t=T} \frac{C_t}{(1+r)^t} - I_0 \tag{2.10}$$

where C_t represents the future values FV or cash inflows at time t, r is the discount rate, and I_0 is the initial investment. By summing the present values of all future cash flows and subtracting the initial investment, NPV provides a clear indication of the net benefit or cost of the investment.

A positive NPV shows that the projected earnings (discounted to present value) exceed the initial investment, suggesting that the investment is likely to be profitable. Conversely, a negative NPV implies that the investment's costs outweigh the benefits, and it may not be a wise financial decision. An NPV of zero indicates that the investment is expected to break even, generating just enough returns to cover the initial costs. Projects or investments are often mutually exclusive, meaning that choosing one option precludes the selection of another. When evaluating such projects, it is advisable to pursue the option with the highest NPV, provided that its NPV is greater than zero.

The NPV represents the maximum amount you would be willing to pay today to secure the future cash flows associated with the investment. However, NPV is only one factor to consider. Organizations also have non-monetary objectives that can influence their decision-making process.

In NPV calculations, discounting transforms future cash flows into their current value, enabling a clear comparison between the initial investment and the expected returns. The power of discounting is particularly evident when comparing investments with different timelines and cash flow patterns. If you expect to receive \$1 million in the future, its present value depends on the discount rate and the time period until you receive it. For example, if the discount rate is 5% and you will receive \$1 million in 10 years, the present value is approximately \$613,913. This demonstrates how discounting reduces the value of future cash flows to reflect their worth in today's terms. Discounting allows us to accurately assess the value of these cash flows by giving more weight to earlier returns, which are more valuable due to their potential earning capacity. This helps investors and managers make more informed decisions by highlighting the true economic value of different investment opportunities.

2.2.2 Internal Rate of Return

Internal rate of return (IRR) is used to evaluate the profitability of an investment or project. It represents the discount rate at which the net present value of all cash flows (both inflows and outflows) from an investment equals zero. In other words, IRR is the rate of return at which the present value of future cash flows exactly offsets the initial investment cost. This makes IRR a useful tool for comparing the potential returns of different investments. To calculate IRR, one must solve for the discount rate that sets the NPV equation to zero:

$$0 = \sum_{t=0}^{t=T} \frac{C_t}{(1 + \text{IRR})^t} - I_0 \tag{2.11}$$

where IRR is the internal rate of return. This equation is typically solved using iterative methods or financial calculators, as it cannot be solved algebraically.

IRR is particularly valuable in capital budgeting, as it aids decision-makers in evaluating the attractiveness of various projects. A project is generally deemed acceptable if its IRR exceeds the required rate of return or the cost of capital. When comparing multiple projects, the one with the highest IRR is often preferred, provided it meets or exceeds the required rate of return.

One of the primary advantages of the IRR is its ability to convey a single percentage figure that summarizes the potential profitability of an investment, thereby facilitating straightforward comparisons among various projects. However, IRR also presents certain limitations. Notably, it assumes that all intermediate cash flows are reinvested at the same rate as the IRR, which may not reflect real-world scenarios. Furthermore, IRR can be misleading for projects with non-conventional cash flows—such as those exhibiting multiple sign changes—resulting in multiple IRR values or failing to yield a meaningful rate of return.

2.2.3 Valuation with Risks

When incorporating risks into BCA, it becomes essential to account for the uncertainty and variability of future outcomes. This involves identifying potential risks, estimating their probabilities, and assessing their impact on both benefits and costs. If the firm faces several possible future cash flow scenarios with probabilities, the criterion for going ahead with the project is the expected NPV—the probability weighted discounted sum of the mutually exclusive cash flow outcomes:

$$\text{NPV} = \sum_{t=0}^{t=T} \frac{\sum_i C_t \times P_i}{(1+r)^t} - I_0 \tag{2.12}$$

where P_i represents the probability of cash flow scenario i taking place.

A critical component of conducting a benefit–cost analysis that incorporates risks is the performance of a sensitivity analysis. This process entails varying the assumptions and parameters within the analysis to ascertain how changes in these factors influence overall results. By exploring various scenarios, decision-makers can identify the variables that exert the most significant impact on outcomes, thereby assessing the robustness of their conclusions. Sensitivity analysis enhances understanding of the range of possible outcomes and the likelihood of different scenarios, offering a more comprehensive perspective on associated risks.

Another important aspect is the use of probabilistic methods, such as Monte Carlo simulations, which is essential for modeling uncertainty in benefit–cost analysis. Monte Carlo simulations generate numerous random samples based on the probability distributions of key variables, facilitating the creation of a spectrum of potential outcomes along with their associated probabilities. By analyzing the distribution of these results, decision-makers gain deeper insights into potential risks, enabling them to make more informed choices.

Overall, incorporating risks into benefit–cost analysis enhances its robustness and reliability. By accounting for uncertainty, performing sensitivity analysis, using probabilistic methods, considering risk aversion, and communicating results transparently, decision-makers can better evaluate the economic efficiency of projects and policies. This comprehensive approach helps in making more informed and resilient decisions in the face of uncertainty.

2.2.4 BCA and the Environment

There are investments that include the values of environmental improvement and environmental deterioration generated by the project. Incorporating environmental values into BCA requires a holistic approach that considers both direct and indirect effects. Direct benefits might include reduced healthcare costs due to improved air quality, while indirect benefits could encompass enhanced ecosystem services, such as water purification and carbon sequestration. On the cost side, direct costs might involve expenses related to pollution control measures, while indirect costs could include long-term impacts on biodiversity and ecosystem health. By capturing these diverse effects, BCA provides a more accurate and comprehensive assessment of the project's overall impact.

Another important aspect of BCA in environmental projects is the consideration of non-market values. Many environmental benefits, such as aesthetic enjoyment and recreational opportunities, do not have market prices but are highly valued by society. Including these non-market values in the analysis ensures that the full range of benefits is recognized and accounted for. This can be achieved through methods like willingness-to-pay surveys, which estimate how much individuals are willing to pay for environmental improvements or to avoid environmental degradation.

In general, the NPV calculation can be revised as follows:

$$\text{NPV} = \text{NPV}' - \text{EC} \tag{2.13}$$

where *NPV′* is the NPV that ignores environmental impacts, and *EC* presents the net value of the project's environmental impacts (can be negative). Only when $\text{NPV}' >$ EC, the project is worthwhile. The above equation also suggests that $\text{EC}^* = \text{NPV}'$ calculates the maximum level of damage for which the project could go ahead.

BCA can help identify trade-offs and synergies between environmental and economic objectives. For instance, a project that generates economic benefits through industrial development might lead to environmental degradation if not properly managed. By explicitly considering these trade-offs, BCA helps decision-makers balance competing objectives and make more informed choices that align with sustainable development goals.

When evaluating the economic impact of a project or policy, it is crucial to consider the size of the population (N) affected. The term EC^* represents the total economic cost or benefit of a project, while N denotes the population size. By dividing EC^* by N, we obtain the per capita economic impact, which provides a more nuanced understanding of how the project affects individuals within the population.

In some cases, even if the total environmental cost or benefit (EC^*) is substantial, the per capita impact (EC^*/N) can be relatively small if the population size is large. This is particularly relevant in densely populated regions or countries with significant populations. For instance, a large-scale infrastructure project might have a high total cost, but when this cost is distributed across millions of residents, the individual financial burden may be minimal. Conversely, the benefits of such a project, when spread across a large population, might also appear modest on a per capita basis.

Moreover, accounting for the size of the population can highlight disparities in how different groups are affected. For example, a project that benefits the entire population might still have varying impacts on different demographic segments. By analyzing EC^*/N, policymakers can identify which groups might bear a disproportionate share of the costs or receive a greater share of the benefits, allowing for more targeted and equitable policy interventions.

However, one of the key challenges in conducting BCA for environmental projects is quantifying the benefits and costs associated with environmental changes. Environmental improvements, such as cleaner air, water, and biodiversity conservation, often provide substantial benefits that are not easily captured in monetary terms. Similarly, environmental deterioration, such as pollution and habitat destruction, can impose significant costs on society. To address this, economists use various valuation techniques, such as contingent valuation, hedonic pricing, and cost–benefit transfer, to estimate the monetary value of environmental benefits and costs.

2.2.5 Is Cost–Benefit Analysis Sufficient?

No economic method can fully capture the intrinsic value of the environment. While BCA uses a single monetary numeraire to quantify the positive and negative impacts of projects or policies, this approach has inherent limitations. The clarity provided by market prices is useful when they accurately reflect the value of goods and services. However, this clarity becomes deceptive when market transactions are partial or absent, as is often the case with environmental goods.

Environmental goods, such as clean air, biodiversity, and ecosystem services, frequently lack market prices because they are not traded in conventional markets. This absence of market transactions makes it challenging to assign a monetary value to these goods. As a result, reliance on pricing rather than comprehensive valuation techniques can lead to the undervaluation of environmental goods. For instance, the benefits of preserving a forest may extend beyond timber value to include carbon sequestration, habitat for wildlife, and recreational opportunities, which are not easily captured by market prices alone.

The question of who should be the arbiter of value is complex. Ideally, the valuation of environmental goods should involve a diverse range of stakeholders, including economists, ecologists, policymakers, and the public. This inclusive approach ensures that multiple perspectives are considered, and the broader societal and ecological significance of environmental goods is recognized. Valuation techniques such as contingent valuation, which surveys individuals' willingness to pay for environmental preservation, and hedonic pricing, which infers values from related market transactions, can provide more comprehensive assessments.

Ultimately, while no economic method can fully "value" the environment, combining various valuation techniques and involving a wide array of stakeholders can help create a more accurate and holistic understanding of the true worth of environmental goods.

2.3 Environmental Evaluation

Environmental evaluation is a systematic process used to assess the environmental impacts of projects, policies, or programs. It aims to ensure that environmental considerations are integrated into decision-making processes, promoting sustainable development. Environmental evaluation involves the assessment of potential environmental impacts resulting from human activities. It is a crucial tool for identifying, predicting, and mitigating negative environmental effects while enhancing positive outcomes. This process helps policymakers, businesses, and communities make informed decisions that balance economic development with environmental protection.

2.3.1 Key Components of Environmental Evaluation

Environmental evaluation consists of several key components, including baseline studies, impact prediction, mitigation measures, public participation, and monitoring and evaluation.

- *Baseline Studies*: Baseline studies involve collecting data on the current state of the environment before a project or policy is implemented. This data provides a reference point against which future changes can be measured. For example, baseline studies for a new industrial facility might include data on air and water quality, soil conditions, and local biodiversity.
- *Impact Prediction*: Impact prediction involves forecasting the potential environmental effects of a proposed project or policy. This step uses scientific methods and models to estimate the magnitude and significance of these impacts. For example, predicting the impact of a new highway on local wildlife populations and air quality.
- *Mitigation Measures*: Mitigation measures are strategies designed to avoid, reduce, or compensate for negative environmental impacts. These measures can include changes to project design, implementation of best management practices, and restoration of affected areas. For example, creating wildlife corridors to allow animals to safely cross a new highway.
- *Public Participation*: Public participation is an essential component of environmental evaluation. It involves engaging stakeholders, including local communities, environmental groups, and government agencies, in the evaluation process. This ensures that diverse perspectives are considered and that the project or policy is socially acceptable. For example, holding public meetings to gather input on a proposed wind farm.
- *Monitoring and Evaluation*: Monitoring and evaluation involve tracking the environmental impacts of a project or policy over time to ensure that mitigation measures are effective and that the project complies with environmental regulations. This step allows for adaptive management, where strategies can be adjusted based on monitoring results, for example, regularly monitoring water quality near a mining site to detect any contamination (Fig. 2.5).

Fig. 2.5 Key environmental evaluation process

2.3.2 Challenges in Environmental Evaluation

Despite its importance, environmental evaluation methods face several challenges that can complicate the assessment process and impact the accuracy and reliability of the results.

Data Availability and Quality: One of the primary challenges is the availability and quality of data. Environmental evaluations require comprehensive and accurate data on various environmental parameters, such as air and water quality, biodiversity, and land use. However, such data may be scarce, outdated, or inconsistent, making it difficult to conduct thorough assessments. Additionally, collecting high-quality data can be resource-intensive and time-consuming.

Complexity of Environmental Systems: Environmental systems are inherently complex, involving numerous interacting components and feedback loops. This complexity makes it challenging to model and predict environmental impacts accurately. Simplifying assumptions are often necessary, but they can lead to oversights and inaccuracies. Capturing the full range of interactions and potential outcomes requires sophisticated modeling techniques and a deep understanding of the system.

Uncertainty and Variability: Environmental evaluations must account for significant uncertainty and variability. Natural systems are subject to fluctuations due to factors such as climate variability, seasonal changes, and human activities. Predicting future conditions and impacts involves considerable uncertainty, which can affect the reliability of the evaluation. Addressing this uncertainty requires robust statistical methods and scenario analysis to explore different possible outcomes.

Stakeholder Engagement: Engaging stakeholders is crucial for the success of environmental evaluations, but it can also be challenging. Different stakeholders, including local communities, government agencies, and environmental organizations, may have diverse perspectives and interests. Balancing these interests and ensuring meaningful participation can be difficult. Effective communication and collaboration are essential to gather input, address concerns, and build consensus.

Political and Economic Pressures: Environmental evaluations often occur in contexts where political and economic pressures can influence the process. Decision-makers may face conflicting priorities, such as economic development versus environmental protection. These pressures can lead to biased assessments or the prioritization of short-term gains over long-term sustainability. Ensuring the integrity and objectivity of the evaluation process is critical in such situations.

Addressing these challenges is essential for conducting robust and reliable environmental evaluations. By improving data collection, enhancing modeling techniques, managing uncertainty, engaging stakeholders effectively, and maintaining objectivity, evaluators can better assess environmental impacts and support informed decision-making for sustainable development.

Example

Hong Kong-Zhuhai-Macao Bridge

The environmental evaluation of the Hong Kong-Zhuhai-Macao Bridge involved a comprehensive assessment due to its unique cross-border nature, linking three different jurisdictions: Hong Kong, Zhuhai, and Macao. Each jurisdiction had its own environmental impact assessment (EIA) requirements, which led to a complex evaluation process. The primary focus was on understanding the potential impacts on the marine ecology of the Pearl River Estuary, where the bridge crosses. Baseline studies were conducted to gather data on the existing environmental conditions, including water quality, marine life, and coastal ecosystems. Impact predictions were made to assess how the construction and operation of the bridge would affect these environmental components. Mitigation measures were proposed to minimize negative impacts, such as habitat restoration and pollution control strategies. Public participation was also a key part of the process, involving consultations with local communities, environmental groups, and government agencies to ensure that the project was environmentally responsible and socially acceptable. Continuous monitoring and evaluation were implemented to track the effectiveness of the mitigation measures and address any unforeseen environmental issues (Fig. 2.6).

Fig. 2.6 Hong Kong-Zhuhai-Macao Bridge

2.4 Valuation of Human Life

Many public programs and projects involve the prevention of loss of life, such as the construction of dams, maintaining roads, installing traffic signs, providing healthcare, and employing firefighters. These initiatives aim to enhance public safety and reduce the risk of fatalities. Resources used to prevent loss of life are scarce, and choices must be made about how to allocate them effectively. For instance, public health campaigns to reduce smoking can save lives by preventing smoking-related diseases. However, these campaigns require funding, which could also be used for other life-saving initiatives, such as improving road safety or increasing access to healthcare. Policymakers must weigh the costs and benefits of different interventions to determine the most efficient use of resources.

However, valuing a life saved (or death averted) in the cost–benefit calculus is a complex and sensitive issue. While life is often considered priceless, various methods are used to estimate the value of a life saved in cost–benefit analyses. These methods, including the Value of a Statistical Life, and Human Capital Method, provide frameworks for quantifying the benefits of life-saving interventions. By considering these values, policymakers can make more informed decisions about resource allocation and prioritize initiatives that offer the greatest overall benefit to society.

One common method for valuing a life saved is derived from individuals' willingness to pay (WTP) to reduce the probability of dying, or the so-called Value of a Statistical Life (VSL). VSL estimates the monetary value that individuals place on reducing their risk of death. For example, if people are willing to pay $100 to reduce their risk of dying by 1 in 10,000, the VSL would be $1 million ($100/(1/10,000)). This method is often used in regulatory impact analyses to assess the benefits of safety regulations, such as those for air quality standards or workplace safety measures.

VSL is widely used in regulatory impact analyses, aiding policymakers in evaluating the benefits of safety regulations, environmental policies, and public health interventions. However, VSL relies on individuals' stated or revealed preferences, which can be influenced by various factors such as risk perception, information availability, and cognitive biases. This subjectivity can lead to variability and uncertainty in VSL estimates. Some critics argue that placing a monetary value on human life is inherently problematic and can lead to ethical dilemmas. Also, VSL estimates can be sensitive to income levels, potentially leading to disparities in the valuation of life across different socioeconomic groups. Higher-income individuals may be willing to pay more for risk reductions, resulting in higher VSL estimates for wealthier populations.

Another approach is the Human Capital Method, which values a life based on the present value of an individual's future earnings. This method considers the economic contributions that a person would have made over their lifetime. For instance, if a young professional is expected to earn $50,000 annually for the next 30 years, the present value of these earnings can be calculated to estimate the economic loss due to premature death. This method is commonly used in legal contexts, such as wrongful death lawsuits, to determine compensation amounts.

This approach is straightforward and relies on readily available economic data. It is commonly used in legal contexts, such as wrongful death lawsuits, to determine compensation amounts, providing a clear and tangible basis for calculating economic losses due to premature death. The Human Capital Method offers a consistent framework for valuing life across different cases, making it easier to apply in various contexts and ensuring comparability of results.

However, this method focuses solely on economic contributions, ignoring non-market values such as the emotional and social impact of a person's life. This narrow scope can lead to an incomplete assessment of the true value of life. It tends to undervalue the lives of individuals who are not in the workforce, such as children, retirees, and homemakers. It also does not account for the potential future earnings of individuals who may not currently be employed.

In summary, both methods have their strengths and limitations in valuing human life. VSL reflects societal preferences and is useful for policy analysis, but it can be subjective and ethically contentious. The Human Capital Method provides an objective and consistent measure based on economic contributions, but it has a limited scope and may undervalue certain individuals. Understanding these pros and cons helps in selecting the appropriate method for different contexts and ensuring a more comprehensive and ethical approach to valuing human life (Table 2.1).

Table 2.1 Comparing VSL and human capital methods

Aspect	VSL method	Human capital method
Basis	Willingness to pay for risk reduction	Present value of future earnings
Reflects	Societal preferences and values	Economic contribution
Policy Relevance	Widely used in regulatory impact analyses	Commonly used in legal contexts
Subjectivity	Influenced by risk perception, information availability, and cognitive biases	Objective and quantifiable
Income Sensitivity	Estimates can vary based on income levels, potentially leading to disparities	Tends to undervalue non-working individuals
Scope	Captures subjective value people place on safety and risk reduction	Focuses solely on economic contributions, ignoring non-market values
Limitations	Variability and uncertainty in estimates due to subjective factors	Limited scope, may undervalue children, retirees, homemakers, and non-employed people

2.5 Integrated Assessment

Integrated assessment (IA) is a method used to analyze complex systems by combining insights from different disciplines. It aims to provide a holistic view of issues that cannot be fully understood through a single disciplinary lens. IA is particularly useful in addressing sustainability challenges, which often involve intricate interactions between environmental, economic, and social factors.

2.5.1 Key Components of Integrated Assessment

Integrated assessment is a multidisciplinary approach that combines knowledge from various fields to analyze complex issues and develop comprehensive solutions. The key components of integrated assessment include system conceptualization, data collection and analysis, model development, scenario analysis, stakeholder engagement, and policy analysis and recommendations.

System Conceptualization: The first step in integrated assessment is system conceptualization, which involves identifying and defining the key components and interactions within the system being studied. This step requires a thorough understanding of the various elements that influence the system, such as environmental, economic, and social factors. By mapping out these components and their relationships, researchers can create a conceptual framework that guides the assessment process. This framework helps in understanding the complexity of the system and provides a basis for further analysis.

Data Collection and Analysis: Accurate and comprehensive data collection is crucial for integrated assessment. This component involves gathering relevant data from various sources, including historical records, scientific studies, and real-time observations. The data collected should cover all aspects of the system, such as environmental conditions, economic indicators, and social metrics. Once the data is collected, it is analyzed to identify trends, correlations, and patterns. This analysis provides insights into the current state of the system and helps in identifying key drivers and pressures.

Model Development: Model development is a critical component of integrated assessment, as it involves creating mathematical or computational models that simulate the behavior of the system. These models integrate data from different disciplines and capture the interactions between various components. By simulating different scenarios, models can predict the potential outcomes of various interventions and policies. This helps in understanding the potential impacts of different strategies and supports evidence-based decision-making.

Scenario Analysis: Scenario analysis is used to explore the potential future states of the system under different conditions. This component involves developing and analyzing scenarios based on varying assumptions, such as changes in policy, technology, or environmental conditions. Scenario analysis helps in understanding the range of possible outcomes and the uncertainties associated with them. It provides a valuable tool for testing the robustness of strategies and identifying potential risks and opportunities.

Stakeholder Engagement: Engaging stakeholders is an essential component of integrated assessment, as it ensures that the perspectives and knowledge of different groups are considered. Stakeholders may include policymakers, industry representatives, community members, and experts from various fields. By involving stakeholders in the assessment process, researchers can gather diverse insights, validate assumptions, and build consensus around potential solutions. Stakeholder engagement also enhances the relevance and acceptability of the assessment outcomes.

Policy Analysis and Recommendations: The final component of integrated assessment is policy analysis and recommendations. This involves evaluating the effectiveness of different policies and strategies based on the insights gained from the assessment. Researchers analyze the potential impacts of various interventions and provide recommendations on the best courses of action. These recommendations are aimed at achieving sustainable and equitable outcomes, addressing the identified issues, and supporting informed decision-making by policymakers.

Example

Scenario Analysis in Climate Change and Agriculture

Imagine we are assessing the impact of climate change on agricultural productivity in a specific region. We develop several scenarios based on different levels of greenhouse gas emissions and climate policies.

- *Scenario 1: Business-as-Usual.* No additional measures are taken to reduce emissions. The model predicts increasing temperatures and changing precipitation patterns, leading to declining crop yields due to heat stress and water scarcity. This scenario highlights the potential consequences of inaction.
- *Scenario 2: Moderate Mitigation.* Moderate efforts are made to reduce emissions through policies like promoting renewable energy and improving energy efficiency. The model predicts a slower rate of temperature increase and more stable precipitation patterns, resulting in less severe impacts on crop yields. This scenario shows the benefits of taking action.
- *Scenario 3: Aggressive Mitigation.* Aggressive measures are implemented to drastically reduce emissions, including widespread adoption of renewable energy and stringent emission regulations.

The model predicts temperature stabilization and improved precipitation patterns, leading to stable or improved crop yields. This scenario demonstrates the positive outcomes of strong climate policies. By comparing these scenarios, policymakers can evaluate the potential impacts of different levels of action and make informed decisions to ensure agricultural productivity and food security in the face of climate change. Scenario analysis provides a valuable tool for exploring future possibilities and preparing for various outcomes (Fig. 2.7).

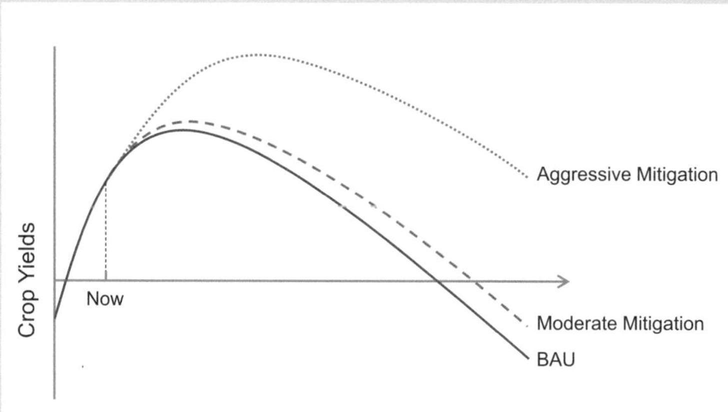

Fig. 2.7 Scenarios for crop yields

2.5.2 Applications of IA in Sustainability

Integrated assessment is widely used in sustainability to address complex issues that require a holistic understanding and multidisciplinary approach. For example, IA is used to evaluate the effectiveness of different strategies for reducing greenhouse gas emissions and adapting to climate change. This includes assessing the impact of renewable energy policies on emissions and economic growth. In water resource management, IA helps in managing water resources sustainably by considering the interactions between water supply, demand, and quality. For example, evaluating the impact of water conservation measures on river ecosystems and agricultural productivity. In biodiversity conservation, IA is used to develop strategies for conserving biodiversity by considering the interactions between ecological, economic, and social factors, for example, assessing the impact of land use changes on species habitats and local communities.

One prominent example of an integrated assessment model is the *Dynamic Integrated Climate-Economy model* (DICE), developed by economist William Nordhaus. The DICE model combines economic and climate systems to explore the relationship between economic growth and climate change. It assists policymakers in evaluating the costs and benefits of various climate policies by simulating the economic impacts of different levels of greenhouse gas emissions and climate change mitigation strategies.

The DICE model includes key elements such as greenhouse gas emissions, atmospheric concentrations, temperature changes, and economic variables like GDP and consumption. By modeling these interactions, the DICE model provides insights into the long-term economic effects of climate change and the potential effectiveness of different policy measures. This model has been widely used in climate policy analysis and has significantly influenced international climate negotiations and policy decisions.

However, it should be noted that IA has its limitations. One of the main challenges in IA is the availability and quality of data. Accurate and comprehensive data is essential for developing reliable models and making informed decisions. IA also involves dealing with complex systems that have many interacting components and uncertainties. This complexity can make it difficult to predict outcomes accurately and develop effective solutions.

2.6 Introduction to System Dynamics

System dynamics is a method for studying and managing complex feedback systems, such as those found in business, economics, and environmental science. It was developed in the 1950s by Professor Jay W. Forrester at the Massachusetts Institute of Technology. The core idea is to use computer simulations to model the interactions within a system and predict how changes in one part of the system can affect the whole. System dynamics is crucial in sustainability analysis because it allows us to understand and model the complex interactions and feedback loops within environmental, social, and economic systems.

2.6.1 *Key Components of System Dynamics*

System dynamics involves key components such as stocks and flows, which describe the accumulation and movement of resources; feedback loops, which either amplify or stabilize changes within the system; delays, which represent the time lags between actions and their effects; and nonlinearities, where small changes can lead to disproportionately large impacts. Together, these elements help in modeling and analyzing the dynamic interactions within systems, making it an essential tool in sustainability economics.

Stocks and Flows: In system dynamics, stocks and flows are fundamental concepts that describe how resources accumulate and change over time. *Stocks* represent the quantities of resources present in a system at any given moment. For instance, the amount of water in a reservoir or the number of trees in a forest are examples of stocks. These stocks are influenced by *flows*, which are the rates at which resources enter or leave the system. For example, the rate of water flowing into or out of a reservoir, or the rate at which trees are planted and harvested in a forest, are flows. Understanding the interplay between stocks and flows is crucial for analyzing how systems evolve and respond to different influences.

Feedback Loops: Feedback loops are mechanisms that either amplify or counteract changes within a system. *Positive feedback loops* enhance changes, leading to exponential growth or decline. For example, in a forest, the more trees there are, the more seeds are produced, which in turn leads to more trees. This self-reinforcing cycle can lead to rapid expansion of the forest. On the other hand, *negative feedback loops* work to stabilize a system by counteracting changes. For instance, in an ecosystem, an increase in the predator population reduces the prey population. As the prey population decreases, the predator population eventually declines as well, maintaining a balance within the ecosystem. These feedback loops are essential for understanding the dynamic behavior of systems.

Delays: Delays are the time lags between actions and their effects within a system. They play a significant role in system dynamics by influencing how quickly a system responds to changes. For example, there is often a delay between planting a tree and growing to maturity. During this period, the effects of planting are not immediately visible, which can complicate efforts to manage and predict system behavior. Recognizing and accounting for delays is important for effective decision-making and long-term planning in sustainability economics.

Nonlinearities: Nonlinearities occur when the relationship between variables in a system is not proportional. This means that small changes in one part of the system can lead to disproportionately large effects elsewhere. For example, a slight increase in pollution might have a negligible impact on a river's ecosystem initially. However, once pollution levels surpass a certain threshold, the ecosystem could collapse, leading to severe consequences. Nonlinearities add complexity to system dynamics, making it challenging to predict outcomes and necessitating careful consideration in sustainability planning.

2.6.2 Building a System Dynamics Model

Building a system dynamics model is a structured process that begins with clearly defining the problem you aim to address. This initial step is crucial as it sets the direction for the entire modeling effort. For instance, you might be interested in managing a forest sustainably to balance timber production and conservation. By articulating the problem, you establish the scope and objectives of your model.

Once the problem is defined, the next step is system conceptualization. This involves identifying the key components of the system, such as stocks, flows, feedback loops, and delays. In the context of a forest management model, stocks could include the number of trees and the amount of timber, while flows might encompass tree growth and harvesting rates. Conceptualizing the system helps in understanding the various elements and their interactions, laying the groundwork for the model.

The third step is model formulation, where you develop mathematical equations to represent the relationships between the components. For example, the growth rate of trees might depend on factors like soil quality and rainfall. These equations form the core of the model, capturing the dynamics of the system. Accurate formulation is essential for the model to reflect real-world behaviors and interactions.

Simulation is the next phase, where you use computer software to simulate the behavior of the system over time. This allows you to test different scenarios and observe how changes in one part of the system affect the whole. Simulation provides valuable insights into the system's dynamics and helps in identifying potential outcomes and trends.

Validation follows simulation, where you compare the model's predictions with real-world data to ensure its accuracy. For instance, you might compare the model's predictions of tree growth with actual measurements from a forest. Validation is a critical step to confirm that the model reliably represents the real system and can be trusted for decision-making.

Finally, the model is used for policy design and testing. This involves using the model to design and test policies that can improve system performance. For example, you might test different forest management strategies to find the one that best balances timber production and conservation. By experimenting with various policies, you can identify effective strategies and make informed decisions to achieve your sustainability goals (Fig. 2.8).

This structured approach to building a system dynamics model ensures that the model is robust, accurate, and useful for analyzing complex systems and making informed decisions. Here are some examples of widely used system dynamics models:

The Limits to Growth: In 1972, a team of researchers at MIT developed "The Limits to Growth," a groundbreaking system dynamics model that examined the long-term consequences of exponential population growth and finite resource supplies. This model was one of the first to use system dynamics to explore global sustainability issues. It simulated interactions between population, industrial growth, food production, and environmental limits, highlighting potential scenarios of economic collapse

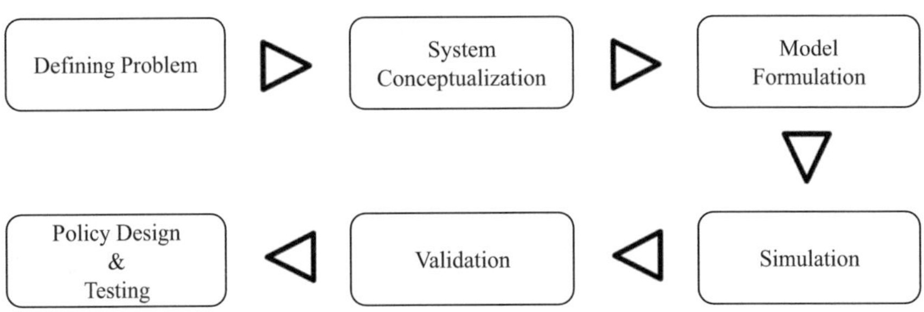

Fig. 2.8 Building a system dynamics model

if growth trends continued unchecked. The model's findings underscored the importance of sustainable development and resource management, sparking widespread debate and influencing environmental policy discussions worldwide.

SIR Model: The SIR model is a widely used system dynamics model in epidemiology that helps understand the spread of infectious diseases. It divides the population into three compartments: susceptible (S), infected (I), and recovered (R). By modeling the rates at which individuals move between these compartments, the SIR model can predict the course of an epidemic and evaluate the effectiveness of intervention strategies such as vaccination and quarantine. This model has been instrumental in public health planning, providing insights into disease dynamics and helping to mitigate the impact of outbreaks.

Urban Dynamics: Urban dynamics, created by Jay Forrester, the founder of system dynamics, is a model that examines the growth and development of cities. It explores the interactions between housing, industry, and services, and how these factors influence urban expansion and decline. By simulating these interactions, the model helps urban planners and policymakers understand the complex dynamics of urban environments. It provides insights into issues such as housing shortages, industrial decline, and service provision, enabling more informed decisions to promote sustainable urban development.

2.6.3 Applications of System Dynamics in Sustainability

System dynamics is a powerful tool in sustainability, enabling the analysis and management of complex problems that involve multiple interacting components and feedback loops. By simulating these interactions, system dynamics models provide valuable insights for policymakers and planners, helping them design effective strategies for sustainable development.

One prominent application of system dynamics is in addressing climate change. These models can simulate the intricate interactions between greenhouse gas emissions, atmospheric concentrations, and global temperatures. By capturing these dynamics, system dynamics models help policymakers understand the potential impacts of different mitigation strategies, such as reducing emissions and enhancing carbon sinks. This enables the design of comprehensive policies aimed at mitigating climate change and promoting environmental sustainability.

In the realm of water resource management, system dynamics models are used to analyze the interactions between water supply, demand, and quality. These models help in designing policies for sustainable water use by optimizing reservoir operations, managing groundwater extraction, and ensuring the equitable distribution of water resources. By simulating various scenarios, system dynamics models provide insights into the long-term impacts of different water management strategies, aiding in the development of sustainable solutions.

Energy systems also benefit from the application of system dynamics. These models can analyze the interactions between energy production, consumption, and environmental impacts. By simulating these dynamics, system dynamics models help in designing sustainable energy policies that promote renewable energy sources, improve energy efficiency, and reduce environmental impacts. This holistic approach enables policymakers to develop strategies that balance energy needs with environmental sustainability.

Urban planning is another area where system dynamics proves invaluable. These models can simulate the interactions between population growth, land use, transportation, and environmental quality. By capturing these complex dynamics, system dynamics models help in designing sustainable urban development policies that reduce urban sprawl, promote public transportation, and enhance the overall quality of urban environments. This comprehensive approach supports the creation of livable, sustainable cities that can accommodate growing populations while minimizing environmental impacts.

Overall, system dynamics provides a robust framework for addressing sustainability challenges, enabling the analysis of complex systems and the design of effective policies for sustainable development.

Reading

Application of System Dynamics in Climate Change Analysis

System dynamics models are highly effective in addressing climate change by simulating the complex interactions between greenhouse gas emissions, atmospheric concentrations, and global temperatures. These models capture feedback loops, such as the melting of polar ice caps, which can accelerate warming, and the increased uptake of carbon dioxide by forests, which helps mitigate climate change. By incorporating delays, system dynamics models highlight the long-term impacts of actions taken today, emphasizing the importance of early intervention.

These models also allow policymakers to evaluate various mitigation strategies, such as reducing fossil fuel consumption, increasing renewable energy production, and implementing carbon capture technologies. By exploring different scenarios and conducting sensitivity analyses, policymakers can identify the most effective approaches to reduce emissions and promote sustainability. Overall, system dynamics provides a comprehensive framework for designing informed and effective climate policies.

2.7 Game Theory and Its Application in Sustainability

Game theory is a significant discipline within both mathematics and economics that examines strategic interactions among rational decision-makers. It offers a systematic framework for analyzing how individuals and organizations arrive at decisions in contexts where the results are influenced not only by their own choices but also by the actions of others. This is particularly important in sustainability, where the actions of one party can significantly impact others, and where collaboration is often necessary to address shared challenges such as climate change, resource management, and environmental protection.

2.7.1 *What Is Game Theory?*

Game theory analyzes situations where multiple players make decisions that affect each other. These situations are called "games," and the players can be individuals, firms, governments, or any entities with decision-making power (Zamir et al., 2020). The goal of game theory is to predict the outcomes of these interactions by understanding the strategies that players use to maximize their payoffs. Several key concepts form the foundation of game theory, including equilibrium, strategies, players, and payoffs,

Players are the decision-makers in a game. Each player has a set of possible actions or strategies they can choose from. For example, in a game involving two firms competing in a market, each firm is a player.

Strategies are the plans of action that players can take. A strategy can be simple, such as choosing a price for a product, or complex, involving multiple steps and contingencies. In the context of sustainability, a strategy might involve a company deciding whether to invest in renewable energy technologies.

Payoffs are the outcomes that players receive based on the combination of strategies chosen by all players. Payoffs can be measured in various ways, such as profits, costs, utility, or environmental impact. For example, the payoff for a firm investing in renewable energy might include reduced energy costs and improved public image.

Equilibrium in game theory denotes a state wherein no player has an incentive to unilaterally change their strategy, leading to a stable outcome for all participants. A key concept within this framework is the Nash equilibrium, named after mathematician John Nash. In a Nash equilibrium, each player's strategy represents the optimal response to the strategies chosen by others, indicating that no player can enhance their payoff by altering their strategy while the strategies of other players remain unchanged.

2.7.2 Prisoner's Dilemma

Prisoner's Dilemma is a fundamental concept in game theory that illustrates the complexities of decision-making in situations where individuals must choose between cooperation and self-interest (Poundstone, 2011). It is a scenario that demonstrates how rational individuals might not cooperate, even if it appears that it is in their best interest to do so.

Imagine two prisoners, A and B, who are accused of committing a crime together. They are interrogated separately and cannot communicate with each other. The authorities offer each prisoner a deal: if one testifies against the other (defects) while the other remains silent (cooperates), the defector will go free, and the silent accomplice will receive a heavy sentence of 3 years in prison. If both prisoners defect, they both receive moderate sentences of 2 years in prison. If both cooperate and remain silent, they receive light sentences of serving 1 year only in prison (Table 2.2).

The dilemma arises because each prisoner must consider the potential decisions of the other. If prisoner A believes that prisoner B will defect, A's best response is to defect as well to avoid the worst outcome. Conversely, if A believes that B will cooperate, A might still choose to defect to gain freedom. This leads to a situation where both prisoners are likely to defect, resulting in a worse outcome for both compared to mutual cooperation.

In the context of sustainability economics, Prisoner's Dilemma can be applied to various environmental and resource management issues. For instance, consider two countries X and Y, each of which faces a choice of whether to abate pollution or not to abate pollution (labeled "*Pollute*"). Pollution abatement is assumed to be a public good so that abatement by either country benefits both. Abatement comes at a cost of 7 to the abater, but confers benefits of 5 to both countries. If both abate, both experience benefits of 10 (and each experiences a cost of 7) (Table 2.3).

Table 2.2 Strategies and payoffs of the two prisoners

	Prisoner B Stays silent (cooperates)	Prisoner B Betrays (defects)
Prisoner A Stays silent (cooperates)	Each serves 1 year	Prisoner A: 3 years Prisoner B: goes free
Prisoner A Betrays (defects)	Prisoner A: goes free Prisoner B: 3 years	Each serves 2 years

Table 2.3 A game of pollute or abate

		Y's strategy	
		Pollute	Abate
X's strategy	Pollute	0, 0	5, −2
	Abate	−2, 5	3, 3

We begin by examining non-cooperative behavior. A player possesses a dominant strategy when one strategy consistently yields a higher payoff than any other, regardless of the choices made by the opposing player. In this context, for player Y, *Pollute* emerges as the optimal choice irrespective of what strategy player X selects. This conclusion holds true when considering player X's perspective as well, where *Pollute* also serves as the dominant strategy. The symmetric structure of the payoff matrix allows for the interchangeability of labels X and Y without loss of information, negating the need for redundant calculations. Consequently, both countries have a dominant strategy represented by *Pollute*. The analysis within non-cooperative game theory thus concludes that the equilibrium solution entails both countries opting not to abate pollution.

If both countries pollute, the environment deteriorates, resulting in severe consequences for everyone. This scenario highlights the tension between individual incentives and collective well-being, demonstrating the need for coordinated efforts and enforceable agreements to achieve sustainable environmental outcomes.

Here, the payoff matrix in environmental pollution mitigation has a structure of payoffs known as a Prisoners' Dilemma. Prisoner's Dilemma highlights the importance of trust, communication, and the establishment of enforceable agreements in achieving sustainable outcomes. It underscores the need for mechanisms that encourage cooperation and discourage self-interested behavior that can lead to collective harm.

2.7.3 Nash Equilibrium

Nash equilibrium, named after mathematician John Nash, is a fundamental concept in game theory that characterizes a situation where no player can enhance their payoff by unilaterally altering their strategy, assuming that the strategies of other players remain unchanged (Holt & Roth, 2004). In this context, it represents a stable state wherein each player's strategy is optimal, taking into account the strategies employed by all other participants. This equilibrium is reached when every player has selected a strategy, and no one can improve their outcome by deviating from their choice while others maintain theirs.

To illustrate this concept, consider the classic scenario of the Prisoner's Dilemma. In this case, the Nash equilibrium occurs when both prisoners choose to defect. Given the other prisoner's strategy, neither A nor B can improve their situation by independently changing their decision. If prisoner A expects prisoner B to defect, A's best response is also to defect to avoid the worst possible outcome. Likewise, if prisoner B anticipates that A will defect, B's optimal response is to defect as well. Consequently, both prisoners end up defecting, despite the fact that mutual cooperation would have yielded a more favorable collective result.

Nash equilibrium is not limited to competitive scenarios; it also applies to cooperative situations. For instance, in environmental agreements, countries might reach a Nash equilibrium where each country's strategy to reduce emissions is optimal, given the strategies of other countries. If any country unilaterally decides to change its emission reduction strategy, it would not achieve a better outcome, assuming other countries stick to their strategies. This equilibrium underscores the importance of mutual expectations and strategic interdependence in achieving stable and sustainable agreements.

However, it is important to note that Nash equilibrium does not always lead to the best collective outcome. In some cases, the equilibrium can result in suboptimal payoffs for all players involved, known as a "Pareto inefficient" outcome. This is often seen in the Prisoner's Dilemma, where the Nash equilibrium leads both players to defect, resulting in a worse outcome than if they had both cooperated. Understanding the Nash equilibrium helps in analyzing strategic interactions and designing mechanisms to encourage better collective outcomes.

2.7.4 Cooperative Games

Cooperative games are a branch of game theory where players can form coalitions and negotiate binding agreements to achieve mutual benefits. Unlike non-cooperative games, where players act independently and often in competition with one another, cooperative games emphasize collaboration and collective strategy. The primary focus is on how groups of players, or coalitions, can work together to maximize their collective payoff and how the gains from cooperation can be fairly distributed among the participants.

Most international environmental challenges involve multiple countries, and global issues typically encompass a larger number of participants. Let's start to think about a game with multiple players. Let N be the number of countries affected by an environmental problem, where $N \geq 2$. For simplicity, we assume that all the N countries are identical. We return to the Prisoner's Dilemma framework, where each unit of pollution abatement incurs a cost of 7 for the abating country while providing benefits of 5 both to that country and to all others. For the case where $N = 10$, the payoff matrix can be described as follows.

Let's consider the perspective of a single country, which we call country i. Table 2.4 illustrates the payoffs for country i from either polluting or abating, depending on the number of other countries that choose to abate. For each scenario, regardless of how many other countries decide to abate, country i achieves a higher payoff by choosing to pollute rather than abate. This demonstrates that, from an individual standpoint, it is rational for country i to opt for pollution.

Since all countries are identical, the same logic applies to any of the N countries. Consequently, the non-cooperative outcome is that none of the 10 countries choose to abate. This situation mirrors the fundamental characteristics of the two-country Prisoner's Dilemma. When each nation acts in its self-interest, they all end up with worse outcomes (0 each) compared to the scenario where all cooperate and abate pollution (43 each). However, the cooperative solution is inherently unstable. Even

Table 2.4 Prisoner's dilemma with ten countries

	Number of abating nations other than i									
	0	1	2	3	4	5	6	7	8	9
Nation i pollutes	0	5	10	15	20	25	30	35	40	45
Nation i abates	−2	3	8	13	18	23	28	33	38	43

if an agreement to abate is reached, any single country can improve its payoff by breaking the agreement and choosing to pollute.

The structure of payoffs plays a crucial role in determining whether cooperation can be sustained among countries. Building on the framework proposed by Barrett (1997), we examine the payoffs associated with different choices in a more generalized manner. Let NBA represents the net benefit to a country if it chooses to abate, and NBP represent the net benefit if it chooses to pollute (not abate). Assume there are N identical countries, with K of them choosing to abate. We define the following payoff-generating functions:

$$\begin{aligned} \text{NBP} &= a + bK \\ \text{NBA} &= c + dK \end{aligned} \tag{2.14}$$

where a, b, c, and d are parameters. By adjusting these parameter values, we can create different payoff matrices. For the specific case involving ten countries, the parameters are set as follows: $a = 0$, $b = 5$, $c = -7$, and $d = 5$. As NBP is always larger than NBA, the consequence of this game is that all countries will pollute.

Now, if the parameter values are adjusted, for instance, by setting the payoff function parameters to $a = 12$, $b = 3$, $c = 0$, and $d = 7$, we can generate an updated figure analogous to Fig. 2.9, leading us to Fig. 2.10. In this scenario, if fewer than three countries choose to cooperate (i.e., abate), none will engage in cooperative behavior, resulting in all countries opting to pollute. Conversely, if three or more countries decide to cooperate, then all participants will follow suit and engage in abatement. This situation yields two potential stable equilibria: either all countries will abate or none will.

To understand why these are stable equilibria, consider the following reasoning. First, assume that no country engages in abatement. In this case, can any individual country enhance its payoff by choosing to abate instead of polluting? The answer is no. Next, if all countries decide to abate, can any single country improve its payoff by opting to pollute? Again, the answer is no. However, no other combination of polluting and abating countries achieves stability. For example, if one country pollutes while two others abate (with seven remaining countries also polluting), the polluting country receives a payoff of 18, whereas the abating countries each receive 14. Here, each of the two abating countries has an incentive to defect and choose pollution, as doing so would elevate their payoff from 14 to 15. This scenario illustrates the instability of mixed strategies.

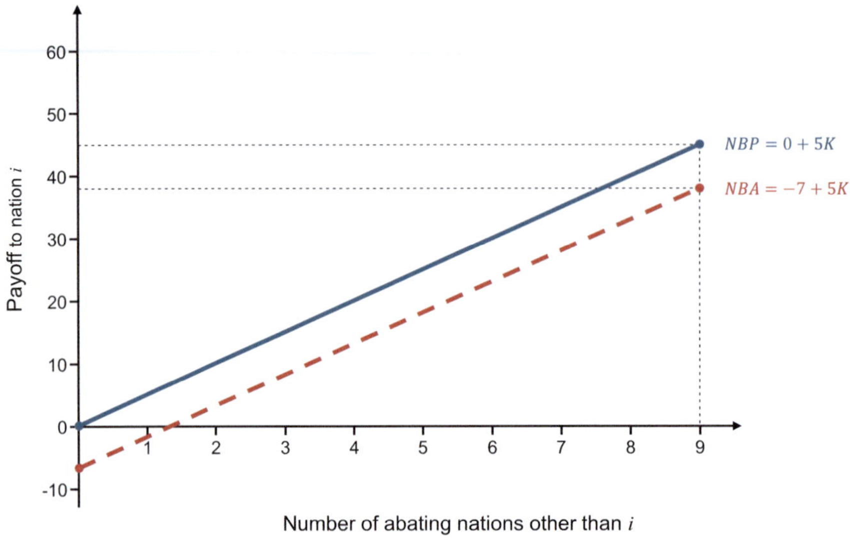

Fig. 2.9 Parallel payoff curves for NBP and NBA

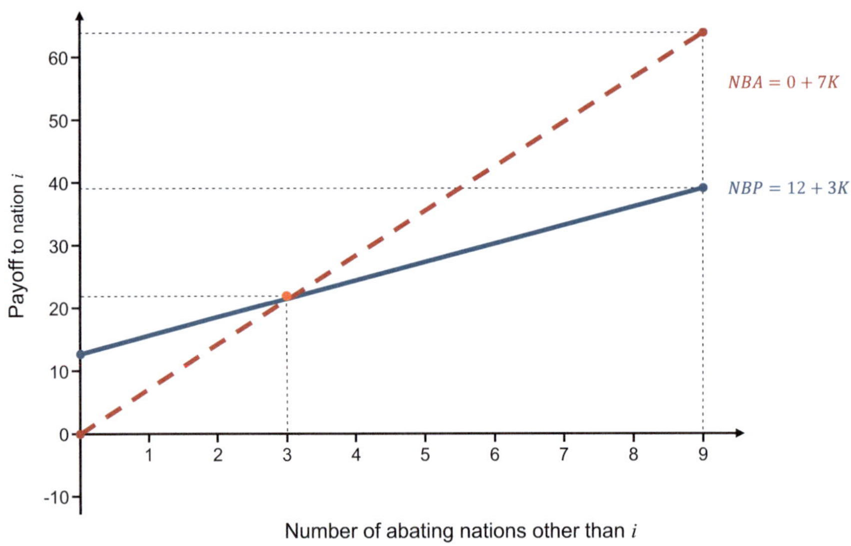

Fig. 2.10 Payoff structure to NBP and NBA with two equilibria

We could further set the parameter values as follows: $a = 0$, $b = 5$, $c = 6$, and $d = 3$. This payoff structure results in an incomplete self-enforcing agreement with three countries choosing to cooperate (abate) and seven opting not to, as shown in Fig. 2.11. It is important to note that the payoff for each cooperating country is lower than that

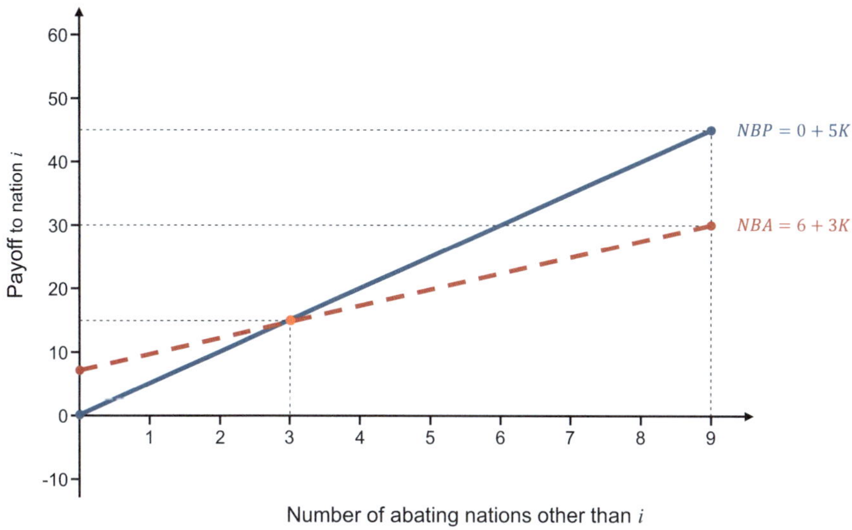

Fig. 2.11 Payoff structure to NBP and NBA with stable equilibrium

for each non-cooperating country, making this scenario similar to a Chicken game. The total payoff for all countries is higher than it would be with no cooperation, but it is still less than what could be achieved with full cooperation. In this sense, the game resembles a partially resolved Prisoner's Dilemma. The key takeaway is that by modifying the payoff structure of a Prisoner's Dilemma, stable cooperation can become feasible. However, this cooperation may still fall short of being complete.

Finally, we can extend the discussion by considering that countries may choose or negotiate their levels of abatement. In a non-cooperative scenario, each country independently determines its level of abatement to maximize its own payoff, without accounting for the effects on others. This means each country determines its abatement level, (z_i), based on the assumption that the abatement levels in other countries, $\bar{z}$, remain constant. Consequently, each nation abates until its marginal benefit from abatement equals its marginal cost.

In contrast, full cooperative behavior involves all N countries jointly deciding on abatement levels to maximize their collective payoff. The cooperative solution entails jointly selecting abatement levels in each country to maximize the collective payoff. This aligns with the principle of efficient provision of a public good. In this context, the marginal abatement cost in each country should equal the aggregated marginal benefits across all beneficiaries of the public good. The full cooperative solution is collectively rational, meaning it maximizes welfare for all N countries considered as a single entity. If a supranational authority existed with the power to enforce decisions aimed at maximizing total net benefits, the result would be the full cooperative solution.

Figure 2.12 is valuable as it highlights the factors that determine two crucial metrics: the differential between full cooperation abatement and non-cooperative

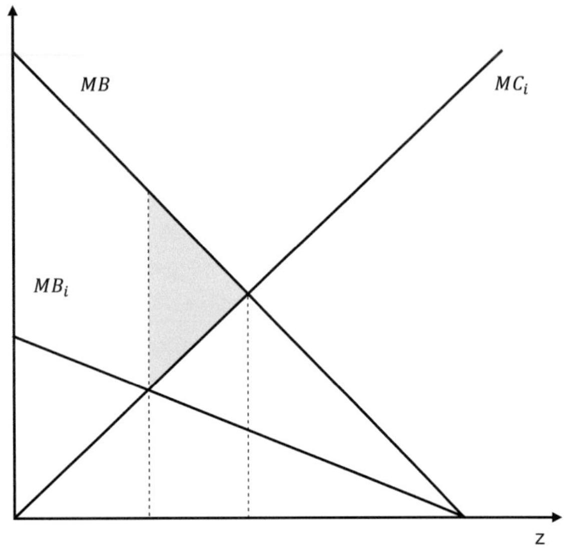

Fig. 2.12 Efficiency gain from full cooperation

abatement, i.e., $z_c - z_N$, and the extent of efficiency gains arising from full cooperation, represented by the shaded triangular area in the diagram. T These metrics are influenced by two primary factors: the relative slopes of the marginal benefit (MB_i) and marginal cost (MC_i) curves, and the number of competing countries, N, which affects the relative slopes of the MB_i and MB curves.

2.7.5 *Sequential Games and Leadership*

So far, we have discussed games in which players make their decisions at the same time, without knowing the choices of the other players. Sequential games are a type of game in game theory where players make decisions one after another, rather than simultaneously. This structure allows players to observe the actions of those who move before them and adjust their strategies accordingly. Sequential games are often represented using game trees, where each node represents a decision point, and branches represent the possible actions a player can take. The outcomes depend on the sequence of choices made by all players involved.

Imagine a situation where country X makes the first move. Country Y then observes X's decision and chooses its own action accordingly. In this scenario, the extensive form of the game is crucial for analyzing the choices made by each country. This is depicted in Fig. 2.13.

The solution to this game can be determined using backward induction. If country X decides to pollute, the best response for country Y is to abate, resulting in a payoff of 5 for X. Conversely, if X chooses to abate, Y's optimal response is to pollute,

Fig. 2.13 Decision tree

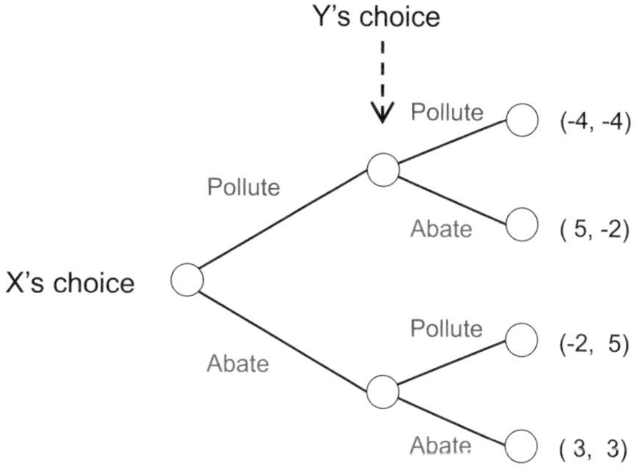

giving *X* a payoff of -2. Knowing *Y*'s best response, *X* will opt to pollute, as this yields a higher payoff of 5 compared to -2 if *X* had chosen to abate.

This example illustrates a broader principle: in sequential games, moving first can sometimes provide a strategic advantage, known as the "first-mover advantage." However, there are other payoff structures where it may be more beneficial to let the other player move first and then capitalize on their decision. In such cases, the second-mover advantage can be more advantageous, depending on the specific dynamics of the game.

Leadership plays a crucial role in sequential games, as the player who moves first can significantly influence the subsequent actions of other players. This first-mover advantage allows the leader to set the tone and direction of the game, potentially steering the outcomes in their favor. In the context of climate change mitigation, leadership is essential for initiating and sustaining global efforts to reduce greenhouse gas emissions.

Consider a scenario where a leading country decides to take bold action on climate change by implementing stringent emission reduction policies and investing heavily in renewable energy. This country, acting as the leader in the sequential game, sets a precedent for other nations. The initial decision to lead in climate change mitigation can create a ripple effect, encouraging other countries to follow suit. The leader's commitment can demonstrate the feasibility and benefits of such actions, reducing the perceived risks for followers.

In real-world scenarios, sequential games and leadership are evident in international climate agreements. For example, the Paris Agreement saw several countries stepping up as leaders, committing to ambitious targets for reducing emissions. These commitments influenced other nations to enhance their own climate actions, creating momentum for global cooperation. Understanding the principles of sequential games and the role of leadership can help countries make more informed and strategic decisions in addressing climate change.

References

Holt, C. A., & Roth, A. E. (2004). The nash equilibrium: A perspective. *Proceedings of the National Academy of Sciences, 101*(12), 3999–4002.
Mishan, E. J., & Quah, E. (2020). *Cost-benefit analysis.* Routledge.
Poundstone, W. (2011). *Prisoner's Dilemma.* Anchor.
Stackelberg, H. V. (2011). *Market structure and equilibrium.* Springer.
Zamir, S., Maschler, M., & Solan, E. (2020). *Game theory.* Cambridge University Press.

Chapter 3
The Economic Origins of Sustainability Challenges

Therein is the tragedy. Each man is locked into a system that compels him to increase his herd without limit – in a world that is limited. Ruin is the destination toward which all men rush, each pursuing his own best interest in a society that believes in the freedom of the commons. Freedom in a commons brings ruin to all.
"The Tragedy of the Commons", Hardin (1968)

3.1 Coase Theorem

3.1.1 Definition of Coase Theorem

Ronald Coase introduced the Coase Theorem in his paper "The Problem of Social Cost" (Coase, 1960), a cornerstone in understanding the economic origins of sustainability challenges. The theorem was a response to conventional economic perspectives on externalities. These views typically advocated for government intervention, like taxation or regulation, to tackle problems such as pollution, noise, and other adverse effects of economic activities. Coase challenged this perspective by proposing that under conditions of well-defined property rights and zero transaction costs, private parties could engage in costless bargaining over the allocation of resources and resolve externalities issues independently. This insight is crucial because it shifts the focus from government intervention to the importance of property rights and voluntary agreements in achieving efficient resource allocation.

The idea was initially inspired by Coase's earlier research on the nature of firms and transaction costs, particularly in his 1937 paper, "The Nature of the Firm" (Coase, 1937). His interest in how business and market function led him to explore the broader implications of transaction costs and property rights. Coase enriched his theorem during his study of radio frequency regulation (Coase, 1959). Competing radio stations broadcasting on the same frequencies caused interference. Regulators

L. Zhang, *Sustainability Economics*, https://doi.org/10.1007/978-3-031-87829-9_3

Fig. 3.1 Ronald Coase. *Source* Coase-Sandor Institute for Law and Economics, University of Chicago Law School

faced the challenge of eliminating this interference and efficiently allocating frequencies. Coase proposed that with clearly defined property rights for frequencies, it wouldn't matter if neighboring stations interfered by using the same frequency band. Additionally, it was irrelevant who received the property rights. Coase reasoned that the station with the potential for increased economic gain from broadcasting would compensate the other station to prevent interference. The Coase Theorem became widely recognized after Coase's discussions with colleagues at the University of Chicago, including George Stigler, who later coined the term "Coase Theorem" to summarize the key insight of Coase's work (Stigler, 1966). This theorem has since become a cornerstone of law and economics, influencing both theoretical frameworks and practical approaches to addressing externalities. In 1991, Coase was awarded the Nobel Prize in Economic Sciences for contributions to the understanding of transaction costs and property rights (Fig. 3.1).

The first notable aspect is the importance of defining property rights. In the Coase Theorem, property rights are the rights to use, transfer, and derive income from a particular resource. These rights must be clearly assigned to either an individual or a group to facilitate negotiation and trade. For instance, if a factory is emitting pollution, defining the property rights involves specifying who has the right to clean air—the factory has the right to emit, or the residents have the right to be free from pollution. When these rights are clearly delineated, they create a framework within which parties can negotiate to mutual advantage.

The second notable aspect is the assumption of zero or very low transaction costs. Transaction costs encompass all the costs associated with buying, selling, and negotiating the terms of a resource exchange. These include the time, effort, and money spent on drafting and enforcing contracts, as well as the costs of information gathering to assess the value of the resource. The Coase Theorem assumes these costs are minimal, allowing parties to negotiate agreements freely without obstacles. In reality, transaction costs can be significant, which may prevent efficient resolution as envisioned by the theorem.

The Coase Theorem significantly impacts environmental policy and sustainability. It proposes that with low transaction costs and well-defined property rights, parties involved in an externality can negotiate a solution that maximizes resource value, irrespective of initial rights assignment. This occurs because the party valuing the resource more will be motivated to purchase the other party's rights. For instance, if a factory pollutes a nearby residential area and residents have the right to clean air, they can negotiate with the factory to lower emissions. If the cost of reducing pollution is less than the compensation the factory would have to pay for the damage, it is in the factory's interest to reduce emissions. This process continues until the resource (in this case, clean air) is allocated to the party that values it most, leading to an efficient outcome.

Critics argue that the Coase Theorem's assumptions are often not met in the real world. First, property rights may be ambiguous. In practice, property rights are not always well-defined or easily enforceable. When these rights are ambiguous, it can be difficult for parties to engage in meaningful negotiations. For example, in the case of air pollution from the factory, it may be unclear who has the right to clean air. This ambiguity can prevent effective bargaining and lead to ongoing conflicts.

Second, transaction costs can be high. These costs include direct expenses for negotiating and drafting agreements, as well as the opportunity costs of time and resources spent in these negotiations. They can involve legal fees, the costs of gathering information, and the potential for disputes and litigation. High transaction costs can render it prohibitively expensive for parties to come to an agreement, even if the agreement would be mutually beneficial.

Third, the theorem does not address the distributional aspects of resource allocation, focusing solely on efficiency. The distributional aspects refer to how the costs and benefits of resource allocation are distributed among different parties. In other words, the theorem does not address who should bear the burden of the externality or who should receive the compensation for it. To illustrate this with the example of air pollution, let's consider the distributional implications. If the factory holds the right to pollute and the residents must compensate the factory to lower emissions, the financial responsibility shifts to the residents. This could be seen as inequitable, especially if the residents are from a low-income community and cannot afford to pay for the pollution reduction. Conversely, if the residents have the right to clean air and the factory must pay for the pollution it emits, the cost is borne by the factory, potentially affecting its profits and the returns to its shareholders.

Despite these criticisms, the Coase Theorem offers a valuable framework for analyzing how clear property rights and low transaction costs are able to enable private solutions to environmental issues.

3.1.2 Example of Coase Theorem

Let's consider an example to demonstrate the application of the Coase Theorem. Imagine a factory is located near a residential neighborhood. The noise generated by the factory's operations disturbs the residents, reducing their quality of life and potentially lowering property values. Assuming low transaction costs and straightforward negotiations, the factory and the residents can achieve a voluntary agreement in line with the Coase Theorem. First, the property rights regarding noise pollution must be clarified. If the residents have a well-defined right to quiet enjoyment of their homes, they are able to either request compensation from the factory for tolerating the noise or invest in noise reduction measures. These measures might include installing sound barriers, upgrading machinery to operate more quietly, or adjusting operating hours to minimize disturbances. Conversely, if the factory owns the property rights, the residents can pay for noise abatement costs.

The Coase Theorem states that as long as the parties can negotiate freely and the initial allocation of property rights is clear, an efficient solution will emerge where the resource (in this case, quiet) is allocated to its highest valued use. However, it's important to note that the theorem does not address the issue of equity—that is, who should bear the cost of the externality. In this example, the residents might be expected to bear the cost of the noise if they are the ones who value quiet more highly, even though they did not create the noise in the first place.

In practice, the Coase Theorem's application is often limited by high transaction costs and the difficulty of defining and enforcing property rights. Residents may have trouble organizing collectively to negotiate with the factory, and the factory may have market power that gives it an advantage in negotiations. Additionally, the legal system may not always be able to clearly define and enforce the property rights necessary for the Coase Theorem to operate effectively.

Case

The Coase Theorem and Air Pollution

Polluter paying: In 2002, American Electric Power purchased all 90 houses in Cheshire, Ohio, prompting all 221 residents to leave due to health concerns over fly ash emissions from the nearby coal-fired Gavin Power Plant. Homeowners received compensation significantly above market value.

Pollutee paying: Japan invested over $500 million in energy efficiency and clean coal projects across seven Asian countries. Due to concerns about sulfur emissions from China affecting Japan, the Cleaner Coal Program was launched to promote desulfuriztion technologies for coal-fired power plants. This program included training, technical assistance, and equipment provision. The fact that desulfurization techniques were not adopted by power plants outside the program indicates that it was the Japanese funding, rather than Chinese concerns about air pollution, that led to the installation of scrubbers.

Source: Deryugina et al. (2021).

Let's consider another example of the theorem in different cases of property rights allocation. In a quiet neighborhood, two neighbors, Sam and Rose, find themselves in a predicament involving Sam's beloved dog, Spot. Sam is a dog lover who cherishes the companionship of his energetic canine, while Rose is a writer who values her peace and quiet to concentrate on her work. The story unfolds as the barking of Spot becomes a point of contention between them, leading to a series of negotiations that exemplify the economic principle known as the Coase Theorem.

In the first scenario, Sam has the legal right to keep Spot, his exuberant dog. The joy Spot brings to Sam is worth $500 to him, but Spot's barking causes Rose a disturbance costing her $800 in lost productivity and peace of mind. The socially efficient outcome, which minimizes the total cost to society, would be for Spot to be removed. However, in a private negotiation, Rose offers Sam $600 to part with Spot, which is more than Sam's valuation and less than Rose's cost. Both parties agree, and the private outcome aligns with the efficient outcome, demonstrating how private bargaining can lead to a socially optimal result.

In the second scenario, Sam's attachment to Spot has deepened, and he now values the dog at $1000. Rose's cost due to Spot's barking remains at $800. The socially efficient outcome is for Spot to stay, as Sam's benefit now exceeds Rose's cost. However, Rose is unwilling to pay more than $800, and Sam refuses to accept less than $1000. No transaction occurs, and Spot stays, which is also the efficient outcome. This case shows that when the valuations change, the private market can still reach an efficient outcome without a transaction, as long as the parties' valuations are respected.

In the third scenario, the legal right to peace and quiet is held by Rose. Sam's benefit from having Spot is $800, and Rose's cost due to the barking is $500. The socially efficient outcome is for Sam to keep Spot, as his benefit is greater than her cost. Sam offers to pay Rose $600 to tolerate Spot's barking, which is more than Rose's cost and less than Sam's benefit. This private transaction results in Spot staying, which matches the efficient outcome. This case illustrates that even when the initial rights are with the party affected by the externality, private negotiations can still lead to an efficient allocation of resources.

These scenarios illustrate a key principle of the Coase Theorem: the private market achieves an efficient outcome regardless of the initial distribution of rights. As long as there are clear property rights and transaction costs are minimal, the parties will negotiate to a consequence that maximizes the total benefit, aligning private incentives with social efficiency. This demonstrates the power of private negotiations in reaching efficient resource allocation without the need for external intervention, as long as the conditions of the Coase Theorem are met.

3.1.3 Resource Allocations

Sustainable development is fundamentally a matter of resource allocation in economics, where efficient resource allocation is crucial for addressing environmental problems. The Coase Theorem offers an approach for recognizing that resources can be allocated efficiently through party negotiations. The dimensions of resource allocation are multifaceted, encompassing both intrageneration and intergeneration considerations.

Intragenerational resource allocation refers to the distribution and use of resources within the current generation. It focuses on ensuring that resources are used efficiently and equitably among different groups and individuals living today. Recall the previous example of factory and air pollution, the negotiation leads to an efficient outcome where pollution is reduced to the level where the marginal cost of further reduction equals the marginal benefit to the residents.

In Fig. 3.2, the red line represents the marginal cost (MC) of reducing pollution. This cost increases as more pollution is reduced because it becomes progressively more expensive to eliminate additional units of pollution. On the other hand, the blue line represents the marginal benefit (MB) to the residents from reducing pollution. This benefit decreases as more pollution is reduced because the initial reductions have the most significant impact on health and well-being. The green dashed lines intersect at the point where the MC curve meets the MB curve. This intersection represents the optimal level of pollution reduction. At this point, the cost of reducing one more unit of pollution equals the benefit gained from that reduction. This balance ensures that resources are used efficiently. Policymakers can use this framework to determine the most efficient level of pollution control. By aiming for the point where MC equals MB, they can ensure that environmental regulations provide the maximum benefit at the lowest cost. The Coase Theorem supports this by suggesting that clear property rights and minimal transaction costs facilitate efficient outcomes. The process also addresses equity by ensuring that those affected by pollution are compensated or that the polluter bears the cost of reducing emissions. This fair distribution of costs and benefits promotes social justice within the current generation.

Intergenerational resource allocation involves the distribution and use of resources across different generations. It focuses on ensuring that the actions of the current generation do not compromise the ability of future generations to meet their needs. The temporal dimension of resource allocation pertains to the trade-offs among short-term and long-term benefits, as well as intergenerational equity. Sustainable development requires that we consider the implications of our resource use not only for the present generation but also for future generations. This involves preserving natural resources, maintaining ecological balance, and ensuring that future generations have access to the same resources and opportunities that we enjoy today.

For example, industrial activities today emit greenhouse gases, contributing to global warming and climate change. These emissions have long-term impacts, affecting the environment and the well-being of future generations. Suppose future

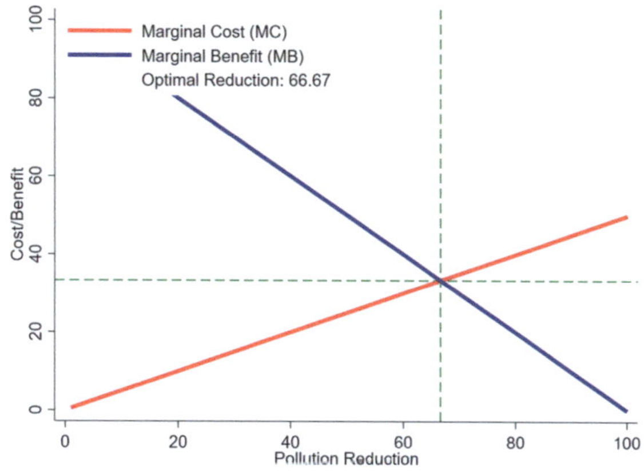

Fig. 3.2 Optimal pollution reduction: where marginal cost equals marginal benefit

generations have the right to a stable climate. Current generations (governments, businesses, and individuals) can negotiate agreements to reduce emissions. For instance, countries might agree to international treaties that set emission reduction targets. These agreements may incorporate systems such as carbon trading, in which nations or firms that cut emissions below objectives can sell carbon credits to those who exceed them. This provides a financial incentive to cut emissions where they are most cost-effective. The negotiation and trading of carbon credits lead to an efficient allocation of resources, where emissions are reduced in the most cost-effective manner. This ensures that the environmental benefits of reduced emissions are maximized while minimizing economic costs. The method also considers equity, ensuring that the costs of lowering emissions are properly distributed among current and future generations. By setting emission reduction targets and allowing for carbon trading, the burden of addressing climate change is distributed in a way that considers the needs of both present and future generations.

Economists study the intertemporal allocation of natural resources using various models and frameworks to understand how resources can be efficiently and sustainably managed over time. Here are some key approaches and concepts they use.

First, dynamic optimization models are used to analyze how resources should be allocated over time to maximize social welfare. These models consider the trade-offs between current and future consumption of resources. For example, the Hotelling Rule is a principle that guides the optimal extraction rate of non-renewable resources, suggesting that the price of these resources should rise at the rate of interest over time (Hotelling, 1931).

Imagine a company that owns an oil field with a total reserve of 1000 barrels of oil. The company wants to maximize its profit from extracting and selling this oil over the next 10 years. According to the Hotelling Rule, the company should consider the present price of oil, the extraction cost, and the interest rate. Let's assume the current market price of oil is \$100 per barrel, the extraction cost is \$20 per barrel, and the interest rate is 5% per year. The Hotelling Rule suggests that the price of

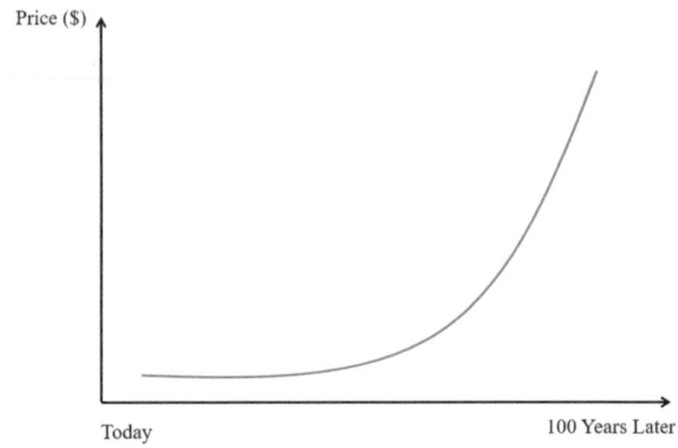

Fig. 3.3 An illustration of the Hotelling rule

oil should rise at the rate of interest over time. This means that if the price of oil is \$100 per barrel today, it should be \$105 per barrel next year (an increase of 5%). In the first year, the company might decide to extract 100 barrels of oil, earning a profit of \$8000 (100 barrels × (\$100–\$20)). In the second year, the price of oil is expected to be \$105 per barrel. The company should adjust its extraction rate so that the marginal profit from extracting an additional barrel of oil equals the interest rate. By following this principle, the company ensures that it is extracting oil at an optimal rate, balancing current profits with future profits, and taking into account the rising price of oil due to the interest rate. We provide a simple illustration of the Hotelling rule in Fig. 3.3.

Second, Overlapping Generations (OLG) models consider multiple generations of consumers and producers, each with their own preferences and constraints. These models help economists understand how resource allocation decisions made by one generation affect future generations.

Consider a simple scenario where a community relies on a shared resource, such as a forest. Each generation consists of young and old individuals. The young generation, during their productive years, has the option to either harvest trees from the forest for immediate economic gain or leave them to grow, benefiting future generations. The old generation, on the other hand, relies on the forest for its environmental services, such as clean air and biodiversity. The decision made by the young generation to harvest or conserve trees affects not only their immediate economic well-being but also the quality of life for the old generation and the availability of the resource for future generations. This model highlights the intergenerational trade-offs inherent in managing natural resources, where current actions have long-term consequences. Sustainable management requires balancing the immediate needs of the current generation with the long-term interests of future generations, ensuring that the resource remains viable and beneficial for all.

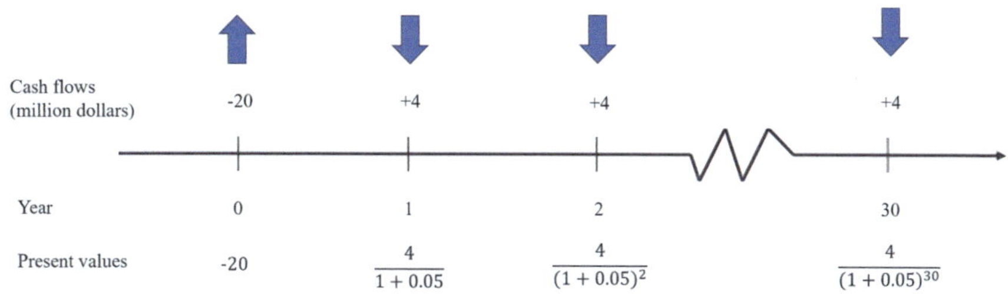

Fig. 3.4 An illustration of the discounting approach

Third, discounting is another method used to compare the value of resources or benefits received at different time. Using a discount rate, economists can determine the present value of future benefits and costs, helping to make decisions that combine short-term and long-term goals. This is crucial for evaluating investments in renewable energy or conservation projects.

Imagine a government debating whether to invest in a new solar power facility, as illustrated in Fig. 3.4. The initial cost of building the plant is significant, but it promises long-term benefits such as reduced carbon emissions and lower electricity costs over time. To decide whether this investment is worthwhile, economists use discounting to compare the present value of these future benefits with the initial costs. Suppose the solar power plant needs an upfront investment of \$20 million. The plant is estimated to save \$4 million per year on electricity bills over the following 30 years. These savings will occur in the future; therefore, their current worth is not equal to their nominal value. To account for this, economists use a discount rate that reflects the value of money over time and the opportunity cost associated with capital.

Let's suppose a 5% discount rate. Using this rate, the present value of the \$4 million annual savings can be calculated. For example, the present value of the savings in the first year is \$4 million divided by $1 + 0.05$, which equals approximately \$3.8 million. For the second year, it would be \$4 million divided by $(1 + 0.05)^2$, and so on. By summing up these present values for each year, the total present value of the savings over 30 years can be determined. If the total present value of the savings exceeds the original investment of \$20 million, the project is considered economically viable. The discounting approach allows decision-makers to compare the immediate costs with the long-term benefits, ensuring that the investment aligns with both short-term financial constraints and long-term sustainability goals. Discounting thus provides a crucial tool for evaluating projects like renewable energy investments, helping to balance immediate expenses with future environmental and economic benefits.

In all, intragenerational considerations focus on the fair and efficient use of resources within the current generation, while intergenerational considerations ensure that future generations are not disadvantaged by today's resource use. This balance is crucial for sustainable development, promoting both economic efficiency and social equity.

3.2 Externalities and Sustainability

3.2.1 *Definition of Market Failure*

In neoclassical economics, market failure arises when a free market's distribution of goods and services is not Pareto optimal, leading to a net loss of economic value. Although the word was first used by economists in 1958 (Francis, 1958), it originated with Victorian philosopher Henry Sidgwick (Medema, 2007).

In the context of sustainability, market failure frequently results in the overuse of natural resources and environmental deterioration. Take overfishing of fisheries as an example, as illustrated in Fig. 3.5. Suppose the red line represents the total cost of fishing effort. As fishing effort increases, the total cost also increases linearly. The blue line represents the total revenue from fishing efforts. Initially, total revenue increases with fishing effort, but after reaching a peak, it starts to decline due to overfishing and reduced fish stocks. The Maximum Economic Yield (MEY) occurs at the point where the difference between total revenue and total cost is maximized. This is the level of fishing effort that provides the highest economic profit. Additionally, the Maximum Sustainable Yield (MSY) occurs at the point where total revenue is maximized. This is the level of fishing effort that provides the highest sustainable catch without depleting the fish stock. Notably, the intersection of the total revenue curve and the total cost curve represents the Economic Breaking Point (EBP). Further increases in fishing efforts beyond the EBP lead to economic losses as costs exceed revenues. Without regulation, individual fishers have an incentive to catch as many fish as possible, which exceeds the fishing effort at MSY, leading to market failure.

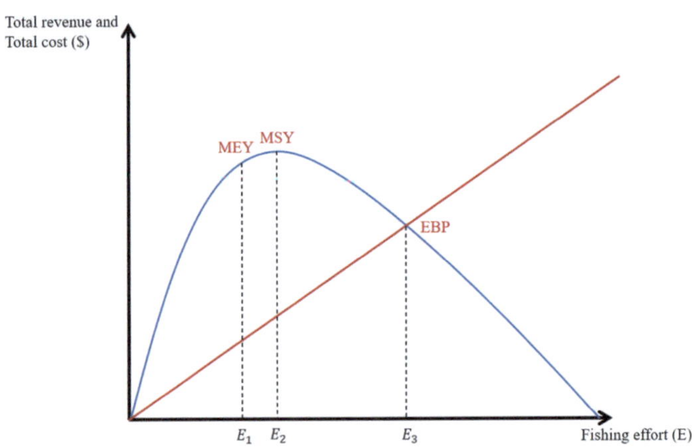

Fig. 3.5 Illustration of overfishing

Divergent views exist among economists regarding the reasons behind market failure. According to conventional economic analysis, there are three basic causes of market failure. First, a small number of companies with substantial market power control or monopolize the market. Second, externalities, external costs or benefits, are produced during the creation of a good or service. Third, the good or service is a public good.

3.2.2 Externalities

Externalities are unintended outcomes of economic activities that impact third parties who do not participate directly in the transaction. The consequences may be either positive or negative. Positive externalities arise when the actions of individuals or firms have a positive impact on third parties that are not represented in market prices. Consider a company that produces vaccines. In Fig. 3.6, the blue curve represents the marginal private cost of the company. Additionally, the orange curve represents the marginal private benefit, while the green curve represents the marginal social benefit of the vaccination. Without externalities, the market equilibrium point is where the marginal cost curve intersects with the marginal private benefit curve (Q^1). However, due to positive externalities, the marginal private benefit is lower than the marginal social benefit, leading to a socially optimal point (Q^*) that is higher than the market equilibrium point (Q^1). This discrepancy results in underproduction ($Q^1 < Q^*$).

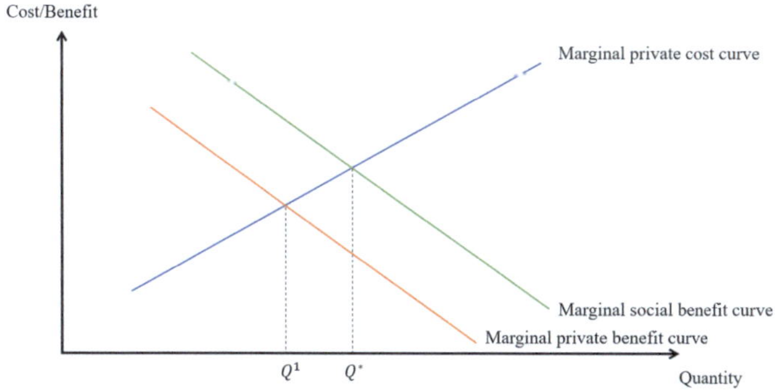

Fig. 3.6 Illustration of positive externalities

Case

Some Examples of Positive Externalities

- Education
  - **Private Benefit**: Individuals gain knowledge and skills that can increase their earning potential and personal development.
  - **Social Benefit**: An educated population can lead to a more informed and productive society, lower crime rates, and greater civic engagement. For example, higher levels of education are associated with increased innovation and economic growth.
- Vaccination
  - **Private Benefit**: Individuals who get vaccinated are protected from certain diseases.
  - **Social Benefit**: When a significant portion of the population is vaccinated, herd immunity is achieved, reducing the spread of infectious diseases and protecting those who cannot be vaccinated. This leads to overall improved public health and reduced healthcare costs.
- Public Parks and Green Spaces
  - **Private Benefit**: Individuals can enjoy recreational activities, exercise, and relaxation in public parks.
  - **Social Benefit**: Green spaces improve air quality, reduce urban heat islands, and provide habitats for wildlife. They also enhance the aesthetic appeal of neighborhoods and can increase property values.
- Research and Development (R&D)
  - **Private Benefit**: Companies that make investments in R&D can create new products and technologies, gaining a competitive edge.
  - **Social Benefit**: Innovations from R&D can result in technological developments that benefit society as a whole, such as medical breakthroughs, improved communication technologies, and more efficient energy sources.
- Public Transportation
  - **Private Benefit**: Individuals using public transportation save on personal transportation costs and avoid traffic congestion.
  - **Social Benefit**: Reduced traffic congestion, lower greenhouse gas emissions, and decreased air pollution benefit the entire community. Public transportation also provides mobility for those who cannot drive, enhancing social inclusion.

These instances illustrate how positive externalities can result in broader societal benefits beyond the immediate advantages to individuals or firms. Encouraging activities with positive externalities can enhance overall social welfare.

Negative externalities, in contrast, are costs placed on third parties without their consent, such as air pollution from a plant that harms surrounding neighbors' health.

In environmental contexts, externalities often manifest as environmental degradation that is not fully accounted for in market transactions. For instance, a company that emits pollution as a byproduct of its production process may not incur the entire expense of the environmental damage or health issues caused by that pollution. This misalignment between private costs and social costs can lead to the overproduction of polluting goods and underinvestment in environmental stewardship. In Fig. 3.7, the blue curve represents the marginal private cost of the company during production. In addition, the orange curve represents the marginal social cost. The green line is the marginal benefit curve for the product. In the absence of externalities, the market equilibrium point is where the marginal private cost intersects with the marginal benefit curve (Q^1). However, due to negative externalities, the marginal social cost is higher than the marginal private cost, leading to a socially optimal point (Q^*) that is lower than the market equilibrium point (Q^1). This discrepancy results in overproduction ($Q^1 > Q^*$).

Greenhouse gas emissions from the burning of fossil fuels are a well-known example of negative externalities in environmental contexts. Figure 3.8 shows the annual global carbon dioxide (CO_2) emissions from 1940 to 2024. These emissions contribute to climate change, which affects ecosystems, agriculture, and human health. These pollutants have a huge social cost, yet they are rarely reflected in the price of gasoline or electricity.

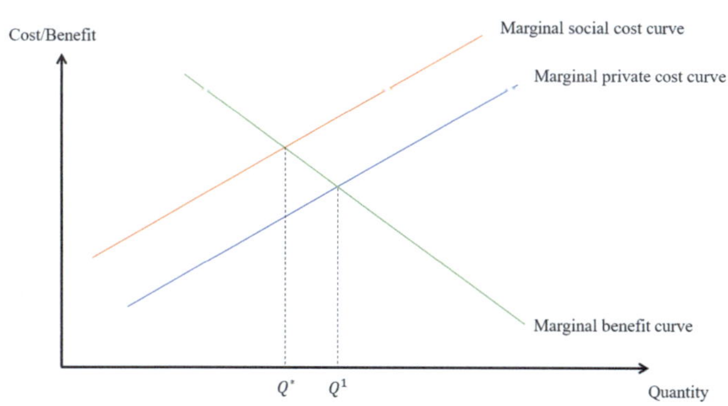

Fig. 3.7 Illustration of negative externalities

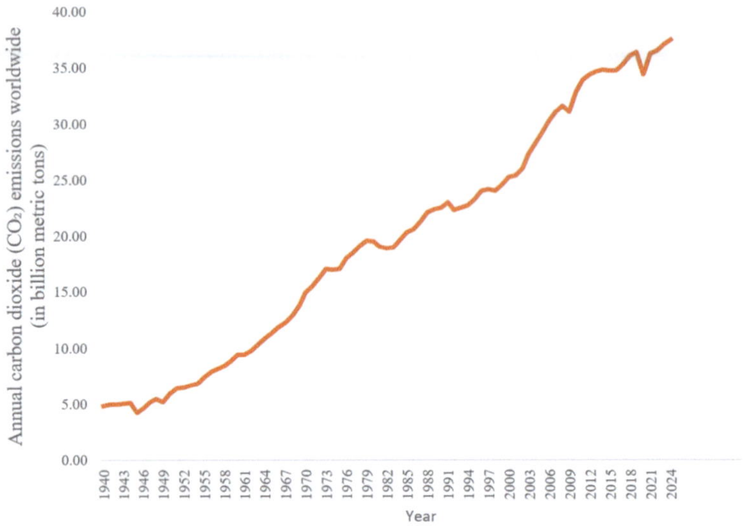

Fig. 3.8 Annual global carbon dioxide (CO_2) emissions from 1940 to 2024. *Notes* Data sources from Global Carbon Project

Deforestation is another example of a negative externality. Figure 3.9 illustrates the global change in forest area in 2015. The clear-cutting of forests for timber or agricultural land results in the loss of biodiversity, disruption of water cycles, and increased carbon dioxide in the atmosphere. While the timber company or farmer may benefit from the immediate economic gain, the broader society and environment bear the long-term costs.

Understanding externalities is crucial for policymakers seeking to promote sustainability. By recognizing that markets often fail to compensate for all of the social costs and benefits of economic activities, policymakers can design interventions to correct these failures.

3.2.3 Impact on Sustainable Development

Negative externalities significantly impede progress toward sustainability by creating a mismatch between private costs and social costs. When a firm emits pollution as a byproduct of production, for example, the immediate cost to the firm may be low, but the broader social cost—including health impacts, environmental degradation, and long-term ecological damage—can be substantial. These social costs are overlooked in the market price of the firm's product, leading to an overproduction of goods that generate negative externalities and an undervaluation of environmental quality. The unchecked release of pollutants into the atmosphere, such as carbon dioxide from burning fossil fuels, contributes to climate change, which poses a severe threat to

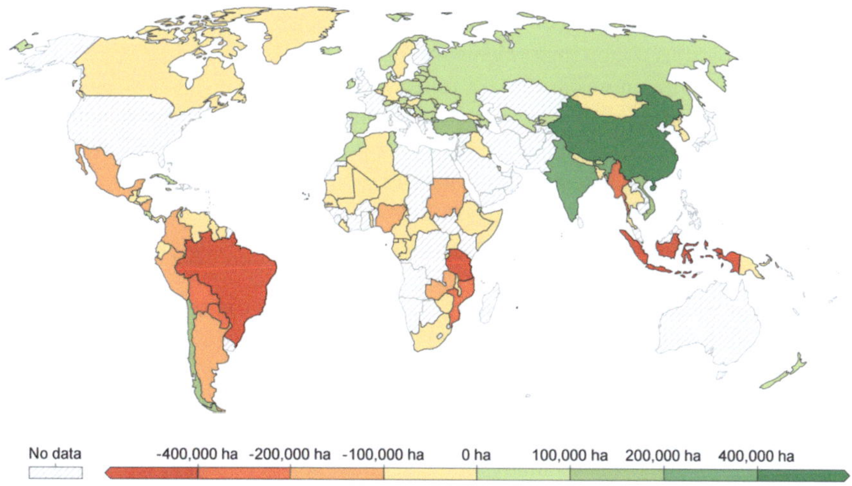

Fig. 3.9 Annual change in forest area, 2015. *Notes* Data sources from UN Food and Agriculture Organization (FAO). The map is plotted by Our World in Data (2024)

global sustainability. This is because climate change can lead to rising sea levels, more frequent and severe weather events, shifts in agricultural productivity, and the loss of biodiversity. These consequences have profound implications for food security, human health, and the overall stability of ecosystems.

On the other hand, positive externalities can promote sustainable practices by providing additional benefits to society. For instance, the planting of trees in urban areas not only improves air quality but also provides shade, reduces urban heat island effects, and enhances the well-being of residents. However, because these benefits are not fully captured by the private entities that plant the trees, there may be an underinvestment in such sustainable activities. Positive externalities play an important role in promoting sustainable practices, yet they are often overlooked. For instance, a company that invests in renewable energy technology may not only reduce its own carbon footprint but also contribute to the development of a more sustainable energy sector, which can have long-term benefits for the environment and the economy.

3.2.4 *Mitigation Strategies*

To address the challenges posed by externalities to sustainable development, various mitigation strategies have been implemented or proposed. These strategies aim to internalize externalities, ensuring that the costs or benefits of economic activities are fully implied in market prices. By doing so, the goal is to align private incentives with social and environmental objectives, thereby promoting more sustainable outcomes. There are numerous policy instruments for internalizing externalities.

Taxes: One of the most direct ways to internalize externalities is through the use of taxes. For instance, a carbon tax imposes a financial burden on entities that emit greenhouse gases, effectively making them pay for the environmental harm they cause. This provides an incentive for firms and individuals to decrease emissions, as they will face higher costs if they do not. Taxes can also be applied to other negative externalities, such as pollution or resource extraction, to discourage overuse and encourage more sustainable practices.

Tradable Permits: Tradable permits' systems, such as cap-and-trade, impose a maximum limit on emissions for a particular pollutant. Companies are given or must purchase permits to emit up to a specific quantity, which they can then exchange with one another. The system creates a market for pollution rights, encouraging companies to cut emissions below their permitted cap, as they can then sell any surplus permits. The financial incentive drives innovation and efficiency in pollution reduction.

Subsidies: Subsidies can be used to promote positive externalities, such as investing in renewable energy or sustainable agriculture. By reducing the cost of sustainable alternatives, subsidies can make these options more competitive and encourage their adoption. Additionally, subsidies can be utilized to fund green technology research and development, fostering innovation that can lead to long-term environmental benefits.

3.3 Public Goods and the Tragedy of the Commons

3.3.1 Public Goods Versus Private Goods

Understanding the distinction between public goods and private goods is crucial for addressing sustainability challenges, as these categories of goods exhibit different characteristics that influence resource allocation and management.

Non-excludability and non-rivalrous consumption are the two key characteristics that define public goods. Non-excludability implies that it is difficult or impractical to prevent individuals from accessing the good once it is provided. For example, clean air and national defense are classic examples of public goods. Once clean air is available, it benefits everyone in the vicinity, and no one can be excluded from enjoying its benefits. Non-rivalrous consumption indicates that one person's use of the good does not diminish its availability to others. For instance, one person enjoying a fireworks display does not prevent others from enjoying it as well. Figure 3.10 indicates that allowing another individual to benefit from a pure public good has no marginal cost. In addition, increasing the level of public good has a positive marginal cost.

Non-rivalry implies that when a number of individuals have a demand curve for a public good, these demand curves are added vertically to create the public

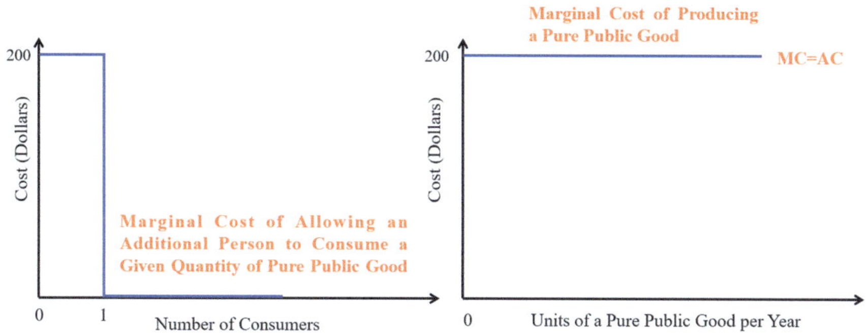

Fig. 3.10 Marginal costs for provision of public goods

good's aggregate demand curve. In contrast, individual demand curves are added horizontally to determine the aggregate demand for a private good. Figures 3.11 and 3.12 provide a graphical interpretation of the optimal provision of both private and public goods. There are two individuals: Kitty and Abby. In the private goods (cake) market, individual marginal benefit curves are summed horizontally to derive the social marginal benefit (i.e., aggregate demand) curve. This means adding the quantities demanded by each individual at each price level. The market equilibrium is achieved where the social marginal cost curve intersects the social marginal benefit curve. In the public goods (missile) market, individual marginal benefit curves are summed vertically to determine the social marginal benefit (i.e., aggregate demand) curve. This involves adding the willingness to pay off each individual at each quantity level. The socially optimal provision of the public good is where the social marginal costs and benefits are equal, known as the Samuelson rule (Samuelson, 1954).

These characteristics lead to a unique challenge known as the "free-rider problem." Because individuals cannot be excluded from using public goods, there is little incentive for them to contribute to their provision. This often results in underinvestment in public goods, as private entities may not find it profitable to produce them. When a community depends on a public park for recreation, for instance, people could decide

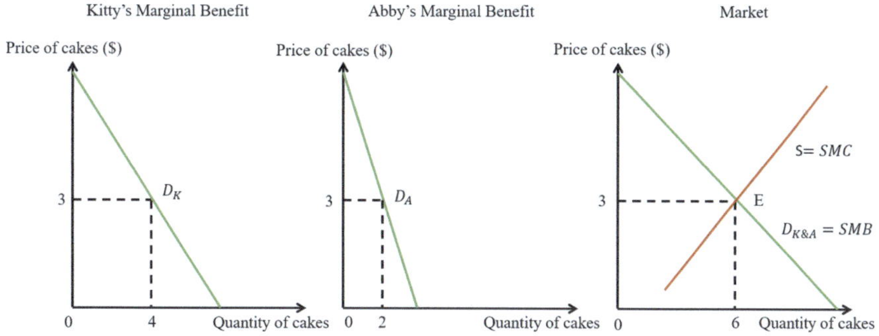

Fig. 3.11 Optimal provision of private goods

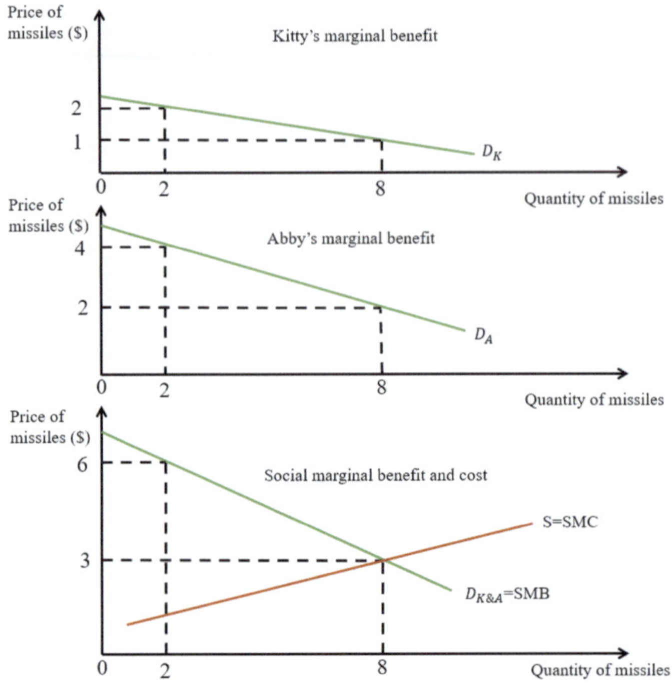

Fig. 3.12 Optimal provision of public goods

not to help maintain it because they believe that others would cover the expenses. As a result, public goods can become underfunded and overused, leading to degradation and diminished quality.

In contrast, private goods are characterized by excludability and rivalry. Excludability means that individuals can be prevented from using the good if they do not pay for it. For example, a sandwich sold in a store is a private good; only those who purchase it can consume it. Rivalrous consumption is the phenomenon wherein the usage of a good by one individual diminishes its availability for others. If one person eats the sandwich, it is no longer available for someone else. This market structure allows for efficient allocation through price mechanisms, as consumers are willing to pay for goods they desire, and producers respond to demand by supplying those goods.

The distinction between public and private goods has significant implications for sustainability. Public goods, such as clean air, biodiversity, and climate stability, require collective action and investment to maintain, yet they are often neglected in market systems due to their non-excludable nature. This neglect can lead to environmental degradation, as the benefits of preserving public goods are not captured in market transactions. To address these challenges, policymakers must implement strategies that encourage the provision and maintenance of public goods. This can include government funding, regulations, and community engagement initiatives aimed at promoting collective responsibility for shared resources. For instance, governments can allocate tax revenues to fund the maintenance of public parks, invest in renewable energy infrastructure, or establish conservation programs to protect biodiversity.

Example

Is Education a public good or a private good?

Education is a service that possesses both private and public goods traits.

As a public good, education serves a socializing purpose that enables us to live in a civic society. It emphasizes the value of cooperating, obeying commands, and abiding by the rules. Furthermore, it equips students with fundamental abilities like following instructions and being on time, which boosts their productivity as employees. Additionally, education aids students in assessing their abilities and selecting suitable careers, raising a country's level of productivity.

On the other hand, education also has characteristics of a private good. There are wide disparities in the quantity and quality of education provided among different school districts. The level of support that parents can give students at home increases with income, and this home support is a crucial factor in children's learning. Furthermore, there is no way to prevent parents who desire more than a standardized quantity and quality of education for their children from purchasing it in the marketplace.

3.3.2 The Tragedy of the Commons

The tragedy of the commons is a compelling concept in environmental economics that illustrates the unsustainable use of shared resources when individual self-interest is not checked by collective action or regulation. This concept was popularized by Garrett Hardin (Hardin, 1968), who used the example of a shared pasture to explain how individuals acting independently according to their own self-interest can deplete a shared resource. The tragedy of the commons describes a situation where individuals, each pursuing their own self-interest, deplete a shared resource. It arises from the fact that individuals have an incentive to use a common resource as much as possible, while the cost of overuse is shared among all users. This leads to overexploitation and, in extreme cases, the collapse of the resource. The term "commons" refers to any shared resource, whether it is a physical entity like a forest or fishery, or an environmental service such as clean air or climate stability.

Sustainability and other environmental challenges are related to the tragedy of the commons. Many resource issues in modern civilization, such as those involving water, forests, fish, and non-renewable energy sources like coal, oil, and gas, can be modeled by the commons dilemma. One common illustration of the tragedy of the commons is climate change. According to this perspective, natural resources have been depleted on Earth, the commons, without taking into account the externalities that affect nearby and future populations. Environmental deterioration is still a result of the combined efforts of individuals, organizations, and governments. Strict limits or alternative remedies are needed to mitigate the long-term effects and approach tipping points, which may negatively affect different industries.

3.3.3 Solutions to the Tragedy of the Commons

Formulating solutions to the tragedy of the commons is a significant challenge. On the one hand, local communities successfully implement some social schemes that work well. On the other hand, various governmental solutions can be considered, such as regulation, privatization, internalizing externalities, and population control.

Non-governmental solutions

- **Cooperation among individuals**: Because collective restraint benefits both the group and the individual, even self-interested people can discover methods to work together (Axelrod & Hamilton, 1981). This cooperation can emerge as people recognize the long-term benefits of sustainable practices. For example, in many traditional societies, communal management of resources has been successful due to strong social norms and mutual monitoring.
- **Community management**: Local communities often develop effective ways to manage common resources (Ostrom et al., 1999). For instance, Swiss farmers have successfully managed a shared commons since 1517, balancing individual and collective benefits. Several key factors contribute to successful community management, including clearly defined boundaries, effective monitoring, graduated sanctions for rule violators, and mechanisms for conflict resolution. These principles have been applied to various resources, such as fisheries, irrigation systems, and forests, demonstrating that local knowledge and social capital are crucial for sustainable resource management.
- **Altruistic Punishment**: Altruistic punishment involves individuals punishing those who do not cooperate, even at a personal cost. This can motivate cooperation and address free-rider problems (Greenwood, 2016). Research has indicated that punishment might serve as a powerful incentive to sustain group collaboration (Fehr & Gächter, 2002). For example, in some traditional societies, social ostracism or fines are used to enforce communal rules and ensure sustainable resource use. This approach relies on strong social norms and a sense of collective responsibility.

Case

Legal Personhood for Nature

Rivers and other natural phenomena can be protected by being given legal rights. According to O'Donnell and Talbot-Jones (2018), the Whanganui River in New Zealand has been given legal personhood, which enables it to be represented in court. This creative method guarantees that the natural entity's interests are taken into account when making decisions. This legal framework seeks to prevent environmental deterioration and encourage sustainable resource usage by acknowledging the inherent value of nature.

The Whanganui River flows 290 kilometers from the center of New Zealand's North Island to the Tasman Sea on the island's lower west coast. As a navigable river, the Crown (the New Zealand government) legally owns the riverbed, but local governments manage the river and catchment. Since 1873, Whanganui Māori have challenged the legislative framework founded on the Treaty of Waitangi, pursuing ownership claims through both statutory and informal avenues. To settle these disputes, the Te Awa Tupua (Whanganui River Claims Settlement) Act 2017 was passed in March 2017, after eight years of negotiations between the Whanganui Iwi (tribe) and the Crown. This Act confers legal personhood on the Whanganui River and its catchment, providing a new governance framework.

This new structure is consistent with the Whanganui Iwi worldview, which sees the river as a living creature that stretches from the mountains to the sea, incorporating both physical and spiritual elements. The Settlement Act recognizes the river and its watershed as a legal entity with all of the rights, functions, obligations, and liabilities of a legal person.

Source: O'Donnell and Talbot-Jones (2018)

Governmental Solutions

- **Regulation**: Governments can impose regulations to limit the use of common resources. Examples include permit systems for activities like fishing, logging, and pollution controls. These regulations help ensure that resource use remains within sustainable limits. For instance, the Clean Air Act in the United States sets limits on air pollution, helping to reduce the negative externalities associated with industrial activities. Similarly, fishing quotas and protected areas help prevent overfishing and preserve marine ecosystems.
- **Privatization**: Converting common resources into private property can incentivize owners to manage them sustainably. However, this approach can also lead to social and economic issues, as seen in historical examples like the Enclosure Acts in England. Privatization must be carefully managed to avoid negative consequences, such as the displacement of local communities or the concentration of resources

in the hands of a few. In some cases, a combination of private and communal ownership can provide a balanced solution, ensuring both efficiency and equity.

- **Internalizing externalities**: Common resources can be managed more effectively if users are held accountable for all the consequences of their use. For example, gasoline taxes can be used to pay for air pollution and road upkeep. Policymakers can promote more sustainable behavior by internalizing the external costs of resource usage, such as pollution or habitat destruction. Examples include carbon pricing schemes that make polluters pay for their environmental impact, like carbon taxes or cap-and-trade schemes, in an effort to lower greenhouse gas emissions.
- **Population control**: Garrett Hardin proposed that overpopulation could be addressed through "mutual coercion, mutually agreed upon," which might involve restricting reproductive rights. While this idea is controversial and has been criticized for its ethical implications, some countries have implemented population control measures to address resource scarcity. For example, China's one-child policy, although now relaxed, was an attempt to curb population growth and reduce pressure on natural resources. However, such policies must be carefully designed to respect human rights and avoid unintended social consequences.

We provide some specific solutions for sustainable resource use with respect to the tragedy of the commons, such as market-based solutions, education and awareness, technological innovation, and collaborative governance.

First, implementing economic incentives, such as taxes, subsidies, or tradable permits, can encourage sustainable resource use by aligning individual incentives with collective well-being. For instance, the Emissions Trading System (ETS) of the European Union is a cap-and-trade scheme that permits companies to purchase and sell emission allowances while imposing a limitation on greenhouse gas emissions. This market-based strategy gives companies the freedom to cut emissions economically while still meeting environmental objectives.

Second, raising awareness about the consequences of overusing common resources can lead to more responsible behavior and support for sustainable practices. Environmental education programs, public awareness campaigns, and community engagement projects can all aid in educating people about the value of resource conservation and sustainable living. For example, campaigns to reduce plastic waste have encouraged people to use reusable bags and bottles, reducing the environmental impact of single-use plastics.

Third, developing and adopting new technologies can help mitigate the tragedy of the commons by improving resource efficiency and reducing environmental impact. For instance, advancements in renewable energy technologies, such as solar and wind power, provide alternatives to fossil fuels and help reduce greenhouse gas emissions. Similarly, precision agriculture techniques can optimize the use of water, fertilizers, and pesticides, minimizing their impact on the environment while maintaining agricultural productivity.

Fourth, effective management of common resources often requires collaboration among multiple stakeholders, including governments, businesses, non-governmental organizations, and local communities. Collaborative governance frameworks, such as multistakeholder partnerships and public–private partnerships, can facilitate cooperation and coordination among different actors. For example, the Global Environment Facility (GEF) brings together governments, international organizations, and civil society to address global environmental challenges, such as biodiversity loss and climate change.

We also offer several examples, including fisheries management, water resource management, forest conservation, and climate change mitigation.

- **Fisheries Management**: As described in Sect. 3.2.1, individual fishers have an incentive to catch as many fish as possible, leading to the depletion of fish stocks. Successful fisheries management often involves a combination of regulatory measures, community-based management, and market-based approaches. For example, the Alaskan halibut fishery uses a system of individual transferable quotas (ITQs), which allocate specific catch limits to individual fishers. This system has helped stabilize fish stocks and improve the economic viability of the fishery.
- **Water Resource Management**: Water scarcity is another example of the tragedy of the commons, where overuse and pollution of water resources can lead to severe environmental and social consequences. Integrated water resource management (IWRM) is an approach that promotes the coordinated development and management of water, land, and related resources. IWRM involves stakeholder participation, cross-sectoral coordination, and the use of economic instruments to promote sustainable water use. For example, the Murray-Darling Basin in Australia uses a combination of water trading, regulatory measures, and community engagement to manage water resources in a sustainable manner.
- **Forest Conservation**: Deforestation and forest degradation are major environmental issues that exemplify the tragedy of the commons. Community-based forest management (CBFM) is an approach that involves local communities in the management and conservation of forests. CBFM has been successful in many countries, such as Nepal and Mexico, where local communities have been given legal rights to manage forest resources. These communities have developed sustainable forest management practices, such as selective logging and reforestation, which help conserve biodiversity and improve livelihoods.
- **Climate Change Mitigation**: Reducing greenhouse gas emissions and lessening the effects of climate change is a worldwide issue that calls for cooperation from everybody. The goal of international agreements like the Paris Agreement is to unite nations in order to establish emission reduction goals and carry out climate action plans. Key instruments for encouraging emission reductions include market-based systems like carbon pricing and emissions trading. Additionally, the shift to a low-carbon economy depends on investments in renewable energy and technological innovation.

Nevertheless, there are several challenges for solutions to the tragedy of the commons, such as equity and justice, institutional capacity, and behavioral change. First, solutions must consider issues of equity and justice, ensuring that the benefits and burdens of resource management are distributed fairly. This is particularly important in the context of global environmental challenges, where vulnerable communities and developing countries often bear the brunt of environmental degradation. In order to advance social and environmental justice and safeguard the rights and interests of marginalized people, policies and interventions must be designed. Second, effective management of common resources requires strong institutions and governance frameworks. Building institutional capacity, particularly in developing countries, is essential for implementing and enforcing sustainable resource management practices. This includes strengthening legal and regulatory frameworks, enhancing monitoring and enforcement mechanisms, and promoting transparency and accountability. Third, addressing the tragedy of the commons requires changing individual and collective behavior. This involves not only raising awareness and providing information but also creating incentives and enabling conditions for sustainable behavior. Behavioral economics and social psychology offer insights into how to design interventions that can influence behavior, such as nudges, social norms, and peer pressure.

References

Axelrod, R., & Hamilton, W. D. (1981). The evolution of cooperation. *Science, 211*(4489), 1390–1396.

Coase, R. H. (1937). The nature of the firm. *Economica, 4*(16), 386–405.

Coase, R. H. (1959). The federal communications commission. *The Journal of Law and Economics, 2*, 1–40.

Coase, R. H. (1960). The problem of social cost. *Journal of Law and Economics*, 1–44.

Fehr, E., & Gächter, S. (2002). Altruistic punishment in humans. *Nature, 415*(6868), 137–140.

Francis, B. (1958). The anatomy of market failure. *Quarterly Journal of Economics, 72*(3), 351–379.

Greenwood, G. W. (2016). Altruistic punishment can help resolve tragedy of the commons social dilemmas. In *2016 IEEE Conference on Computational Intelligence and Games.*

Hardin, G. (1968). The tragedy of the commons. *Science, 162*(3859), 1243–1248.

Hotelling, H. (1931). The economics of exhaustible resources. *Journal of Political Economy, 39*(2), 137–175.

Medema, S. G. (2007). The hesitant hand: mill, sidgwick, and the evolution of the theory of market failure. *History of Political Economy, 39*(3).

Our World in Data. (2024). *Forests and deforestation.* https://ourworldindata.org/deforestation

Ostrom, E., Burger, J., Field, C. B., Norgaard, R. B., & Policansky, D. (1999). Revisiting the commons: Local lessons, global challenges. *Science, 284*(5412), 278–282.

Samuelson, P. A. (1954). The pure theory of public expenditure. *The Review of Economics and Statistics, 36*(4), 387–389.

Stigler, G. (1966). *Theory of price* (3rd edn.). Macmillan Library Reference.

Part II
Energy-Related Sustainability Issues

Chapter 4
Energy Supply

We know we'll run out of dead dinosaurs to mine for fuel and have to use sustainable energy eventually, so why not go renewable now and avoid increasing risk of climate catastrophe? Betting that science is wrong and oil companies are right is the dumbest experiment in history by far.
Elon Musk, Founder, Tesla

4.1 Energy Resources

4.1.1 Types of Energy Resources

Energy is ubiquitous. From giant machines like airplanes and trains to small gadgets like bulbs and watches, the modern lifestyle is sustained by consuming various forms of energy. The broader scope of "energy" includes various other forms. For example, in Physics, kinetic energy is defined as the energy an object possesses due to its motion. Sustainable Economics, however, focuses on the sustainability issues associated with particular energy forms that are identified as resources and hold economic value, such as coal, oil, and hydropower. We can directly perceive our consumption of energy products through activities such as receiving gasoline station receipts, electricity bills, and natural gas invoices. We also indirectly utilize and consume energy because any modern industrial processes and products, such as smelting, manufacturing, and transportation, rely on energy use. Energy is such a crucial resource that modern life without it is unimaginable. Achieving sustainable utilization of energy resources is a key goal of sustainable economics. Energy may come from renewable or non-renewable resources. Non-renewable energy resources refer to coal, petroleum (oil), natural gas, propane, and nuclear fuel (Fig. 4.1). The designation of these resources as "non-renewable" stems from their finite availability on Earth and the irreversible nature of their utilization process. For example, crude oil is derived from ancient marine organisms and plants that underwent a lengthy natural process spanning millions of years for its creation, rendering its quick replenishment improbable.

L. Zhang, *Sustainability Economics*, https://doi.org/10.1007/978-3-031-87829-9_4

Fig. 4.1 Non-renewable energy

After extraction, crude oil is refined to produce aviation fuel, gasoline, diesel, and various other products, which are then combusted for their intended purposes. Oil, coal, and natural gas are also known as fossil fuels. Non-renewable energy resources are depletable and will eventually run out. Since the *Industrial Revolution*, fossil fuels have been extensively used, leading to pollution and carbon emission issues, posing a significant threat to sustainable development. Nuclear energy is produced through nuclear fission or fusion processes, which consume nuclear fuels, such as uranium and plutonium. Thus, nuclear is also classified as non-renewable energy. While the use of nuclear fuel does not result in air or water pollution, there is a potential risk of nuclear leaks.

Reading

Carbon Dioxide Emissions Factors of Different Fuels

When comparing the carbon emissions associated with different fossil fuels, it is evident that not all energy sources contribute equally to greenhouse gas emissions. Figure 4.2 illustrates the carbon intensity of various fossil fuels, emphasizing the importance of understanding these differences in the context of climate change mitigation strategies. For instance, while coal has a higher carbon intensity compared to natural gas and oil, the latter two still contribute significantly to overall emissions. Recognizing and addressing these variations in carbon output is crucial for developing informed policies aimed at transitioning toward a more sustainable energy landscape.

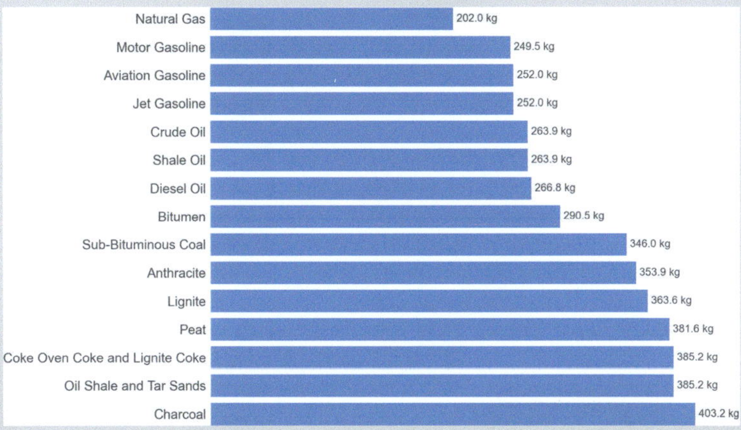

Fig. 4.2 Carbon dioxide emissions factors of different fuels. *Note* Adapted from Our World in Data (2024a)

Renewable energy resources, such as wind, solar, hydropower, and geothermal energy, are inexhaustible sources in nature (Fig. 4.3). We can convert these energy sources into utilizable kinetic or electrical energy with specific devices. For instance, traditional windmills can transform wind energy into mechanical energy for grinding grains, while modern wind turbines can generate electricity. Solar panels directly convert solar energy into power. Hydroelectric and tidal power stations convert the kinetic energy of rivers and oceans into electricity. Utilizing renewable energy sources emits no pollution or greenhouse gases directly. However, the devices used to harness these energy sources may have certain environmental impacts, such as noise from large wind turbines, resource consumption and emissions from manufacturing solar panels, and the potential disruption of fish migration caused by hydroelectric dams.

You may have noticed that some common forms of energy are not classified as either renewable or non-renewable energy sources, such as electricity. Another way of categorizing energy sources may help us better understand the characteristics of these different energy forms. More generally, we can classify different forms of energy into primary energy and secondary energy based on the degree of processing applied to the energy. Primary energy is the energy form that exists directly in nature without any processing, such as coal, crude oil, and natural gas. Secondary energy is the product of refining primary energy resources. For example, gasoline and diesel are secondary energy forms as they are refined from crude oil. Hydrogen is a form of secondary energy because we can produce hydrogen by splitting water. Secondary energy is a product of the energy industry and is not widely found in nature. Electricity is also secondary energy. Although lighting is quite common in nature, unfortunately, we have not yet found a way to collect and store lightning as an utilizable resource. Electricity is one of the most familiar forms of energy for us. In modern society, a significant amount of energy is converted into electricity and distributed through the

Fig. 4.3 Renewable energy

ubiquitous grid to households to power appliances and devices. Thus, electricity is sometimes referred to as "tertiary energy."

4.1.2 Energy Production: A Global Trend

The twentieth century witnessed a shift in the world's primary sources of energy. Coal fueled the Industrial Revolution and was the main energy source for producing steam, electricity, and steel. Until the end of the first half of the twentieth century, coal remained the primary energy source globally. The modern oil industry began in the late nineteenth century and experienced exponential growth thereafter. This period also saw the rapid rise of the oil refining industry, leading to the success of industrialists such as John D. Rockefeller, who founded the Standard Oil Company in 1870 and at one point controlled 90% of the American oil market. Figure 4.4 illustrates the trends in annual global coal, oil, and gas production since the twentieth century. All energy units are standardized to terawatt-hours (TWh). Following World War II, global energy demand surged. By the 1960s, oil surpassed coal to become the world's primary source of energy, a position it maintains to this day.

It is noteworthy that coal production in the twenty-first century has exhibited a significant polarized growth trend, as seen in Fig. 4.5. Coal combustion, such as in coal-fired power plants for electricity generation, is a major source of air pollution in many countries, emitting pollutants like sulfur dioxide, nitrogen oxides, and particulate matter. Consequently, coal has gained a notorious reputation, leading to the phasing-out of coal combustion in some high-income countries. *The Great Smog of London* in December 1952 was one of the most severe environmental disasters in

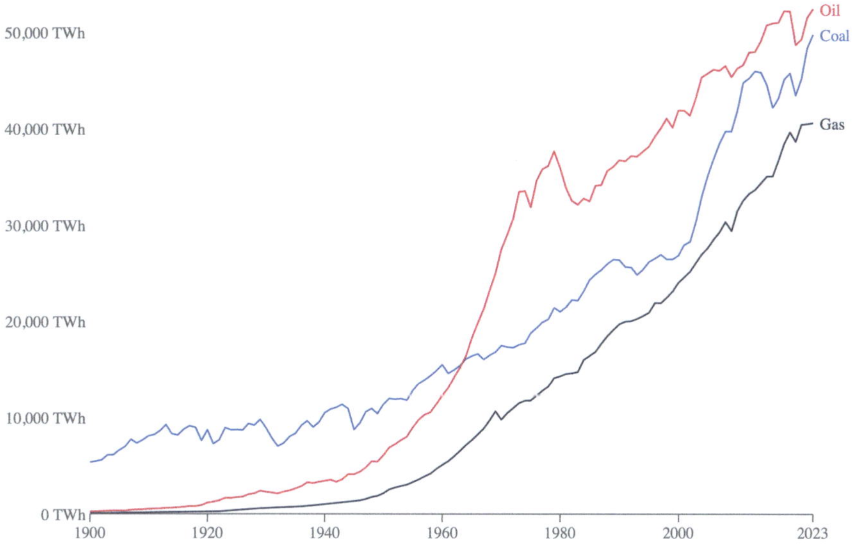

Fig. 4.4 Global trends in non-renewable energy production. *Data source* EIA (2024)

British history, with thousands of deaths attributed to air pollution caused by coal combustion. Following this event, the UK revised several regulations, such as the Clean Air Act of 1956, to reduce air pollution from coal combustion. Germany's coal production has experienced a significant decline since the 1990s, while the United States has seen a similar trend since the 2010s.

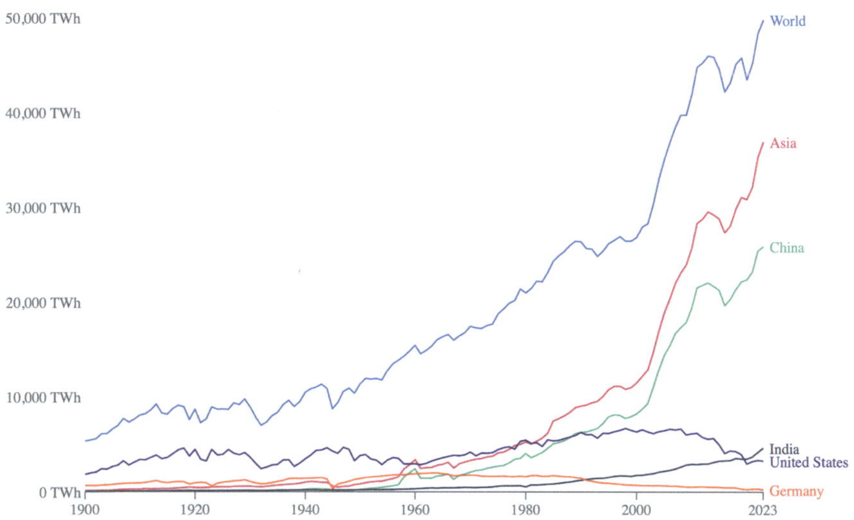

Fig. 4.5 Trends in global annual coal production. *Data source* EIA (2024)

However, in the twenty-first century, developing countries' economies and energy demands, particularly in Asian countries like China and India, have been rapidly growing. The abundance of coal resources, lower prices, and lower technological barriers have made coal the primary energy source driving the industrial growth of these nations. Consequently, we have witnessed a resurgence in coal production in the 2000s. Despite numerous studies confirming that air pollution from coal combustion leads to health issues and shortens the lives of residents, unfortunately, we are still observing the "pollute first, clean up later" approach in some emerging economies (Lelieveld et al., 2015). In the subsequent chapters, we will take a closer look at the nexus of economic growth, energy, and sustainable development.

Oil is the most crucial energy source in the world today. However, the distribution of oil resources globally is uneven. Figure 4.6 illustrates the oil production in different countries in 2023. The United States, Saudi Arabia, and Russia are the top three oil-producing countries, with each country accounting for over 10% of the global oil production. OPEC is a permanent intergovernmental organization established at the Baghdad Conference in September 1960. Countries such as Saudi Arabia, Iran, Iraq, the United Arab Emirates, and Kuwait are among the 12 member countries of this organization. OPEC coordinates its member countries' oil production, policies, and prices. OPEC's oil production represents 38% of global production and significantly impacts global oil prices. Despite significant demand for oil in some countries, limited reserves necessitate importing oil from other countries, such as China and India.

Natural gas is a common gaseous energy source primarily composed of methane, often found in the same geological formations as crude oil and extracted alongside crude oil. In addition to natural gas, other common gaseous energy sources include liquefied petroleum gas (LPG) and biogas. LPG is a byproduct of petroleum refining,

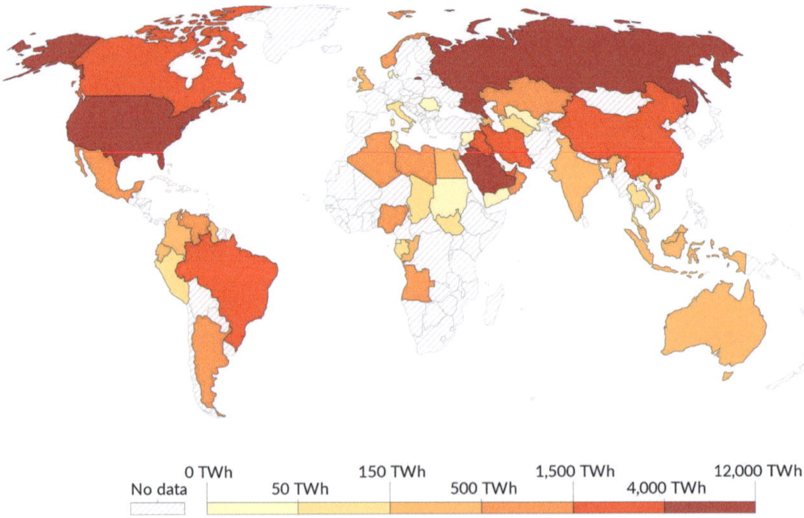

Fig. 4.6 Global oil production in 2023. *Note* The map is plotted by Our World in Data (2024b)

while biogas can be produced through the fermentation of agricultural waste, urban waste, and sludge, making it a renewable energy source. Gas can be transported and stored through pipelines, liquefaction, or compression, providing high flexibility in usage and transportation. Gas is widely used in power generation, heating, cooking, and industrial production. The main components of these gaseous energy sources are hydrocarbons like methane, with the combustion byproducts being carbon dioxide and water. Compared to coal and oil, the combustion of gaseous energy sources produces less carbon dioxide and harmful waste, resulting in a smaller environmental impact. Therefore, gaseous energy sources like natural gas are also referred to as "transition fuels." In many countries, replacing coal and oil with gas is considered a clean energy policy.

Figure 4.7 shows that, after 2000, the supply of renewable energy has grown rapidly. Hydropower is the most significant clean energy source, providing over 4000 TWh of energy annually (10% of global gas production). Emphasis on sustainability, policy incentives, and the emergence of new technologies have accelerated solar and wind energy utilization in the past two decades. Large-scale solar and wind farms have quickly been established around the world and have provided a significant amount of green and clean energy for the global economy. From the energy supply perspective, the growth rate of renewable energy far exceeds that of fossil fuels, which is encouraging and signifies a profound energy transition rapidly becoming a reality. However, we must be reminded that the pace of global warming depends on the stock of greenhouse gases. The continued growth of global fossil fuels implies the need for greater achievements in the transition to renewable energy to achieve net-zero greenhouse gas emissions ultimately.

4.2 Energy Reserves and Extractions

4.2.1 Definition

Not all energy resources are accessible. *Energy resources* refer to the total estimated amount of energy that can be extracted from a source, including some undiscovered, hypothetical, or inaccessible resources. Energy resources encompass the energy sources assumed to exist or technically impossible to extract or economically unfeasible to extract. *Energy reserves* are defined as discovered energy sources of known quantity that can be extracted and commercialized. Therefore, energy reserves provide a more conservative measure of the value of energy assets based on current technology and market conditions.

The World Petroleum Council classifies reserves into three categories based on the probability of extraction (Etherington & Ritter, 2008):

- Proved (P): probability of extraction > 90%,
- Proved and probable (2P): probability of extraction > 50%,
- Proved, probable, and possible (3P): probability of extraction > 10%.

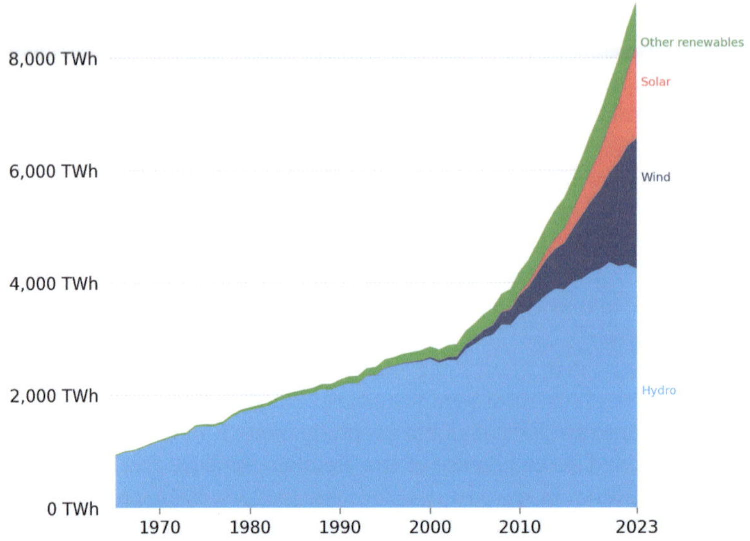

Fig. 4.7 Global production of renewable energy. *Note* Adapted from Our World in Data (2024c), original data from EIA (2024)

There exist other classification systems used to categorize resources and reserves. For example, the Petroleum Resources Management System (PRMS) divides petroleum resources into four categories: Reserves, Contingent Resources, Prospective Resources, and Unrecoverable Resources. These concepts may be useful in different contexts. Energy economists and geologists sometimes lean toward using the concept of "resource" rather than "reserve." Due to technological advancements or changes in energy prices, some resources may be reclassified as reserves. This means that changes in reserves are influenced by more factors than just the energy source deposited underground. For prudent reporting, the U.S. Securities and Exchange Commission (SEC) requires companies to disclose their holdings of energy assets using P-reserves as the standard.

4.2.2 Global Distribution of Energy Reserves

Energy reserves, determined by the energy sources deposited underground, are entirely a "roll of the dice." Fig. 4.8 illustrates the global distribution of three resources: coal, oil, and natural gas. As revealed by global energy production, the distribution of global energy reserves is highly uneven. North America, Northeast Asia (Russia and China), and Australia have abundant fossil energy reserves. The United States has the largest coal reserves in the world, reaching 2.3 trillion tons. The *Appalachian Basin* in the eastern United States is the largest coal field in the world. China, the world's largest coal producer, has over 1 trillion tons of coal reserves (ranking third globally). In contrast, the coal reserves in Africa and South America are relatively small.

Venezuela in South America holds the world's largest oil reserves. According to the Energy Information Administration (EIA), Venezuela's proven oil reserves reach 304 billion barrels, accounting for 18% of global oil reserves (see Table 4.1), surpassing Saudi Arabia, which holds about 17% of global oil reserves. However, Venezuela's oil is predominantly heavy crude and is less favored in the international crude oil market. Moreover, these heavy crude oil reserves are buried deep, leading to high extraction and refining costs. Due to a lack of advanced extraction technologies and equipment, Venezuela's oil production ranked only 20th globally in 2023, with a daily output representing only about 6.4% of the United States. Apart from Venezuela and the Middle East, North America, North Africa, Russia, and China also possess relatively abundant oil reserves. Europe and the majority of Africa exhibit relatively sparse oil reserves. Natural gas deposits often accompany crude oil. Once again, natural gas reserves in Africa and Europe are relatively scarce, while Russia, North America, and the Middle East enjoy substantial natural gas reserves.

Unlike fossil fuels, the "reserve" of renewable energy is not underground but "above ground." In fact, we are more used to using "power potential" to describe the distribution of solar, wind, and hydroelectric resources. Figure 4.9 provides an example of a photovoltaic power potential map produced. Photovoltaic panels can directly convert solar energy into electricity. Therefore, photovoltaic power potential is mainly influenced by factors such as sunlight duration and intensity (latitude), terrain, climate, and more. According to data from the International Energy Agency (IEA), Africa possesses 60% of the world's best solar energy resources, yet its current solar power generation capacity accounts for only 1% globally, indicating significant untapped solar energy development potential.

Wind power potential forecasting is particularly crucial for decision-making in the construction of wind farms. In recent years, with rapid advancements in wind field observation and prediction technologies, the accuracy of wind power intensity estimates has significantly improved, leading to a rapid increase in the utilization rates of wind power in some countries or regions. In addition to the number of windy days, the magnitude and stability of the wind are also crucial factors. Due to their more open nature, offshore wind farms (wind power plants built offshore not far from the land) can install taller and larger turbines, operating within a wider range of wind speeds. Wind in the sea area faces no obstructions, resulting in higher stability. Therefore, in the wind power potential map in Fig. 4.10, offshore regions are highlighted as areas with the highest power intensity.

4.2.3 Energy Extraction and Reserve-to-Production Ratio

The reserve-to-production ratio (RP ratio) is the ratio between reserves and annual production, measured in years. It estimates the number of years that the reserve of resources will continue to be productive based on current production rates.

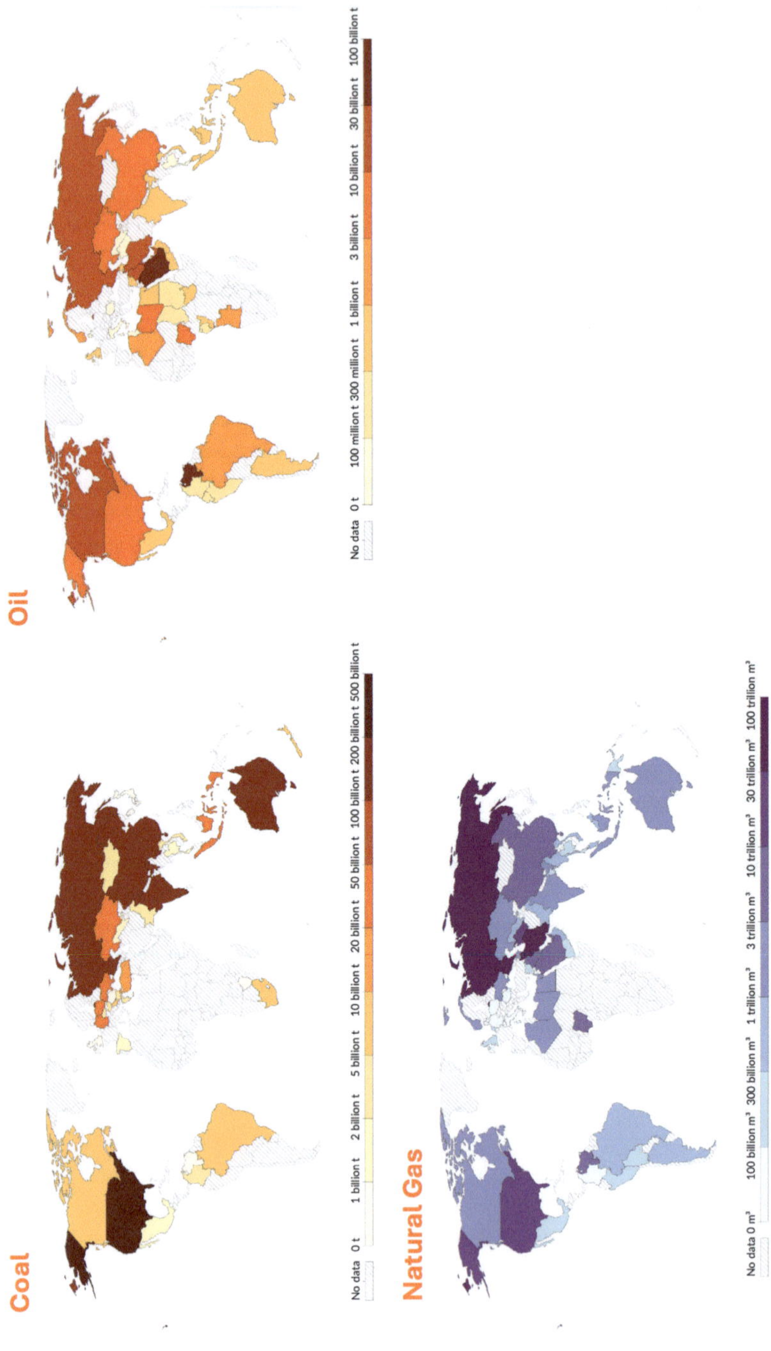

Fig. 4.8 Global distribution of coal, oil, and gas reserves. *Note* Adapted from the maps plotted by Our World in Data (2024d)

Table 4.1 Top 20 countries with the largest oil reserves

Country	Proven reserves (millions of barrels)	Country	Proven reserves (millions of barrels)
Venezuela	304,000	Nigeria	37,000
Saudi Arabia	267,000	Kazakhstan	30,000
Iran	209,000	Somalia	30,000
Canada	170,000	China	26,000
Iraq	145,000	Qatar	25,000
UAE	113,000	Brazil	13,000
Kuwait	102,000	Algeria	12,000
Russia	80,000	Guyana	11,000
United States	74,000	Namibia	11,000
Libya	50,000	Ecuador	8,300

Data source EIA (2024)

$$RP_t = \frac{\text{Reserves}_t}{\text{Production}_t} \tag{4.1}$$

Are fossil fuel reserves about to run out soon? Well, it depends. Table 4.2 displays the RP ratio of the top 20 countries with the largest oil production. Based on the 2023 oil production rates, the oil reserves of the United States are projected to be depleted before 2040. China's oil reserves are estimated to run out in 17 years. Mexico is expected to have no extractable oil in less than 10 years. Meanwhile, countries like Canada, Iraq, Iran, and Kuwait have RP ratios exceeding 90 years. However, this is just one aspect of the reality. Indeed, the RP ratio is influenced by many factors and changes annually.

If you observe the trend in global oil RP ratios since 1980, the RP ratios have even increased (see Fig. 4.11). In 1980, the global oil RP ratio was approximately 30 years. This implied that by 2010, global oil reserves would be depleted. However, by 2020, global oil reserves were still sufficient to sustain supply for close to 60 years at the extraction rate of that year, despite the daily oil production in 2020 being more than doubled than that of 1980.

Why is this the case? There may be several reasons:

- *New reserves*: Oil and gas explorers and other extractors are constantly identifying new natural resources. As the annual extraction of fossil fuels increases, more resources and reserves are also being discovered.
- *Technological advances and market conditions*: Energy reserves are also influenced by extraction technologies, equipment, and energy prices. With the emergence of new technologies, more energy resources are being reclassified as reserves due to technological or economic feasibility. The application of new exploration technologies has accelerated the discovery of new reserves. For example, 3D and 4D seismic imaging, utilizing advanced seismic instruments

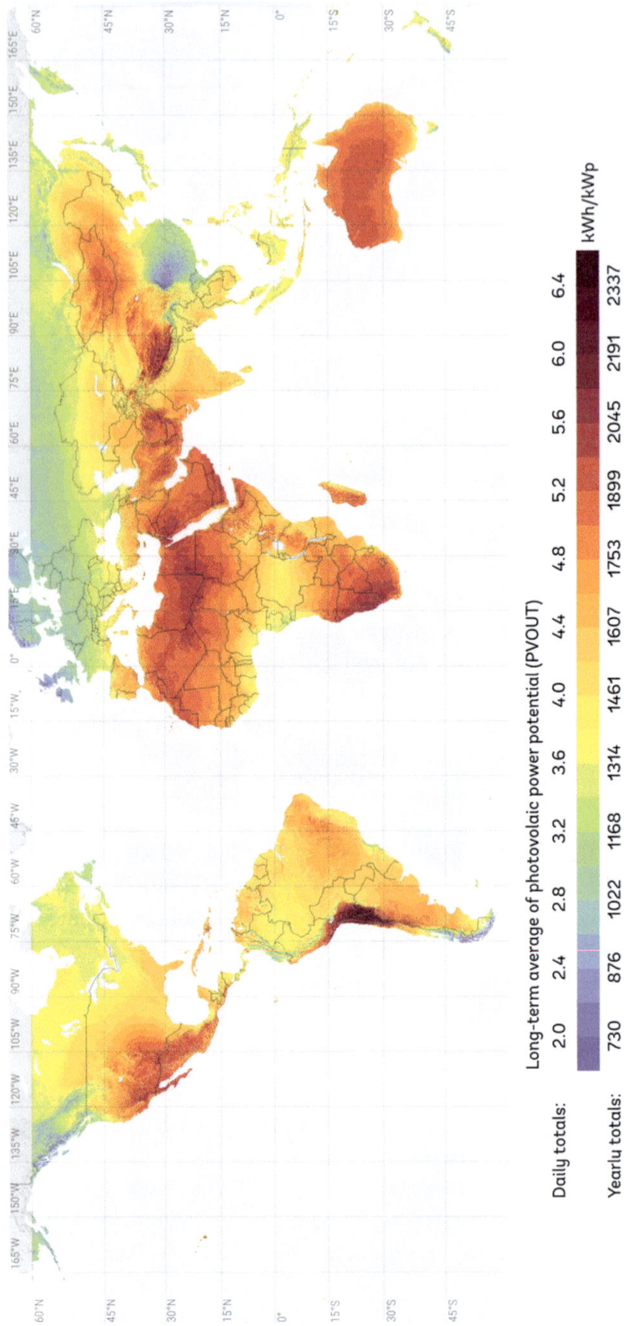

Fig. 4.9 Global distribution of photovoltaic power potential. *Note* The map is obtained from Solargis (2019)

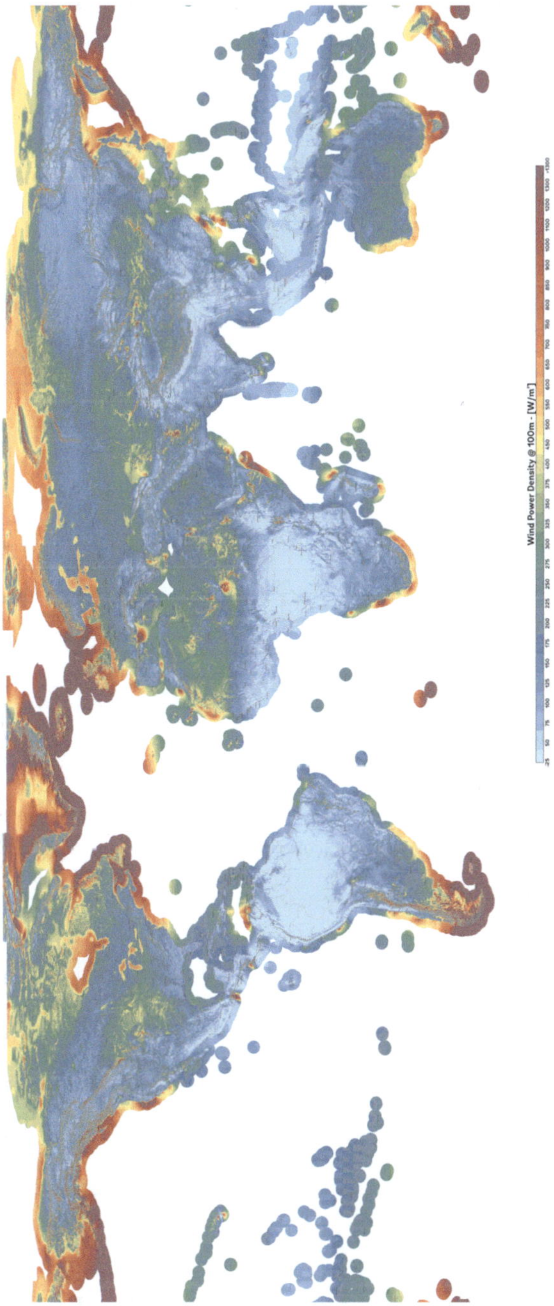

Fig. 4.10 Global wind power intensity potential. *Note* The map is obtained from the Technical University of Denmark (2020)

Table 4.2 RP ratio of the top 20 countries with the largest oil production

Country/ Region	Proven reserves (millions of barrels)	Oil production Dec 2023 (millions of bbl/day)	RP ratio
United States	74,000	13.295	15
Russia	80,000	10.126	22
Saudi Arabia	267,000	8.95	79
Canada	170,000	4.971	94
Iraq	145,000	4.375	91
China	26,000	4.172	17
Iran	209,000	4.084	140
Brazil	13,000	3.585	10
UAE	113,000	3.3	81
Kuwait	102,000	2.644	106
Norway	8,100	2.079	11
Mexico	6,000	1.903	8.4
Kazakhstan	30,000	1.897	43
Nigeria	37,000	1.573	64
Qatar	25,000	1.322	52
Libya	50,000	1.248	109.8
Angola	7,800	1.235	17
Algeria	12,000	1.16	28
Oman	5,400	1.04	14
Venezuela	304,000	0.852	978

Data source EIA (2024).

and computer simulations, help scientists identify newly proven reserves located miles below the seabed floor.

4.2.4 Energy Peaks

Peak oil is a critical point at which global oil extraction reaches its maximum level and subsequently begins an irreversible decline, symbolizing the depletion of oil reserves. Peak oil is also known as "*Hubbert's Peak.*" At the 1956 American Petroleum Institute (API) meeting, American geophysicist King Hubbert (1903–1989) presented his prediction of the peak of oil extraction in the United States. He forecasted that oil extraction in the United States would peak in 1970 and rapidly decline thereafter (see Fig. 4.12). If peak oil does occur, in the absence of sufficient alternative energy sources, fossil fuel prices may experience significant fluctuations and even trigger public panic.

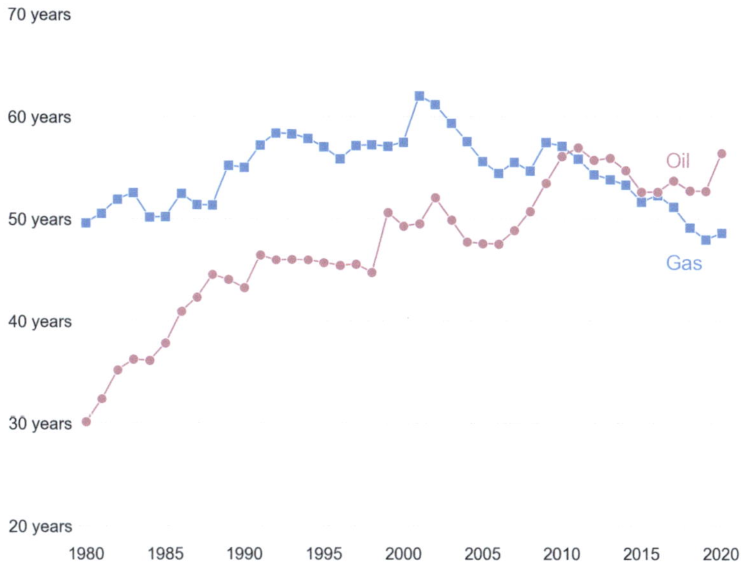

Fig. 4.11 RP ratio of global oil and gas from 1980 to 2020. *Note* Data source includes EIA (2024)

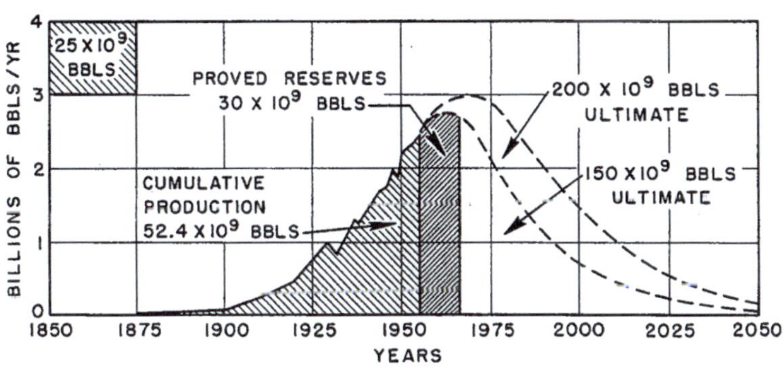

Fig. 4.12 Hubbert's, 1956 peak oil graph. *Source* Hubbert (1956)

The American people quickly realized that Hubbert's prophecy became a reality in the 1970s and 1980s (see Fig. 4.13). Although the discovery of Alaskan oil delayed the decline in U.S. oil production in the 1980s, the decrease in production was evident until just before 2008. However, after 2008, breakthroughs in horizontal drilling and hydraulic fracturing technologies led to a significant reduction in production costs for U.S. shale oil and gas, resulting in a sustained increase in production afterward. U.S. shale oil production has grown from less than 500,000 barrels per day in 2008 to the current scale of 4 million barrels per day, with the share of shale oil production in U.S. crude oil production expected to gradually increase in the future.

Fig. 4.13 Oil production in the United States since 1900. *Data source* EIA (2024)

The peak of fossil fuel extraction will eventually arrive, sooner or later. However, we may not have to wait for this peak to endure the catastrophic consequences of pollution and global warming resulting from the cumulative burning of fossil fuels. Therefore, some scholars argue that we should define the peak of fossil fuels based on consumption rather than production to better track and predict trends in carbon emissions.

4.3 Energy Storage and Grid

4.3.1 Energy Storage

Energy storage is a technology that stores energy for later use. Specifically, it involves converting electrical energy, which is difficult to store directly, into other forms of energy for storage.

Two main factors have driven the development and application of energy storage technologies. The first is the rapid increase in the demand for powering portable devices and emergency energy since the twenty-first century, leading to the rapid advancement of energy storage technologies such as batteries. The second is the rapid growth of renewable energy sources. Burning fossil fuels is controllable, allowing for smoothing power supply by adjusting fuel use. In contrast, the supply of wind, solar, and hydropower is determined by natural factors, resulting in supply instability. The instability of renewable energy sources is a key factor limiting their complete replacement of non-renewable energy sources. Energy storage technologies are crucial in overcoming this challenge.

The main technological pathways for energy storage include:

- *Electrochemical energy storage*, such as batteries. Batteries are common in daily life, such as button batteries and dry batteries. The batteries can also be huge. For instance, large data centers may be equipped with massive battery arrays to form uninterruptible power supply (UPS) systems to prevent power outages. The types of batteries include lead-acid batteries, nickel-metal hydride batteries, and lithium-ion batteries.
- *Gravity energy storage*, including pumped hydro storage and solid weight storage. Pumped hydro storage is the primary large-scale energy storage technology. It uses electricity during low-demand, low-price periods (such as at night) to pump water to a higher elevation, converting and storing electrical energy as gravitational potential energy. During high-demand, high-price periods, the stored water is released to generate electricity using hydropower turbines. Pumped hydro storage facilities are often paired with hydropower stations. Pumped hydro storage is a feasible large-scale, high-efficiency energy storage solution, with energy conversion efficiencies generally exceeding 70%. Gravity energy storage can also be achieved using solid weights by raising them to a height with an engine and then releasing them to drive the engine reversely to produce electricity.
- *Thermal energy storage.* This technology involves converting electrical energy into heat and storing it in thermal storage materials, which can then be converted back into electrical energy when needed.
- *Hydrogen energy.* Hydrogen storage technology involves storing energy in the form of hydrogen gas. The most common method of hydrogen production is the electrolysis of water, with the produced hydrogen gas being converted back into electricity through fuel cells. However, due to significant energy losses in electrolyzing, the actual efficiency of hydrogen storage is lower compared to other energy storage technologies.

Energy storage generates private profits, surplus, and increased societal welfare. Electricity demand exhibits significant cyclical variations. For example, electricity demand is generally higher during midday and lower during midnight; demand is higher in summer and winter (for cooling or heating) and lower in spring and autumn. Energy storage provides a strategy for intertemporal arbitrage, allowing energy storage capacity owners to purchase and store electricity when prices are low and sell and supply electricity when prices are high. This arbitrage, technically, provides stability to the electricity supply. Such arbitrage also stabilizes electricity prices (buying electricity at low prices raises prices; selling electricity at high prices lowers prices). Additionally, without energy storage systems, additional supply capacity must be provided to meet peak electricity demand. However, these for-peak-only capacities remain idle during non-peak periods and are costly. Energy storage systems offer a more economically efficient solution for addressing peak electricity demand, reducing peak and average electricity prices.

4.3.2 Electricity Grid

The electricity grid (or *electricity network*) is the power transmission network that connects power generators to electricity consumers (Fig. 4.14). Figure 4.15 illustrates how the power grid delivers electricity to customers. The electricity generated by power plants is first stepped up in voltage before being connected to the grid. The loss of power during the transmission of electricity through the grid is known as *line loss*. According to Ohm's Law, line loss is inversely proportional to voltage. Therefore, voltage is increased before long-distance transmission.

The grid then transmits electricity over long distances using alternating current (with a small portion in direct current for special purposes). Finally, substations lower the voltage of the grid to the voltage required by users and distribute it to customers. This part of the grid is also known as the distribution network. The grid is responsible for the transmission and distribution of energy. Without it, electricity systems and markets would be impossible to function. As a matter of fact, most power outages are not caused by power plant failures but by faults in the grid's transmission lines. As a

Fig. 4.14 Electric poles and transmission lines

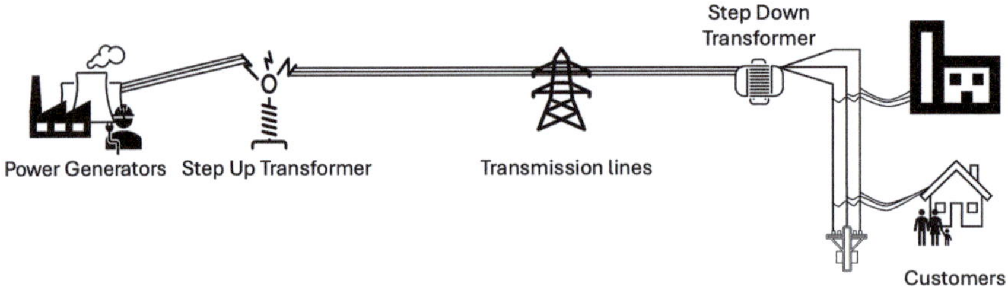

Fig. 4.15 Electricity grid system

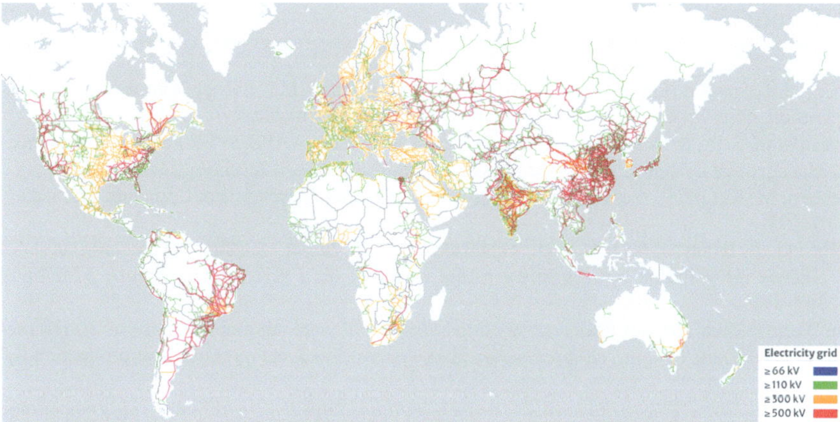

Fig. 4.16 Global power grid. *Note* The map is obtained from the Global Wind Atlas 3.3 (2024)

critical modern infrastructure, the density, efficiency, and reliability of the electricity grid are also considered metrics for regional economic development, particularly in developing countries. Figure 4.16 displays the distribution of contemporary global transmission networks. The electricity grid functions like a dense network of veins across the globe, continuously supplying energy to its customers.

Electricity travels at the speed of light in the grid. Therefore, electricity production and consumption must be synchronized. Modern grids also bear the responsibility of grid load management. When electricity demand (or electricity load) increases, the grid must dispatch power generators to increase power supply. Conversely, when the electricity load decreases, and the power supply exceeds demand, the grid must promptly regulate power generators to reduce power output. On a grid, we always have:

$$\text{Generation} = \text{Load}$$

Curtailment is a key challenge faced by the development of renewable energy, such as wind, solar, and hydropower. Let's take wind power as an example. When the grid's load is lower than the electricity supply, indicating no need for additional power, the wind may continue to blow. In this scenario, wind power generation may be curtailed as the power generation must equal the load, leading to the wastage of wind resources. Accelerating the development of energy storage technologies may help alleviate curtailment issues and accelerate the growth of renewable energy.

References

Energy Information Administration (EIA). (2024). *Opendata - U.S. Energy Information Administration.* https://www.eia.gov/opendata/index.php#bulk-downloads

Etherington, J. R., & Ritter, J. E. (2008). The 2007 SPE/WPC/AAPG/SPEE Petroleum Resources Management System (PRMS). *Journal of Canadian Petroleum Technology*, *47*(08).

Hubbert, M. K. (1956). *Nuclear energy and fossil fuels.* Shell Development Company. Publication No. 95.

Lelieveld, J., Evans, J. S., Fnais, M., Giannadaki, D., & Pozzer, A. (2015). The contribution of outdoor air pollution sources to premature mortality on a global scale. *Nature, 525*(7569), 367–371.

Our World in Data. (2024a). *Coal reserves.* https://ourworldindata.org/grapher/coal-proved-reserves

Our World in Data. (2024b). *Oil production.* https://ourworldindata.org/grapher/oil-production-by-country

Our World in Data. (2024c). *Renewable energy generation.* https://ourworldindata.org/grapher/modern-renewable-energy-consumption

Our World in Data. (2024d). *Carbon dioxide emissions factor*. https://ourworldindata.org/grapher/carbon-dioxide-emissions-factor

Solargis. (2019). *Global solar atlas.* The web-based application is developed and operated by the company Solargis s.r.o. on behalf of the World Bank Group, utilizing Solargis data, with funding provided by the Energy Sector Management Assistance Program. https://globalsolaratlas.info

Technical University of Denmark. (2020). *Global wind atlas version 3.3.* https://globalwindatlas.info

Chapter 5
Energy Demand

Not enjoyment, and not sorrow,
Is our destined end or way;
But to act, that each to-morrow
Find us farther than to-day.
"A Psalm of Life", Henry Wadsworth Longfellow

5.1 Energy Demand and Development

The way energy is utilized determines the way humans live. Before the Industrial Revolution, the vast majority of the world's population relied on farming, herding, or fishing, with little difference from a thousand years ago. When droughts, floods, frost, or plagues occurred, people could only huddle in low, mud, or stone-built houses, trembling and praying for god's intervention. During the Little Ice Age, spanning from the 16th to the eighteenth century, thousands of European farmers scraped a living and often had to bear hunger. In 1815, the eruption of the Mount Tambora volcano in Southeast Asia led to the infamous "Year Without a Summer." The cold climate caused widespread famine, and countless poor farmers perished in the cold and hunger. Human survival, even the rise and fall of wars and dynasties, depended on unpredictable natural factors such as the weather.

In a sense, humanity should be thankful for fossil fuels, despite the negative connotations we have extensively discussed in this book regarding pollution and carbon emissions. The massive use of fossil fuels has empowered humanity to conquer nature. In the late eighteenth century, during the First Industrial Revolution, the invention of steam power and the extensive use of coal as a resource propelled industrialization, leading to advancements in coal mining and transportation technologies. In the Second Industrial Revolution, the discovery of oil and the application of electricity became key drivers of industrial development. The invention of the internal-combustion engine empowered the development of transportation, with oil replacing coal as a primary energy source. Renewable energy emerged during the third Industrial Revolution. Undeniably, the large-scale application of fossil fuels

L. Zhang, *Sustainability Economics*, https://doi.org/10.1007/978-3-031-87829-9_5

Fig. 5.1 Medieval plowing with oxen. *Note* From an illustration in Myers (1905)

has been the most significant force driving the Industrial Revolutions and human development over the past 200 years and has profoundly changed people's lifestyles.

The energy revolution has liberated humanity. In the Middle Ages, nearly all the population had to toil in the fields all day to provide barely enough food for the community to survive. Figure 5.1 illustrates the scene of farmers in the Middle Ages using livestock to power agricultural tools in the fields. Today, engines spin, and in highly modernized and automated factories, one person can manage and operate an entire factory of machinery, creating a myriad of products. Of course, all engines require energy to drive them. Nevertheless, a significant amount of labor has been liberated from the agricultural grind, allowing individuals to engage in enjoyable work, life, and leisure activities, such as reading a book on sustainable economics.

The standard of living of individuals and the wealth of a nation is directly correlated with the country's per capita energy consumption. Wealthier citizens use transportation more frequently and travel to distant places more often. To maintain comfort, people need to control the temperature of their homes within a comfy range, which requires energy for heating or cooling. Wealthier individuals also pursue fashion and gourmet food more often. Producing these customized clothing and food requires more energy consumption. Put simply, residents of wealthier countries own and use more appliances for convenience and consume more energy.

Gross Domestic Product (GDP) is an indicator that measures the total economic activity of a country or region, representing the total value of all final goods and services produced within a country or region during a specific period (usually a year). Figure 5.2 illustrates the relationship between global energy consumption and the trend of global energy consumption since 1800. This "hockey stick" graph provides us with two fundamental pieces of information. Firstly, after the Industrial Revolution, global energy consumption and the economy experienced rapid growth, particularly after the mid-twentieth century. Secondly, the growth of energy consumption and the economy occurred simultaneously.

When considering horizontal comparisons, we often scale the GDP metric by population for an apple-to-apple comparison. Figure 5.3 plots the per capita GDP and per capita energy consumption of different countries in 2022 on the same graph. The GDP values in the data of this figure have been adjusted to account for the different purchasing power of the US dollar in different countries. An evident conclusion is that per capita energy consumption is highly correlated with a country's level of

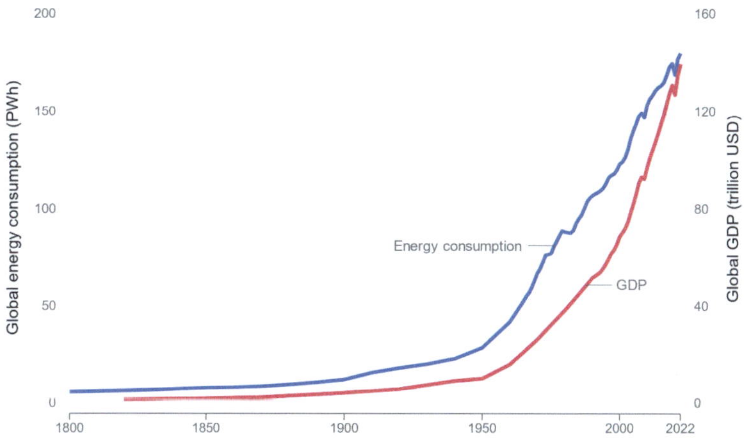

Fig. 5.2 Global primary energy consumption and GDP since 1800. *Data source* EIA (2024)

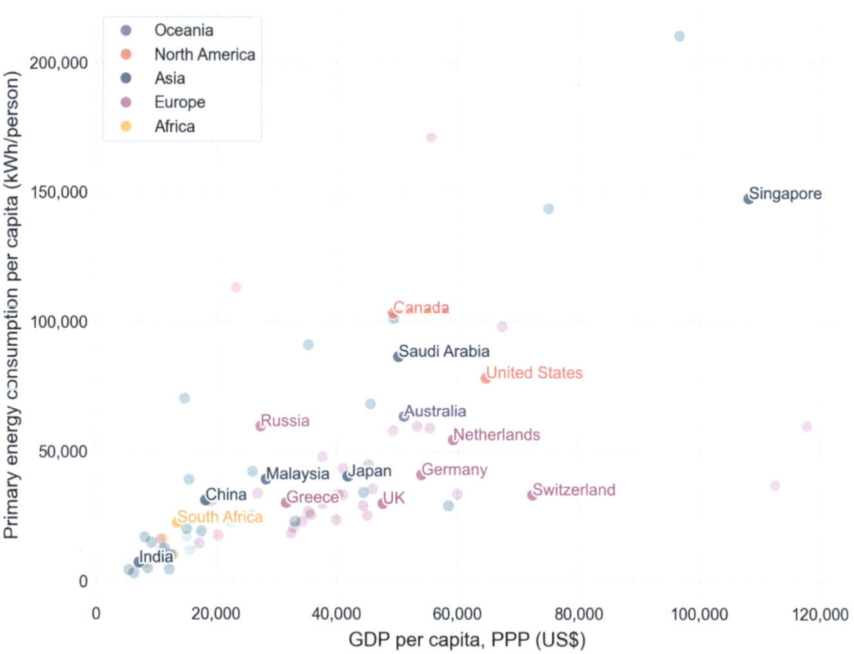

Fig. 5.3 Energy use per person versus GDP per capita in 2022. *Data source* EIA (2024)

economic development. Countries with higher incomes also exhibit higher energy consumption intensity. OECD countries, such as the United States, European Union countries, and Australia, completed industrialization earlier and entered the fleet of developed countries. These countries have higher per capita incomes and energy consumption levels. Many African and Asian countries, identified in purple and blue on this graph, are located in the lower left corner, indicating lower levels of economic development, lower per capita incomes, and lower per capita energy consumption.

Case

Economic Growth and Energy Demand: China 2000–2022

Analyzing the case of China from 2000 to 2022 provides a compelling illustration of the strong correlation between energy consumption and economic growth at a national level over time (Fig. 5.4). During this period, China's economy experienced remarkable growth, with its GDP expanding significantly each year. Concurrently, energy demand and consumption soared, driven by industrialization and urbanization efforts. China's rapid industrial development necessitated a firm reliance on coal and other fossil fuels to meet its burgeoning energy needs. The data clearly reveals that as energy consumption increased, so did economic outputs, reinforcing the notion that a robust energy supply is essential for sustained economic performance. This case study underscores the importance of energy resources in supporting national economic trajectories, demonstrating that even in the context of a single country over a given period, energy consumption remains a critical driver of economic growth.

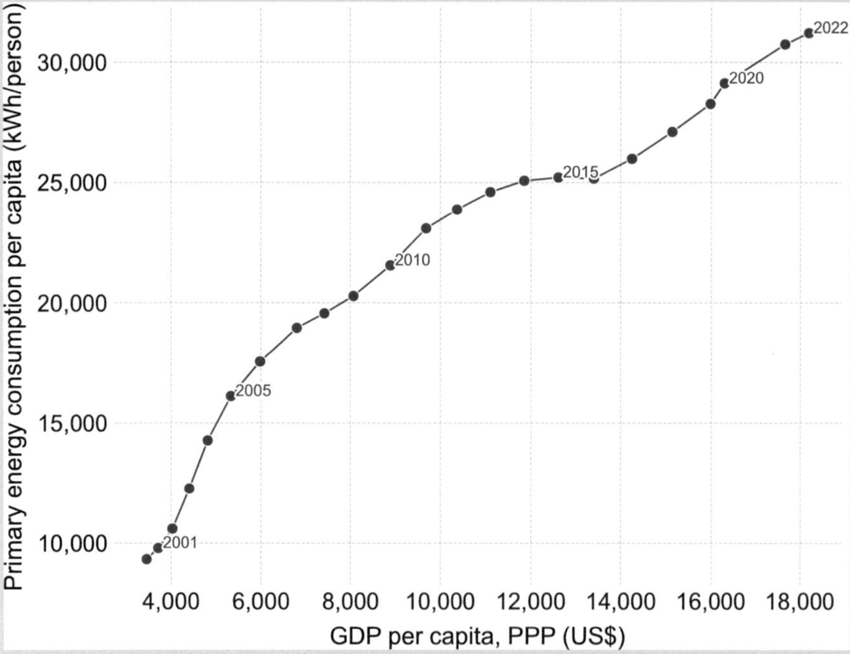

Fig. 5.4 GDP per capita and primary energy consumption per capita in China, 2000–2022. *Data source* World Bank (2024)

In recent years, the advancement of space-borne remote sensing technology has enabled us to measure surface nighttime light intensity at a very fine scale globally (e.g., 1 km by 1 km). Nighttime light remote sensing provides a unique opportunity to directly observe human activities from space. It is a common understanding that in regions with higher levels of economic development and affluence, there tends to be more public lighting at night and lights from transportation vehicles, even when observed from space. Therefore, nighttime light data is widely used for mapping urban areas and estimating population and real GDP. GDP accounting relies on government statistical systems. Sometimes, this accounting system may be influenced by political situations, leading to biases and distorting the measure of GDP. Economists are keenly aware that nighttime light serves as an indicator of regional development that is difficult to manipulate. Nighttime light data can also serve as a proxy for regional energy consumption intensity.

Figure 5.5 presents the results from National Aeronautics and Space Administration (NASA's) remote sensing of global surface nighttime light in 2012 from space. Cities in the United States, Europe, Western Asia, and Southeast Asia exhibit the most intense nighttime light, reflecting high economic activity in these regions. Therefore, it is important to emphasize that, despite the persistent calls in sustainable economics to conserve energy and reduce emissions, this does not imply that we should reduce energy usage overall at the expense of economic development and people's quality of life. In fact, as we are attempting to illustrate here, consuming more energy is generally indicative of higher levels of economic development. Of course, we do not differentiate between the sources of energy here. Therefore, sustainable development does not mean development without using energy; rather, the idea is that we should rely more on clean energy sources.

In addition to its pivotal role in economic development, energy consumption profoundly influences life expectancy and overall human development. As depicted

Fig. 5.5 Light at night. *Note* NASA Earth Observatory images by Joshua Stevens, using Suomi NPP VIIRS data from Miguel Román, NASA GSFC

in Fig. 5.6, there exists a significant positive correlation between energy consumption and life expectancy at birth across various countries. This relationship can be attributed to the fact that higher energy consumption typically leads to improved living standards and access to essential services such as healthcare, sanitation, and nutrition. For instance, countries that invest in energy infrastructure are better equipped to provide reliable electricity, which is crucial for hospitals and medical facilities. This enhanced access to healthcare services subsequently reduces mortality rates and increases longevity. Indeed, burning fossil fuels for energy can lead to air and water pollution, which may adversely affect local health. However, sustainable development should not adopt a simplistic and counterproductive perspective, as it is evident that energy availability significantly enhances residents' well-being. We will discuss this trade-off later in the book.

The Human Development Index (HDI), which encompasses life expectancy, education, and per capita income, also reflects the interconnectedness of energy consumption and human welfare. Each dimension of HDI is assessed through specific indicators. These indicators are then normalized and aggregated to generate the overall HDI, placing them on a scale from 0 to 1. By providing a comprehensive view of human development, the HDI serves as a valuable tool for comparing well-being across nations. As nations progress in their energy consumption patterns, they often experience parallel advancements in HDI. Increased energy availability enables educational institutions to function more effectively, facilitating better learning environments and outcomes. Education, in turn, nurtures a proficient workforce capable of propelling economic advancement, thereby establishing a virtuous cycle where energy consumption not only propels economic advancement but also enriches human development.

The relationship between energy consumption and human development is multifaceted and integral to understanding sustainable economics. By ensuring an adequate energy supply, nations can enhance public health, educational opportunities, and overall quality of life, thus promoting a more sustainable and equitable future. Policymakers must recognize this synergy and prioritize investments in sustainable energy systems that promote both economic stability and human well-being.

5.2 Primary Energy Demand

5.2.1 Global Trends

Since the Industrial Revolution, fossil fuels (mainly coal and oil) have been the primary energy sources driving human society. Currently, coal accounts for approximately 25% of global primary energy consumption, while oil represents about 30%. The shale gas revolution starting from the twenty-first century significantly enhanced the technology and efficiency of natural gas extraction, leading to a rapid increase in its share of primary energy consumption, reaching around 22%. Nuclear energy

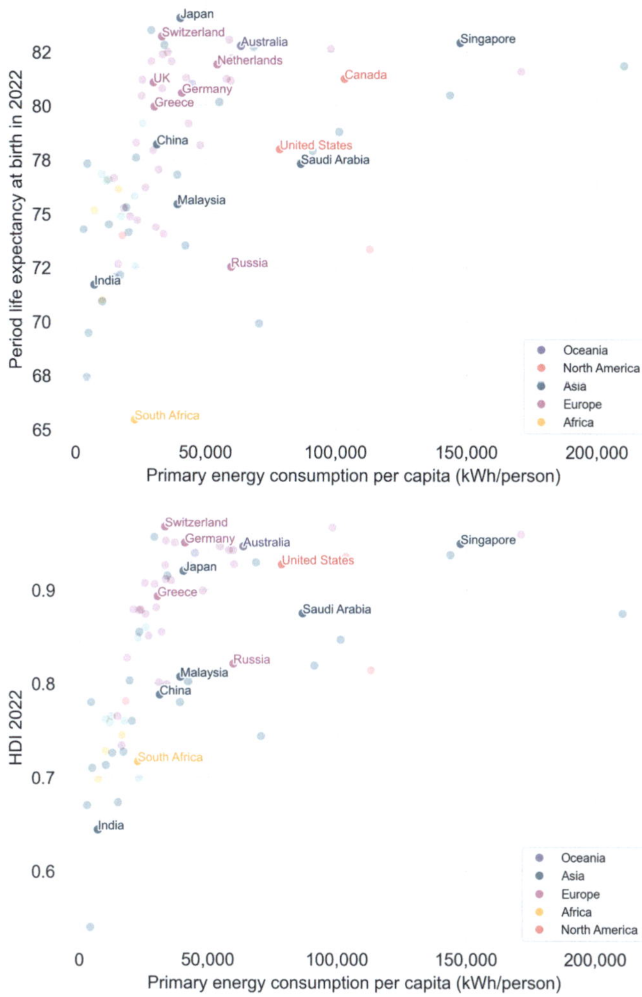

Fig. 5.6 Energy use per person versus life expectancy at birth and HDI, 2022. *Note* Adapted from Our World in Data (2024a, 2024b)

was once favored for its low cost and pollution-free advantages. However, in the past decade, its share in primary energy consumption has declined, possibly due to the panic triggered by the Fukushima nuclear disaster. Today, nuclear energy constitutes less than 4% of primary energy consumption. Additionally, nearly 6% of energy still comes from traditional biomass, such as wood burning. In developing countries, particularly in rural areas, these traditional biomass energy sources remain the main means for cooking, heating water, and warming homes.

Renewable energy is the direction of consumption development; however, its share in the current energy structure remains relatively low. Hydropower accounts for about 6% of primary energy consumption, only one-fifth of oil's share. The combined share of wind and solar energy in primary energy consumption is around 5%. The metrics convey two key messages. On one hand, global energy continues to depend significantly on fossil fuels. As energy demand continues to rise, our annual consumption of fossil fuels is increasing. Despite ongoing efforts to create new energy reserves through advanced exploration and extraction technologies, we may soon face the reality of energy depletion. Notably, European countries do not possess abundant fossil energy resources and are unlikely to in the future. However, an analysis of energy consumption across European countries reveals that they still heavily depend on fossil fuels. Forecasts indicate that by 2050, Europe's proven oil reserves will be reduced to about 14%, coal reserves to 72%, and natural gas reserves to 18%. At that point, European countries will either need to find alternative energy sources or import fossil fuels from other countries and regions at significantly higher prices.

On the other hand, the potential for developing renewable energy is immense. As global awareness of environmental issues and the impacts of climate change intensifies, there is a growing commitment among nations to transition toward cleaner energy sources. This change is propelled by technological progress that has greatly reduced the costs of producing renewable energy, especially from solar and wind sources. According to Fig. 5.7, since 2019, the installation of solar panels and wind turbines has increased dramatically, resulting in unprecedented contributions to the energy mix. As a result, countries are increasingly integrating renewable resources into their energy portfolios, reducing their dependence on fossil fuels and mitigating greenhouse gas emissions. Government policies and incentives play a crucial role in promoting the adoption of renewable energy. Many nations have implemented subsidies, tax breaks, and renewable energy targets that encourage investment in new technologies and infrastructure. International agreements and conferences, such as the Paris Agreement and the Conference of the Parties (COP), further bolster these efforts by setting ambitious climate goals. Such collaborative global frameworks inspire nations to innovate and diversify their energy sources while fostering the growth of a worldwide clean energy technology market. As these trends continue, the possibility for renewable and clean energy to become the backbone of a sustainable energy future grows increasingly tangible, promising to reshape the global energy landscape in the years to come. Of course, whether these prospects can become a reality still depends on sustained investment in new energy sources and breakthroughs in technological innovation (An & Zhang, 2023).

Reading

The Paris Agreement and the 2 °C Goal

The Paris Agreement, adopted on December 12, 2015, during the 21st Conference of the Parties (COP21) to the UNFCC in Paris, represents a landmark commitment by nearly every nation to combat climate change and its impacts. This groundbreaking accord emerged from the urgent need to tackle the growing threats of global warming, which has far-reaching implications for ecosystems, economies, and human health. The Paris Agreement aims to limit global temperature rise to well below 2 °C above pre-industrial levels and strives for a 1.5-degree limit. It provides a framework for countries to strengthen their climate actions and resilience.

One of the key features of the Paris Agreement is its flexible, nationally determined contributions (NDCs) mechanism. Under this framework, each country is required to set and communicate its own climate targets, reflecting its unique circumstances, capabilities, and levels of development. This approach encourages nations to take ownership of their climate strategies and fosters a spirit of collaboration and accountability. The agreement highlights the need for transparency and accountability, urging countries to track and report their progress in meeting their NDCs. This system seeks to build trust among nations and ensure that collective efforts align with the agreement's overarching goals.

Moreover, the Paris Agreement recognizes the critical role of financial and technological support in facilitating climate action, particularly for developing countries that may lack the resources to implement ambitious climate strategies. The agreement calls for enhanced financial flows to support climate resilience and adaptation efforts, as well as the transfer of technology to help nations reduce GHG emissions. The Paris Agreement seeks to build a sustainable and resilient future by caring for the needs of vulnerable populations and promoting equitable development. It underscores the connection between climate action, social equity, and economic progress.

Fig. 5.7 Global primary energy consumption by source since 1800. *Data source* EIA (2024)

5.2.2 Regional Particulars

In examining the regional particulars of energy consumption, significant disparities emerge across different global regions, reflecting their unique energy profiles and resource endowments. Figure 5.8 illustrates the structure of primary energy consumption in various regions for the year 2022. The Middle East, for instance, demonstrates a heavy reliance on natural gas, which comprises nearly 52% of its total energy use. Oil is significant, making up about 45%. The region's energy profile is shaped by its vast hydrocarbon reserves, which satisfy domestic needs and establish it as a major force in global energy markets. However, the limited diversification into renewable energy sources, with negligible contributions from wind, solar, and other renewables, suggests a potential vulnerability to shifts in global energy trends and a pressing need for diversification strategies to enhance energy resilience.

In contrast, Asia presents a more diversified energy mix, though it still leans toward fossil fuels, particularly coal, which constitutes over 40% of the region's energy consumption. Natural gas is also prominent, providing about 18% of energy needs. However, the growth of wind and solar energy, although currently small, reflects an emerging trend toward sustainable energy solutions. This gradual shift reflects Asia's awareness of the importance of diversifying energy sources, which is motivated by environmental concerns and the pursuit of energy security. Over the past 30 years, Asia has experienced the fastest growth in energy consumption compared to other regions, accompanied by substantial economic progress. This rapid increase has also created significant pressure on the region to pursue green development, especially as global demands for sustainability continue to intensify.

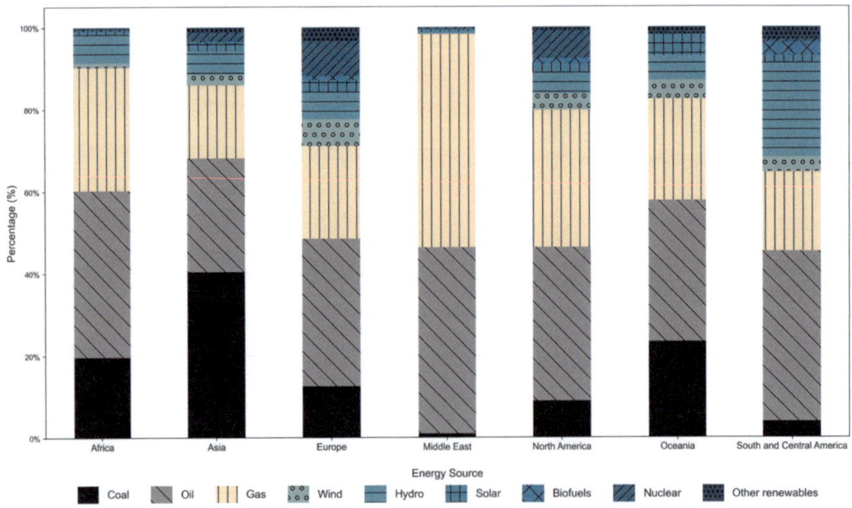

Fig. 5.8 Primary energy consumption by source and regions, 2022. *Data source* EIA (2024)

Europe and North America present a more balanced energy consumption profile, with a notable emphasis on renewable energy sources. In Europe, the share of renewables, including wind and hydro, reaches about 10%, reflecting the region's commitment to sustainability and climate goals. North America, while still reliant on fossil fuels, shows a growing trend toward cleaner energy, with renewables accounting for approximately 8% of its energy mix. This shift is indicative of broader efforts to reduce carbon emissions and transition toward a more sustainable energy future. The varying degrees of renewable energy integration across these regions underscore the importance of tailored energy policies that consider local resources, economic conditions, and environmental goals to foster a more sustainable global energy landscape.

5.3 Energy Consumption and Environment

5.3.1 Environmental Impact

Burning fossil fuels significantly contributes to air pollution, leading to various negative impacts on human health and the environment. Emissions from burning coal, oil, and natural gas release harmful pollutants such as particulate matter (PM), nitrogen oxides (NO_x), and sulfur dioxide (SO_2). The combustion of fossil fuels for electricity generation accounts for around one-seventh of human exposure to harmful air pollutants, such as fine $PM_{2.5}$ (particulate matter with an aerodynamic diameter of 2.5 μm or less), and about 40% of CO_2 emissions that contribute to climate change (Lelieveld et al., 2015). These pollutants are linked to respiratory diseases, cardiovascular issues, and other serious health conditions, disproportionately affecting vulnerable populations (Brunekreef & Holgate, 2002). We will discuss the issue of pollution related to fossil fuels in detail later in this book.

On the other hand, the transition to renewable energy sources, while generally more environmentally friendly, is not without its own challenges. For example, wind power can kill birds, lead to noise pollution, and habitat disruption, raising concerns among local communities. The manufacturing process of solar panels can itself lead to air pollution and GHG emissions, while also posing a threat to the ozone layer. However, it is important to approach these environmental impacts with a balanced perspective. While renewable energy facilities may have some drawbacks, they are essential for reducing carbon emissions and addressing climate change issues. Therefore, rather than abandoning renewable energy initiatives due to these concerns, we should focus on technological innovations that can mitigate their environmental effects, ensuring a sustainable path forward.

The combustion of fossil fuels is a primary source of carbon dioxide, significantly contributing to global GHG concentrations in the atmosphere. Beyond the emissions produced during combustion, the extraction, transportation, and post-extraction phase of coal also release other potent GHGs, such as methane (CH_4), which has a much higher heat-trapping ability than CO_2 over a shorter time frame (Liu et al., 2024). This dual contribution—both from burning fossil fuels and from leakage during extraction processes—exacerbates the greenhouse effect, leading to rising global temperatures and alarming climatic changes. As human activities continue to emit these GHGs at unprecedented levels, the associated risks of global warming intensify, resulting in severe environmental impacts, such as more frequent and severe weather events, sea-level rise, and biodiversity loss. Addressing these emissions is paramount for fostering a sustainable future and mitigating the harmful effects of climate change.

Figure 5.9 shows the global temperature anomaly since 1900. The temperature anomaly is calculated by subtracting the average global temperature from 1961 to 1990 from the annual global mean temperature. A continuous period of 30 years or more is often referred to as a climate state. Thus, the temperature anomaly measures the degree to which temperatures deviate from this climate state. Clearly, global temperatures have risen significantly since 1900. At the same time, annual carbon emissions from fossil fuel combustion have increased at the exact same pace. Concurrently, the frequency of global climatological and meteorological extreme events—such as droughts, wildfires, extreme temperatures, fog, and storms—has also risen sharply. Human activities have driven greenhouse gas emissions, leading to global climate change and an increasing frequency of extreme events.

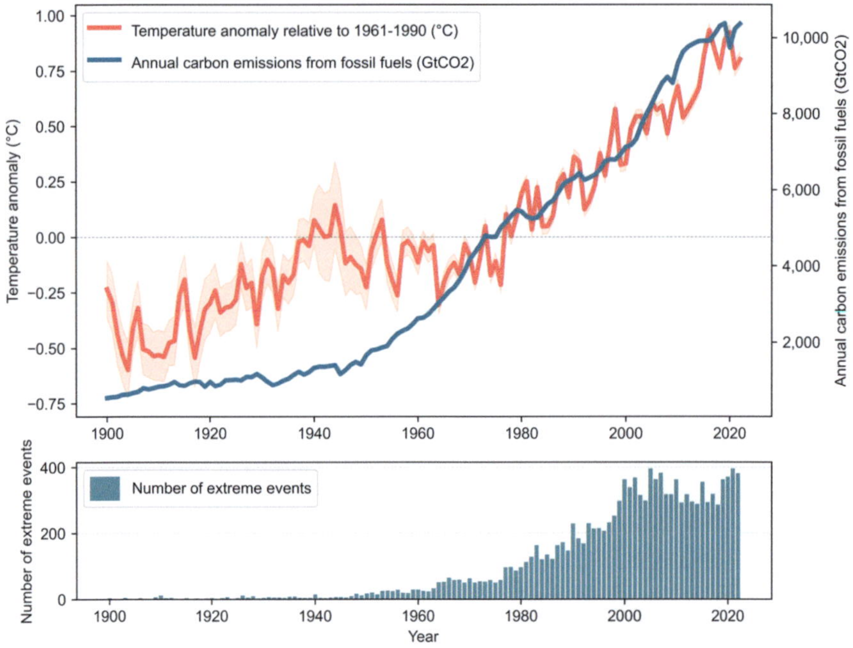

Fig. 5.9 Temperature and extreme events increase due to carbon emissions from fossil fuels. *Data sources* EM-DAT (2024), Global Carbon Budget (2024), and Morice et al. (2020)

5.3.2 *Environmental Kuznets Curve*

Does economic growth necessarily result in environmental degradation? The Environmental Kuznets Curve (EKC) hypothesis proposes an inverted U-shaped relationship between environmental degradation and per capita income (Kuznets, 1955). In the early stages of economic growth, as income rises, pollution and environmental degradation tend to increase. However, once a certain income level is reached, further economic growth can lead to environmental improvements. The EKC implies that wealthier societies are better positioned to invest in advanced technologies and implement rigorous environmental regulations, resulting in reduced pollution levels. The EKC demonstrates how environmental quality evolves with economic development, underscoring the complex interaction between economic activity and environmental pressures.

Reading

Simon Kuznets and the EKC Theory

Simon Kuznets (see Fig. 5.10) was a Russian-born American economist and statistician, renowned for his pioneering work in economic growth and development. Born on April 30, 1901, in Pinsk, which is now part of Belarus, Kuznets emigrated to the United States in 1922. Kuznets made significant contributions to the field of economics, particularly through his development of methods to measure national income and his analysis of economic growth patterns.

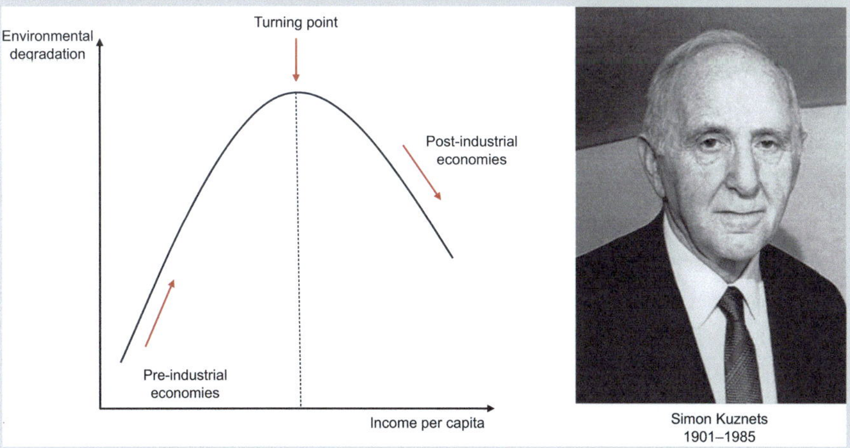

Fig. 5.10 Simon Kuznets and the EKC

In 1971, Simon Kuznets was awarded the Nobel Memorial Prize in Economic Sciences for his "empirically founded interpretation of economic growth, which has led to new and deepened insight into the economic and social structure and process of development." His work provided a comprehensive understanding of how economies grow and evolve, influencing both economic theory and policy.

The concept of the EKC is derived from the work of Simon Kuznets in 1955 (Kuznets, 1955). The application of this concept to environmental issues began in the early 1990s. The first empirical studies on the EKC were conducted by Grossman and Krueger in 1991, who examined the environmental impacts of the North American Free Trade Agreement (NAFTA) (Grossman & Krueger, 1991). This was followed by the World Bank's 1992 World Development Report, which further popularized the EKC hypothesis. The term "Environmental Kuznets Curve" was first coined by Panayotou in 1993 (Panayotou, 1993). Since then, the EKC has been a focal point of numerous studies exploring the relationship between economic growth and environmental quality.

From the perspective of the EKC theory, economic growth and environmental degradation initially increase simultaneously due to several factors. In the early stages of economic development and industrialization, the primary focus is on increasing production and economic output, often at the expense of environmental quality. This phase is characterized by the "scale effect," where the expansion of economic activities leads to higher consumption of natural resources and increased emissions of pollutants (Stern, 2004). As industries grow, they tend to use more energy and raw materials, resulting in greater environmental degradation. Additionally, the lack of stringent environmental regulations and the prioritization of economic growth over environmental protection further exacerbate the situation. People and governments in developing economies may also have limited resources and incentives to invest in cleaner technologies or enforce environmental regulations, leading to higher levels of pollution.

As an economy continues to grow and reaches a higher level of income per capita, the relationship between economic growth and environmental degradation begins to change. This shift can be explained by several factors that come into play after a certain income threshold is crossed. One of the key factors is the "composition effect," where the structure of the economy changes from heavy, polluting industries to cleaner, service-oriented industries. As economies develop, there is a natural progression toward sectors that are less resource-intensive and generate fewer emissions.

Another significant factor is the "technique effect," which involves the adoption of cleaner technologies and more efficient production processes. Higher income levels enable both governments and businesses to invest in advanced technologies that reduce pollution. Increased public awareness and demand for a cleaner environment also play a crucial role. As people become wealthier, they place a higher value on environmental quality and push for stricter environmental regulations and better enforcement (Dinda, 2004).

Furthermore, globalization and international trade can facilitate the transfer of cleaner technologies and best practices from developed to developing countries, accelerating the reduction in environmental degradation. The combined impact of these factors leads to a *decoupling* of economic growth from environmental degradation, resulting in the observed decline in pollution levels after a certain point in the EKC.

The EKC offers a clear and compelling theory to explain the relationship between economic growth and environmental degradation. This theory suggests an optimistic outlook, indicating that environmental issues may improve once economic development reaches a certain level. However, is reality truly aligned with the EKC theory? Since its introduction, the EKC theory has faced significant scrutiny (Dinda, 2004). Here are several key criticisms:

- *Inconsistent Evidence.* Empirical studies on the EKC have produced mixed and often conflicting results. While some studies find evidence supporting the EKC for specific pollutants and regions, others do not (Apergis & Ozturk, 2015).
- *Generalization Issues.* The EKC hypothesis cannot be generalized for all types of environmental degradation. For instance, GHG often shows a monotonically increasing relationship with income. Some evidence indicates that economic growth can only reduce emissions of certain pollutants at extremely high-income levels (e.g., particulate matter in the air) (Stern, 2017).
- *Impact of Trade and Globalization.* EKC does not adequately account for the effects of international trade and globalization. The *displacement hypothesis* suggests that developed countries may reduce their pollution by outsourcing dirty industries to developing countries, which can distort the true relationship between income and environmental quality from a global perspective. This leads to a "*race to the bottom*" scenario where developing countries may lower environmental standards to attract foreign investment.
- *Turning Points Variability.* The income level at which environmental degradation starts to decline (the turning point) varies widely across studies and pollutants. This variability makes it difficult to draw general conclusions about the EKC and its applicability to different contexts.
- *Heterogeneity Across Countries.* The assumption that all countries follow a similar EKC trajectory is problematic. Countries differ significantly in terms of their economic structures, regulatory frameworks, and environmental conditions. Pooling data from diverse countries without accounting for these differences can lead to incorrect inferences.

5.4 Energy Future

5.4.1 Key Drivers of Energy Future

Transitioning from our exploration of energy demand, economic development, and their environmental impacts, the future of energy stands as a vital theme in sustainable economics. Several factors will shape this future, influencing everything from supply and demand dynamics to technological advancements and regulatory frameworks (see Fig. 5.11). A clear understanding of these elements is crucial for formulating effective strategies that address energy challenges and opportunities.

Supply challenges. As global energy needs continue to grow, accessing reliable and sufficient energy sources becomes increasingly complex. The depletion of easily accessible fossil fuels, coupled with geopolitical tensions, raises concerns over the reliability of energy supplies. The move to renewable energy, though promising, encounters challenges related to resource availability and infrastructure development. Overcoming these supply issues is crucial for a stable energy future.

Demand growth. As populations expand and economies develop, energy consumption is projected to keep rising substantially. This increased demand places pressure on existing energy systems, making it crucial to explore sustainable solutions that can meet future needs without exacerbating environmental degradation. Strategies that promote energy efficiency and alternative energy sources can reduce the impact of this rising demand. Developed countries have made notable achievements in these areas, but there is still a long way to go. For instance, the UK has managed to

Fig. 5.11 Key drivers of energy future

reduce overall energy demand and significantly phase out coal. In addition to population control, industrial transformation, and efficiency improvements, the transfer of carbon emissions to other countries has likely played a significant role (Shapiro, 2016). Many developing countries still have considerable economic growth potential. To improve the living standards of people in the Global South, a greater demand for energy is assured.

Environmental impacts. The reliance on fossil fuels contributes significantly to climate change and local air quality issues. Environmental impacts present a key challenge in shaping the future of energy. On the one hand, addressing local environmental issues, such as air, water, and soil pollution, is crucial as these problems directly affect public health and community welfare. Thus, there is a strong motivation for governments and communities to mitigate these local pollutants, given their immediate effects on citizens' quality of life. On the other hand, climate change and GHG emissions, also directly related to the energy nexus, represent a global challenge with substantial externalities. The effects of global climate change are unevenly and inequitably distributed, often burdening regions and countries that contribute a smaller proportion to GHG emissions (Sultana, 2022). This discrepancy complicates efforts to formulate fair and effective policies. Recognizing and addressing this duality—local environmental concerns alongside global climate issues—will be vital in creating a sustainable and clean energy future that protects community health and the planet's climate.

Availability and security. A just energy transition is essential to guarantee that all communities have equal access to energy, particularly amid energy poverty. This requires implementing policies that address disparities in energy access, providing marginalized populations with opportunities for participation in the energy market. Simultaneously, the energy security challenge must be tackled to guarantee that energy supplies remain stable and resilient against disruptions. However, we must confront the reality that natural disasters, technological failures, geopolitical conflicts, and occasional regional hot wars can draw us further away from these goals. Striking a balance between promoting equitable access to energy and safeguarding supply security is crucial. Effective strategies must recognize the interconnectedness of all related issues, ensuring that efforts to transition to sustainable energy sources do not compromise the reliability and affordability of energy for all, particularly vulnerable communities. By collectively addressing availability and security, we can foster a more inclusive and resilient energy future.

Technology and policy. Innovations in energy production, storage, and distribution can enhance efficiency and open up new avenues for sustainable practices. Policy frameworks that support renewable energy development and the shift from fossil fuels are also essential. Joint efforts among governments, enterprises, and communities are essential to build an energy future that is sustainable, secure, and economically viable. However, it is essential to recognize that there is no silver bullet for the energy transition. Table 5.1 lists examples of energy technologies under different categories. While many promising technological pathways exist, significant breakthroughs often prove to be "high-hanging fruit," which is difficult to achieve and requires substantial

Table 5.1 Energy technologies

Categories	Energy technologies
Primary Energy Sources	Light crude, Heavy oil, Tar sands, Wet gas, Tight gas, Nuclear, Coal, Solar, Wind, Biomass, Hydro, Geothermal, …
Extraction & Conversion Technologies	Exploration, Deeper water, Arctic, LNG, Refining, Differentiated fuels, Advantaged chemicals, Gasification, Syngas conversion, Photovoltaics, Bio-enzyimatics, Hydrogen production, …
End Use Technologies	ICEs, Adv. Batteries, Hybridization, Fuel cells, Hydrogen storage, Gas turbines, Building efficiency, Urban infrastructure, Systems design, Appliances, Retail technologies, Evs, …

investment and innovation. This reality extends to policy as well; effective frameworks are necessary to guide the transition, but simply implementing new policies is not a guaranteed success. More effort must be directed toward preventing the catastrophic consequences of technological and policy failures.

5.4.2 Bottom-Up Approach

Understanding the specific pathways and strategies for achieving a sustainable energy future becomes increasingly important. Therefore, certain analytical methods are essential for our systematic analysis and for gaining insights.

The bottom-up approach, often referred to as microeconomic process analysis, focuses on individual components or entities within the energy system to understand their behavior and impact on overall performance. This approach is particularly useful for achieving detailed analyses that require insights into specific sectors, technologies, or consumer behaviors. In the context of energy demand forecasting, the bottom-up method allows for a granular assessment of how various factors—such as demographics, technological advancements, and economic conditions—affect energy consumption patterns. By analyzing these details and nuances, researchers can create more accurate and tailored energy demand models that reflect real-world conditions.

The bottom-up approach typically involves several key steps. First, data collection is conducted at the individual or household level, encompassing aspects like energy usage, appliance efficiency, and lifestyle choices. Next, this data is utilized to develop models that simulate the energy consumption of these entities under varying conditions. Scenarios are then created to evaluate future energy demand based on different variables, such as policy changes or economic shifts. The results are aggregated to provide an overall picture of energy demand, allowing policymakers and stakeholders to understand and anticipate future trends.

The demand for energy $E(t)$ in a period of t is the sum of energy demand of all energy-using capital $\mathrm{Cap}(t)$. Through simple mathematical transformations, we can decompose energy consumption into energy efficiency, energy intensity, and working hours.

$$\begin{aligned} E(t) &= \sum_i \mathrm{Cap}_i(t) \\ &= \sum_i \underbrace{\mathrm{eff}_i(t)}_{\text{energy efficiency}} \times \underbrace{w_i(t)}_{\text{working time during period } t} \\ &= \sum_i \mathrm{eff}_i(t) \times \frac{w_i(t)}{t} \times t \\ &= \sum_i \mathrm{eff}_i(t) \times \underbrace{v_i(t)}_{\text{intensity of use}} \times t \end{aligned} \tag{5.1}$$

The age structure of appliances is also an important factor to consider in the bottom-up analysis. The efficiency of appliances can decline as they age, impacting overall energy consumption. Older appliances often use more energy due to decreased performance and outdated technology, which can lead to higher operational costs and increased energy demand. Accounting for the age distribution of appliances can inform strategies for promoting energy-efficient upgrades and optimizing energy usage, ultimately contributing to more accurate energy demand forecasts. Taking household appliances as an example, the overall process analysis structure is demonstrated in Fig. 5.12.

While the bottom-up approach offers detailed insights and the ability to capture real-world factors that influence energy demand, it also has limitations. First, the method can be resource-intensive, requiring extensive data collection and analysis, which may not always be feasible. Second, combining all the insights from individual components into a coherent overall picture can be challenging, particularly if various assumptions need to be made. Besides, certain macroeconomic factors may be overlooked in such analyses. Regardless, the bottom-up approach is particularly

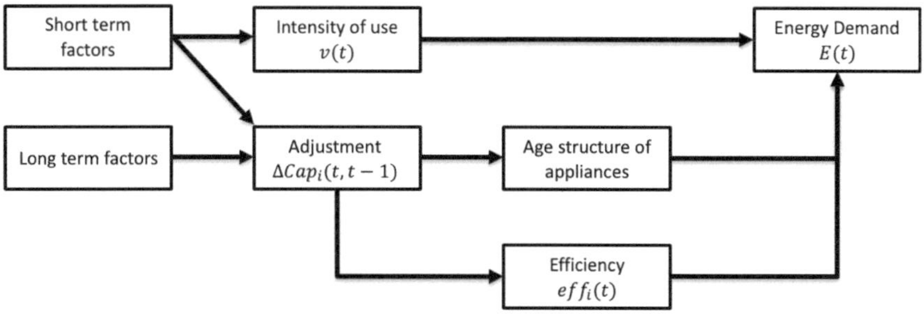

Fig. 5.12 Process analysis of the household appliances

suitable for scenarios where detailed understanding at the micro level is crucial, such as in urban planning, energy efficiency programs, and the integration of renewable energy technologies.

5.4.3 Top-Down Approach

Applying bottom-up models for energy demand analysis has to meet considerable micro-data requirements. Various components of energy demand are typically associated with key macroeconomic variables like GDP, population, and relative energy prices. You may wonder: "Why not model energy demand directly as a function of these macro variables?"

The top-down approach in energy demand forecasting involves analyzing aggregate data at a broader level, such as national or regional energy consumption patterns, to draw conclusions about overall energy demand. This method typically utilizes historical data, macroeconomic indicators, and statistical models to identify trends and project future demand. The primary objective of the top-down approach is to provide a macro-level understanding of energy usage and facilitate the identification of overarching patterns and relationships, allowing policymakers and energy planners to assess overall requirements and allocate resources effectively.

The IPAT model is a wildly used framework for understanding the impact of human activity on the environment, specifically focusing on the relationship between population growth (P), affluence (A), and technology (T). This identity can be easily proved, where the GDP per capita indicates affluence and the environmental impact per unit of GDP is a proxy for the technologies.

$$
\begin{aligned}
I &= I \\
&= P \times \frac{I}{P} \\
&= P \times \underbrace{\frac{\text{GDP}}{P}}_{\text{Affluence}} \times \underbrace{\frac{I}{\text{GDP}}}_{\text{Technology}} \\
&= P \times A \times T
\end{aligned}
\tag{5.2}
$$

While the IPAT equation can be used for any environmental impact analysis, the *Kaya identity*, a concrete form of the IPAT model, specifically articulates the relationship between carbon emissions and the factors affecting energy consumption.

$$
F = P \times \frac{G}{P} \times \frac{E}{G} \times \frac{F}{E}, \tag{5.3}
$$

where F is global CO_2 emissions from human activities, P, G, and E indicate global population, GDP, and energy consumption, respectively. In this identity, $G/$

P, *E*/*G*, and *F*/*E* are proxies for affluence, energy intensity, and emission intensity, respectively.

Table 5.2 presents the results of the IPAT decomposition analysis for greenhouse gas emissions in several countries. For instance, China's GHG emissions rose from 10,510.8 Mt CO_2e in 2010 to 13,706.1 Mt CO_2e in 2020, which can be attributed to increases in both population and affluence, with per capita income increasing from \$5.6 thousand to \$10.4 thousand. However, the CO_2e intensity per dollar (T) decreased from 1.4 to 0.9 kg, reflecting a shift toward cleaner technologies. India also saw a significant increase in GHG emissions, driven by population growth and rising affluence. In contrast, countries like France, Germany, and the United States managed to reduce their GHG emissions despite slight increases in population and affluence by significantly lowering their emission intensity.

After collecting data and constructing models such as the IPAT and Kaya identities, a typical top-down analysis would also incorporate scenario development and analysis to explore the potential impacts of varying factors on the environment. This analytical approach allows researchers to simulate multiple future scenarios and assess the effects of different policies or technological changes. Discussing the implications and feasibility of these policies is essential, providing decision-makers with actionable insights that support the achievement of sustainability goals.

In reality, no one can predict the future accurately. Therefore, the essence of scenario analysis lies not in precise future predictions, but in exploring, based on empirical models, how our variables of interest, such as environmental impact, might evolve if certain macro factors develop along the paths outlined in the scenarios. This allows policymakers to assess the potential benefits and costs of adopting different policies.

5.4.4 Energy Demand and Household Utility Function

This section will provide an example to illustrate how to incorporate household energy demand and consumption into a microeconomic model. This model can enhance our understanding of energy demand and energy consumption decision-making within the residential sector.

We begin with the household utility function. When considering the energy consumption decisions of households, we discuss their consumption of energy services separately. Energy services refer to products or services that require energy for proper use or access, such as cooking, laundry with a washing machine, and lighting. Therefore, we consider a household's utility as dependent on the consumption of two types of goods: ordinary goods (unrelated to energy) and energy services, as described in Eq. 5.4:

$$U = U(C, \mathrm{ES}) \tag{5.4}$$

Table 5.2 IPAT analysis, 2010–2020

2010				
	GHG emissions (Mt CO_2e)	P (million)	A (1000 US$ per capita)	T (kg CO_2e per US$)
China	10,510.8	1337.7	5.6	1.4
India	2584.5	1240.6	1.2	1.7
Brazil	1398.2	196.4	8.7	0.8
France	471.3	65.0	35.7	0.2
Germany	923.3	81.8	37.8	0.3
Indonesia	987.4	244.0	2.7	1.5
Italy	461.2	59.3	32.0	0.2
Japan	1256.1	128.1	32.9	0.3
Russian	1490.3	142.8	8.8	1.2
US	5887.7	309.3	52.8	0.4
UAE	214.3	8.5	34.5	0.7
Australia	642.6	22.0	53.7	0.5
2020				
	GHG emissions (Mt CO_2e)	P (million)	A (1000 US$ per capita)	T (kg CO_2e per US$)
China	13,706.1	1411.1	10.4	0.9
India	3208.9	1396.4	1.8	1.3
Brazil	1643.1	213.2	8.3	0.9
France	386.7	67.6	35.8	0.2
Germany	736.7	83.2	41.6	0.2
Indonesia	2029.5	271.9	3.8	2.0
Italy	334.2	59.4	29.4	0.2
Japan	1109.0	126.3	34.6	0.3
Russian	1826.9	145.2	9.8	1.3
US	4795.8	331.5	59.4	0.2
UAE	262.8	9.3	41.2	0.7
Australia	515.5	25.6	58.2	0.3

Data source World Bank (2024)

U: Household Utility
C: Consumables not related to energy
ES: Energy services.

In this framework, the energy services (ES) produced by households primarily involve two inputs—electricity and appliances (or so-called capital stock). The function for producing electricity services is often referred to as the household production function. A production function typically describes a firm's technology to transform inputs into goods or services. Here, the household production function illustrates how households convert electricity and other inputs into energy services. Of course, other inputs, such as labor, are often involved in obtaining energy services. For simplicity, we will consider only electricity and appliances as the two inputs, namely:

$$\text{ES} = f(E, A), \tag{5.5}$$

ES: Energy services
E: Energy
A: Appliances.

If we substitute ES from Eqs. 5.5 into 5.4, we then have the household utility function expressed as:

$$U = U(C, \text{ES}(E, A)). \tag{5.6}$$

Next, we need to consider the final aspect of household decision-making, which is the household budget constraint. Households aim to minimize the cost of achieving utility given a set of prices and income. This optimization problem is expressed as follows:

$$\begin{aligned} &\max U = U(C, \text{ES}(E, A)), \\ &\text{s.t. } C(P_E, P_A, \text{ES}) + P_C C \leq Y, \end{aligned} \tag{5.7}$$

P_E: Energy price
P_A: Appliance price
P_C: Price of other goods
Y: Household income.

At this point, a household's decision involves maximizing utility given a certain level of income or budget. This encompasses two decision-making aspects:

- Choosing the optimal combination of energy services (ES) and other goods (C) to maximize utility (U) within the constraints of income (Y).
- Selecting the optimal combination of inputs to minimize production costs while achieving the desired level of energy services.

5.5 Demand-Side Management

In the final section of this chapter, we will explore several policies related to energy demand-side management (DSM). DSM refers to strategies and initiatives aimed at altering consumer electricity usage patterns to enhance efficiency, reduce peak demand, and lower overall energy costs. By actively engaging consumers in energy consumption decisions, DSM promotes practices that can lead to more sustainable energy use. Generally, achieving DSM requires simultaneous efforts in load control management and the automation of remote metering and billing. Here, we primarily discuss key policies that assist in load control. These policies aim to optimize energy usage and demand, alleviate peak demand, and enhance overall efficiency, thereby supporting sustainable development. Specifically, we will focus on *Time-of-Use* (TOU) pricing and *Critical-Peak Pricing* (CPP), analyzing their mechanisms and impacts on energy demand management.

TOU pricing is regarded as an effective policy tool to enhance consumer awareness and increase household price elasticity, thereby reducing peak electricity demand and improving market efficiency. TOU pricing sets electricity prices based on the time of day, reflecting variations in electricity demand and supply. The total demand for electricity exhibits strong patterns. For instance, midday and dinnertime may represent peak energy usage periods, while the demand from midnight to early morning is typically at its lowest. The primary goal of TOU is to encourage consumers to shift their electricity usage to off-peak hours when demand is lower (e.g., 6 p.m. to 7 p.m., certainly not 6 p.m. to 11 p.m.!), thereby reducing strain during peak periods. When prices vary according to consumption times, users might adjust their habits—such as scheduling energy-intensive tasks like laundry for lower cost periods. Studies have shown that TOU has effectively reduced peak demand in several regions, resulting in significant savings on infrastructure investments needed for new power generation capacity. A study on the TOU policy reform in Spain (Fig. 5.13) found that TOU pricing reduced household electricity consumption by up to 9% during peak periods and permanently altered household electricity usage habits (Enrich et al., 2024).

Critical-peak pricing, on the other hand, is a pricing strategy focused specifically on addressing extreme peak demand periods. Compared to TOU, CPP emphasizes short-term demand adjustments rather than a continuous pricing system. Therefore, CPP is often implemented on a temporary basis; when needed, the electricity management authorities issue notifications to implement CPP and specify the duration of the temporary pricing. The effectiveness of CPP is evident in its ability to significantly curtail electricity demand during peak hours, thereby safeguarding grid stability and reducing reliance on costly and often polluting backup generation sources.

Electricity, as a commodity, operates under standard price mechanisms; when prices rise, consumers generally ponder reducing their usage. However, because electricity is a necessity, its price elasticity tends to be relatively low. In other words, people are less likely to forgo the benefits associated with electricity consumption simply to save a small amount on their utility bills. This situation is further complicated for firms where adjusting production schedules can be prohibitively expensive. Short-term changes in production planning often incur significant costs and logistical challenges, making companies less responsive to electricity price fluctuations. Indeed, empirical studies indicate that businesses may show little reaction to TOU policies. This limited responsiveness can stem from the essential nature of continuous power supply for operations, which creates a tension between cost-saving measures and the need to maintain productivity. As a result, while TOU and CPP schemes can be beneficial for overall demand management, the actual behavioral shifts may be

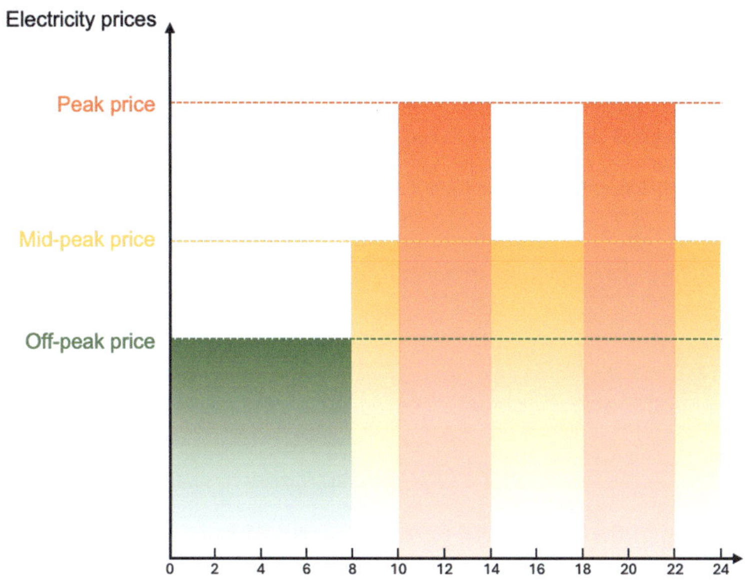

Fig. 5.13 TOU pricing program introduced in Spain

more gradual and less pronounced than anticipated, highlighting the complexities involved in effectively implementing such policies.

References

An, Y., & Zhang, L. (2023). The thirst for power: The impacts of water availability on electricity generation in China. *The Energy Journal, 44*(2), 250–450.

Apergis, N., & Ozturk, I. (2015). Testing environmental Kuznets curve hypothesis in Asian countries. *Ecological Indicators, 52*, 16–22.

Brunekreef, B., & Holgate, S. T. (2002). Air pollution and health. *The Lancet, 360*(9341), 1233–1242.

Dinda, S. (2004). Environmental Kuznets curve hypothesis: A survey. *Ecological Economics, 49*(4), 431–455.

EM-DAT. (2024). *The international disasters database*. https://www.emdat.be

Energy Information Administration (EIA). (2024). *Opendata—U.S. Energy Information Administration*. https://www.eia.gov/opendata/index.php#bulk-downloads

Enrich, J., Li, R., Mizrahi, A., & Reguant, M. (2024). Measuring the impact of time-of-use pricing on electricity consumption: Evidence from Spain. *Journal of Environmental Economics and Management, 123*, 102901.

Global Carbon Budget. (2024). *Global carbon budget*. https://globalcarbonbudget.org

Grossman, G. M., & Krueger, A. B. (1991). *Environmental impacts of the North American Free Trade Agreement*. National Bureau of Economic Research, Paper No. 3914.

Kuznets, S. (1955). Economic growth and income inequality. *American Economic Review, 45*, 1–28.

Lelieveld, J., Evans, J. S., Fnais, M., Giannadaki, D., & Pozzer, A. (2015). The contribution of outdoor air pollution sources to premature mortality on a global scale. *Nature, 525*(7569), 367–371.

Liu, Q., Teng, F., Nielsen, C. P., Zhang, Y., & Wu, L. (2024). Large methane mitigation potential through prioritized closure of gas-rich coal mines. *Nature Climate Change, 14*, 652–658.

Morice, C. P., Kennedy, J. J., Rayner, N. A., Winn, J. P., Hogan, E., Killick, R. E., Dunn, R. J. H., Osborn, T. J., Jones, P. D., & Simpson, I. R. (2020). An updated assessment of near-surface temperature change from 1850: The HadCRUT5 dataset. *Journal of Geophysical Research: Atmospheres, 126*(3), e2019JD032361.

Myers, P., van, N. (1905). In mediaeval and modern history. In *Growth of nations*. Ginn & Company.

Our World in Data. (2024a). *Life expectancy*. https://ourworldindata.org/grapher/life-expectancy

Our World in Data. (2024b). *Human development index*. https://ourworldindata.org/grapher/human-development-index

Panayotou, T. (1993). *Empirical tests and policy analysis of environmental degradation at different stages of economic development*. International Labour Organization, ILO-WEP 2-22/WP. 238.

Shapiro, J. S. (2016). Trade costs, CO_2, and the environment. *American Economic Journal: Economic Policy, 8*(4), 220–254.

Stern, D. I. (2004). The rise and fall of the environmental Kuznets curve. *World Development, 32*(8), 1419–1439.

Stern, D. I. (2017). The environmental Kuznets curve after 25 years. *Journal of Bioeconomics, 19*(1), 7–28.

Sultana, F. (2022). The unbearable heaviness of climate coloniality. *Political Geography, 99*(99), 102638.

World Bank. (2024). *World development indicators*. https://databank.worldbank.org/source/world-development-indicators

Chapter 6
Energy Efficiency

The crew who burn from stem to stern
To win the foremost place,
The crew to row, the boat to go
The eight to win the race.
"A Racing Eight", James Lister Cuthbertson

6.1 Energy Productivity and Efficiency

Will technological advancement be the ultimate solution to our energy challenges? In the previous two chapters, we examined the energy supply crises and the pressures for transition in energy consumption. We have reached a point where, from both bottom-up and top-down perspectives, technological progress can significantly alleviate energy supply and demand crises. From the supply side, technological advancement allows us to discover more energy reserves and deliver energy at lower prices. On the demand side, technology enhances the efficiency of devices, enabling us to accomplish more with less power consumed. Remarkable!

However, the reality is likely to be more complex than we anticipate. In this chapter, we will explore critical topics related to energy productivity and efficiency. Indeed, the future of our energy landscape heavily relies on technological advancements and energy efficiency. Yet, we must also consider factors such as the rebound effect and technology lock-in, and how they influence the evolution of energy systems.

6.1.1 Energy Productivity

In economics, the concepts of productivity and efficiency, while related, have distinct meanings. Productivity refers to the output produced per unit of input utilized in a production process. It is a broad metric that encompasses various types of inputs,

L. Zhang, *Sustainability Economics*, https://doi.org/10.1007/978-3-031-87829-9_6

including labor, capital, and materials, and measures the overall effectiveness of these inputs in generating goods or services. Efficiency, on the other hand, specifically examines how well resources are used to achieve a particular output. It focuses on minimizing waste and maximizing output from given inputs. Essentially, while productivity assesses the quantity of output relative to input, efficiency is concerned with the quality of resource use and the degree to which inputs are optimized to produce that output.

Productivity typically compares output to input, measuring how effectively resources are transformed into goods and services. In the context of energy, energy productivity is defined as the amount of economic output generated per unit of energy consumed. It serves as a vital indicator of how efficiently an economy utilizes energy resources. Energy productivity can also be measured using other units related to the output generated from energy consumption, such as:

$$\frac{\text{cooling space}\left(\text{m}^3\right)}{\text{energy input}}, \frac{\text{electricity (kWh)}}{\text{energy input}}, \frac{\text{steel (ton)}}{\text{energy input}}, \frac{\text{GDP}(\$)}{\text{energy input}}, \ldots$$

The ratio of a country's gross domestic product (GDP) to its energy consumption over a specific period is an important indicator that effectively measures the value-added, or wealth, generated per unit of energy consumed in a nation or region. We may be more familiar with its reciprocal—energy intensity, as discussed in Chap. 5—which indicates the amount of energy required to produce one unit of economic output.

$$\text{Energy Productivity} = \frac{\text{GDP}}{\text{Energy Use}} = \frac{1}{\text{Energy Intensity}} \tag{6.1}$$

Many factors can influence macro energy productivity. Technological advancement plays a significant role, particularly those aimed at reducing energy usage. For instance, more advanced engines can utilize fuel more efficiently to generate power, producing less waste heat; energy-efficient lighting bulbs can provide illumination with less electricity; and new materials and designs can significantly lower the energy demands of buildings. However, it is important to recognize that more "advanced" technologies may not solely focus on energy efficiency. In reality, technological progress may involve sacrificing a degree of energy efficiency in favor of gains in other areas, such as automation.

A reassuring piece of news is that global energy productivity has been rising at a nearly steady rate. Figure 6.1 illustrates the trend of energy productivity at the global level from 1990 to 2022, which has seen significant improvements. The metric in the chart is measured in US$/kWh and represents the ratio of global GDP to primary energy consumption. The global GDP is the sum of all countries' GDP in 2017 international dollars, while the major energy consumption is standardized to be measured in terms of electricity. For instance, the global energy productivity in 1990 was 0.48, meaning that for every kWh of energy consumed, the world can

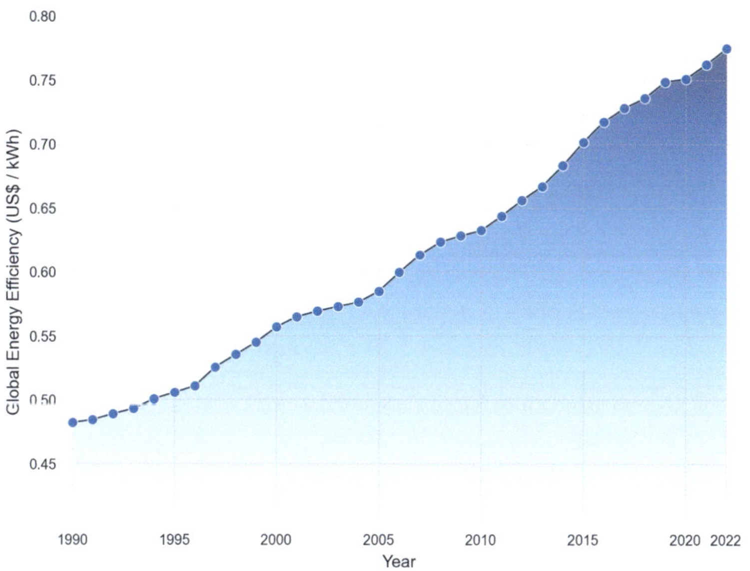

Fig. 6.1 Global energy productivity trends. *Data source* EIA (2024) and World Bank (2024)

generate $0.48 in terms of added value. Three decades later, the wealth generated from the same amount of energy consumption has doubled.

Furthermore, as the global focus on sustainable development has intensified, policy initiatives and technological innovations have further accelerated gains in energy productivity. In the 2010s, as the global economy gradually recovered from the 2008 Financial Crisis and GDP steadily increased, energy consumption per unit declined. Despite the impact of the COVID-19 pandemic in 2019 and 2020, which led to a great shock to both energy supply and demand, energy productivity remained at a high level, demonstrating the potential and accomplishments in achieving sustainable development goals worldwide.

There are significant disparities among different countries. Table 6.1 utilizes country-specific data from the World Bank (2024), highlighting the substantial differences in energy productivity improvements across various nations and regions, despite the dataset only covering up to 2014. For instance, Brazil's energy productivity declined from 13.44 in 1990 to 12.66 in 2014, reflecting a decrease of 5.81%. This trend highlights the challenges Brazil faces in energy management and technology application, particularly the conflict between economic development and environmental sustainability. In contrast, China's energy productivity experienced significant growth during the same period, rising from 2.15 to 6.16, an increase of 187% (Filippini & Zhang, 2016; Zhang & Adom, 2018). This progress is attributed to strong government policies, active investment in renewable energy, and modernization of industrial processes (Jaunky & Zhang, 2016).

Table 6.1 Changes in energy productivity in major countries from 1990 to 2014

Country Name	Energy productivity[a] 1990	Energy productivity 2014	Change (%)
Brazil	13.44	12.66	– 5.81
China	2.15	6.16	187.00
France	10.32	14.12	36.80
Germany	9.65	15.78	63.55
India	6.23	9.67	55.19
Indonesia	9.06	12.06	33.03
Japan	9.98	12.36	23.85
Russian Federation	5.04	7.46	48.10
South Africa	4.85	5.54	14.33
United Kingdom	9.89	18.40	85.98
United States	5.78	9.08	56.94

Note [a]US$ per kg of oil equivalent. The GDP is standardized using constant 2021 PPP. *Data source* World Bank (2024)

6.1.2 Energy Efficiency

Energy efficiency, often referred to as energy utilization efficiency, focuses on the effectiveness with which energy is converted into useful work or output. It represents a specific aspect of productivity related to energy consumption. While energy productivity emphasizes overall output relative to energy input, energy efficiency indicates the performance of individual systems and processes (see Fig. 6.2). Despite their differences, these two concepts are frequently used interchangeably, as improvements in energy efficiency can lead to enhanced energy productivity.

There are many approaches for the measurement of energy efficiency. In thermodynamics, energy efficiency, or the thermodynamic efficiency factor, is defined as the ratio of effective energy output to the energy input of a system or process. By calculating one minus the thermodynamic efficiency factor, one can measure the extent of energy waste in that system or process. Here, both the numerator and the denominator are measured in units of energy, and thus, the efficiency value is necessarily less than 1.

$$\omega = \frac{\text{useful energy output}}{\text{energy input}} \tag{6.2}$$

Energy efficiency metrics provide an important way to assess the performance of energy consumption systems or processes. However, single-factor efficiency indicators, which focus solely on the relationship between energy input and final output, have limitations. These indicators may overlook the complementary or substitutive roles of energy with other inputs, such as labor and capital. For example, in a factory,

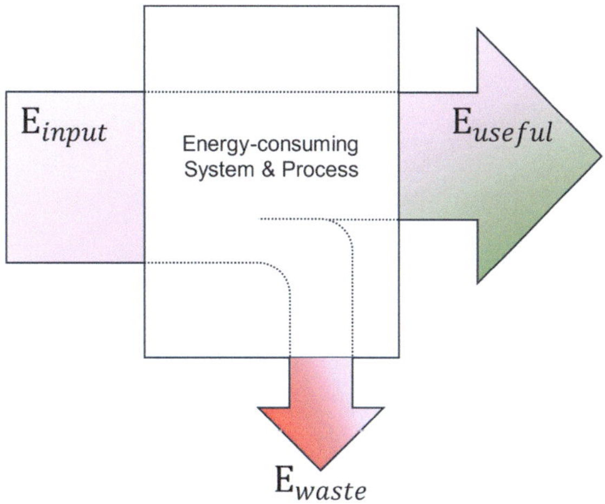

Fig. 6.2 Input, output, and waste of an energy-consuming system

one worker operating a Model-A machine consumes 100 kWh and assembles 100 mobile phones per day. After a technological upgrade, the same worker can operate a Model-A-Pro machine, consuming 250 kWh to assemble 200 mobile phones per day. From a single-factor productivity perspective, energy efficiency decreases as the electricity consumed per unit of output increases when using the pro version machine. However, labor productivity increases twofold when considering the output per worker. Therefore, multi-factor productivity or efficiency approaches have been developed to quantify such system-wide technology changes.

In practice and research, the methods used for measuring efficiency include the Solow Residual, Growth Accounting, and Frontier Analysis. Among these, Frontier Analysis is the most widely used approach. Here, we briefly outline the concepts, and the process related to measuring technical efficiency using Frontier Analysis.

From an economic perspective, the essence of the production process for a given product is a process of transforming inputs into outputs. This process is abstracted and quantified by economists as the production function, or "technology".

$$y = f(\text{energy inputs, other inputs}) \tag{6.3}$$

The production possibilities curve (PPC) represents the minimum input required to produce a given output. Given a specific quantity and combination of inputs, the maximum output is determined by technological and physical constraints. For instance, a company using solar energy to break down water into hydrogen and oxygen faces a physical limit on the amount of hydrogen produced from a given amount of water. In Panel A of Fig. 6.3, the PPC divides the area of all input combinations into two parts: the possible region and the impossible region. Point *A* lies on the PPC, indicating that Company *A* can produce output *Y* using the input combination (i, e_A). Clearly, Company *B*, located within the possible region, can also achieve output *Y*, but it requires more energy input than Company *A*. This suggests that

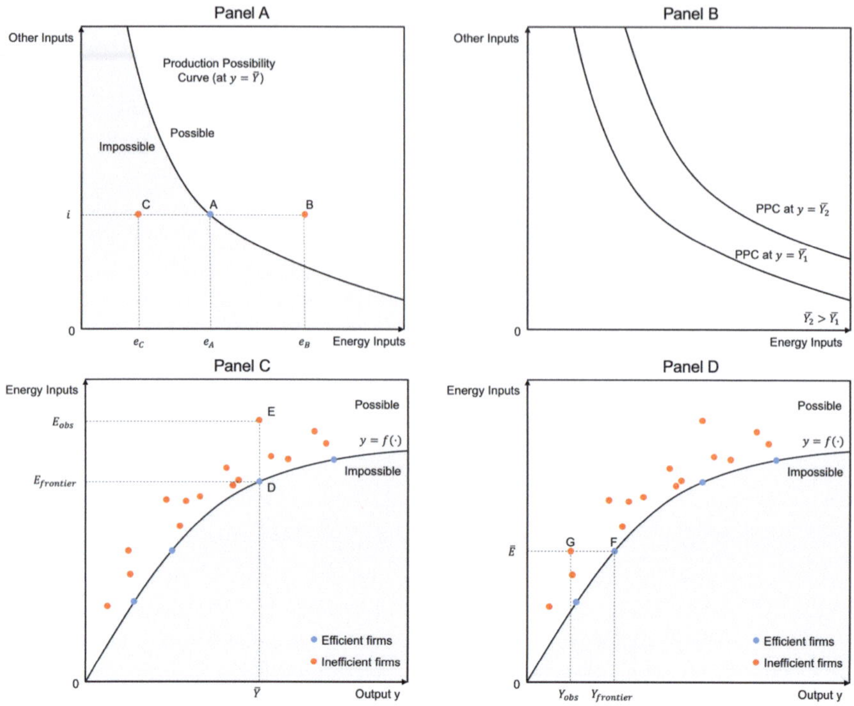

Fig. 6.3 Production frontier and technical efficiency

Company B is less efficient than Company A. Point C serves as an example of the impossible region. With other inputs held constant, no company can produce $\overline{Y}$ units of output using less energy than Company A. As shown in Panel B of Fig. 6.3, any two production possibilities curves do not intersect.

Panel C of Fig. 6.3 illustrates a method for defining and calculating energy efficiency. In this figure, the vertical axis represents energy inputs, while the horizontal axis indicates output. The curve that intersects the origin is known as the "frontier" (or "frontier surface" when multiple inputs are concerned). This frontier is a PPC and is estimated based on all observed input–output combinations within the sample. The frontier possesses all the characteristics of a PPC. All firms located on the frontier are considered efficient. In Panel C of Fig. 6.3, input-oriented technical efficiency is defined. Clearly, at the given output level $\overline{Y}$, the firm at point E is less efficient than the firm at point D on the frontier. Therefore, we can define the energy efficiency of any firm as:

$$\mathrm{EF}_{\mathrm{input}} = \frac{E_{\mathrm{frontier}}}{E_{\mathrm{obs}}} \leq 1. \tag{6.4}$$

Thus, energy efficiency EF measures the distance of a given input–output combination from the frontier and is a value that never exceeds 1. It equals 1 only when the input–output combination lies on the frontier. Similarly, the out-oriented energy efficiency can be defined as:

$$\mathrm{EF}_{\mathrm{output}} = \frac{Y_{\mathrm{obs}}}{Y_{\mathrm{frontier}}} \leq 1. \tag{6.5}$$

Input-oriented energy efficiency emphasizes minimizing energy input for a specified output target. In contrast to input-oriented energy efficiency, output-oriented energy efficiency highlights maximizing output given a certain level of energy input.

The input–output combinations of different decision-making units (DMUs) can be observed, while the position of the frontier cannot be directly measured. Therefore, estimating the frontier is crucial. In practice, methods such as Data Envelopment Analysis (DEA), Stochastic Frontier Analysis (SFA), and some non parametric approaches are used to estimate the frontier.

Energy efficiency offers a new perspective on energy issues, particularly concerning energy policy. In practice, the targets of energy and climate policies typically are defined by "reducing energy intensity" while "improving energy efficiency" as the recommended means to achieve these goals. Defining policy objectives based on energy intensity is highly feasible. However, measuring energy efficiency requires more data collection and involves more complex mathematical modeling. In contrast, calculating energy intensity is much less costly. Yet, the concept of energy efficiency aligns better with the expectations of governments, the public, and economists regarding energy transition, while energy intensity can be easily manipulated. Some local governments may consider temporarily shutting down large, energy-intensive enterprises to meet the energy intensity-oriented targets. However, these enterprises might be allowed to resume operations after the assessment period to maintain local tax revenues. Such strategic behavior not only harms businesses and the market economy but also impedes energy transition (Zhang, 2017).

Case

The Energy Efficiency in the UK

The energy transition in the UK has been marked by significant changes aimed at reducing carbon emissions and promoting sustainability. Over the past few decades, the UK has shifted its energy mix, moving away from coal and increasing the share of renewable energy sources like wind, solar, and biomass. Robust government policies, including subsidies for renewable energy projects and investments in energy efficiency, have supported this transition.

Since 2005, the UK has even seen a decline in primary energy consumption (the red line in Fig. 6.4). This reduction can be attributed to a combination of factors, including population control, the restructuring of the economy, the outward migration of energy-dependent industries following the 2008 crisis, and the shift to renewable energy sources. The improvement in energy efficiency has clearly played a significant role in this trend. Over the past half-century, from the 1970s to the 2020s, the UK's energy efficiency has quadrupled. This substantial increase reflects advancements in technology, changes in consumption behaviors, and effective energy policies aimed at promoting sustainability.

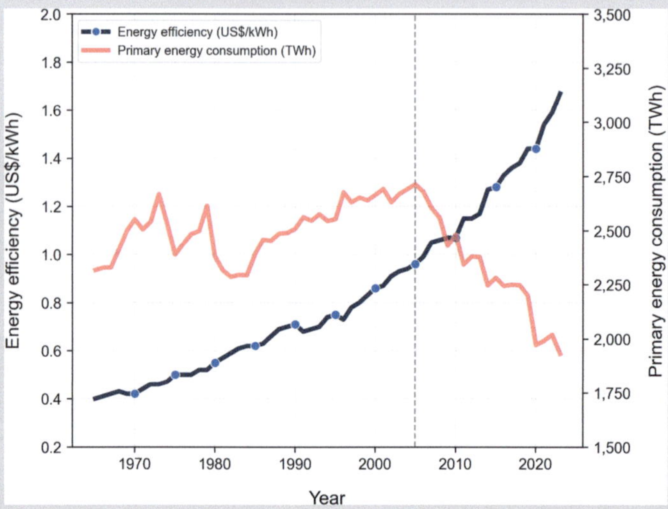

Fig. 6.4 Energy efficiency and primary energy consumption in UK, 1970–2022. *Data source* U.S. EIA (2024)

6.2 Green Technology and Re-directed Innovation

6.2.1 Emergence of Green Technologies

The term "green technology" has become a common phrase in everyday life. You may find this term appearing in over half of product advertisements nowadays! It is a broad concept that refers to technologies and innovations aimed at reducing or eliminating negative environmental impacts. These technologies focus not only on energy efficiency and resource conservation but also on pollution reduction, waste management, and the use of renewable energy sources. The "green" in green technology symbolizes environmental protection and sustainability, representing a commitment to respecting and safeguarding the natural environment. It emphasizes achieving a balance between economic development and environmental protection, thereby promoting sustainable societal growth. A closely related but narrower concept is clean technology. Clean technology specifically refers to technologies aiming at improving efficiency and reducing environmental impact.

Determining a specific origin for green technology is challenging. However, it is undeniable that green technology is trending. The Dutch invented the windmill in the thirteenth century to harness wind energy for grinding grain. In the early nineteenth century, scholars began to recognize the adverse effects of coal pollution and advocated for improvements in production processes to reduce pollution. In the United States, during World War II, more than 400,000 volunteers collected metal, paper, rubber, and other materials for contributing to the war, raising awareness of resource conservation and reuse, marking a significant milestone in the history of green technology.

In the past fifty years, issues related to resources, pollution, and carbon emissions have gained recognition from governments, appearing in their policy documents. A recent example is the Infrastructure Investment and Jobs Act, signed by President Biden in November 2021, which allocates $1.2 trillion to support the development of green technologies. This act is considered the largest investment in clean energy transmission and electric vehicle infrastructure in human history and includes funding to convert thousands of schools and transit buses to electric power, as well as the establishment of a new agency to build a resilient, clean grid. Ultimately, the pursuit of green development and growing awareness within society have driven policies, innovations, and applications related to green technology.

When you turn on the television, you will find that advertisers label nearly all products with the term "green technology." The concept of green technology is indeed extensive. Generally, green technology may exhibit one or more of the following characteristics: (1) reduced resource input, such as energy savings; (2) lower environmental impact; (3) recycling and reuse; (4) higher efficiency; (5) decreased reliance on fossil fuels and increased use of green energy. The application of green technologies spans multiple sectors, including energy, construction, agriculture, manufacturing, transportation, and waste management. For instance, the energy sector significantly reduces reliance on fossil fuels by developing renewable energy technologies such as solar and wind power. In the construction sector, the use of efficient insulation materials and smart energy management systems effectively lowers energy consumption and enhances residential comfort. The promotion of these technologies not only aids in environmental protection but also fosters sustainable development.

The number of patent filings is often regarded as a key indicator of innovation and technological advancement. Figure 6.5 illustrates the annual filing numbers of green technology patents in energy sectors such as solar and wind power since 2000. From 2000 to 2010, the number of green technology patents rose exponentially. In the following decade, annual filings remained steady at around 60,000 to 70,000 per year. This significant increase in green patents reflects the global emphasis on and demand for sustainable development, highlighting the crucial role of green technology in energy transition and green growth.

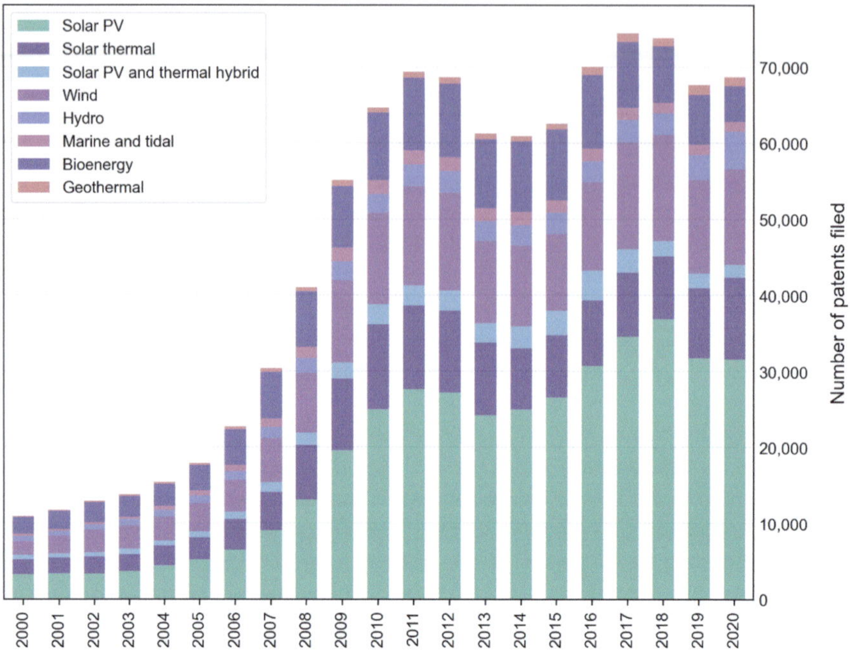

Fig. 6.5 Emergence of green patents. *Note* Adapted from Our World in Data (2024)

Figure 6.5 also reflects the concept of redirected innovation. This emerging concept refers to how businesses and researchers shift their innovation focus toward green technologies, facing urgent climate challenges. Of course, policy and market incentives may also drive innovators to adjust their priorities. For example, temperature fluctuations caused by global climate change can negatively affect the US agriculture sector. In response, the agricultural sector is redirecting its innovation efforts by allocating more research and development resources to the creation of heat- and drought-resistant seeds to mitigate the impacts of climate change (Moscona & Sastry, 2023).

In summary, the emergence of green technologies, particularly their development over the past 20 years, has significantly enhanced energy efficiency and offers a promising outlook for a clean, sustainable energy future. Increasing efforts are being directed toward the advancement of green technology, including government policies and investments, growing capital from entrepreneurs and innovators, and heightened academic research in this field. Despite these initiatives, there is a call for even greater efforts, as the urgency of resource depletion and climate change suggests that we may not be doing enough.

6.2.2 Negative Carbon Technologies

Negative carbon technology (NCT) refers to technologies capable of removing carbon dioxide from the atmosphere and either permanently storing it or converting it into other forms. Carbon that is removed from the atmosphere and is not expected to return for a significant period is referred to as a carbon sink. As a significant aspect of green technology, it warrants dedicated discussion in this section. Other green technologies aiming at improving energy efficiency or promoting recycling can only minimize the incremental GHG emissions released into the atmosphere, lacking the ability to remove existing carbon dioxide. NCT technologies are important from several perspectives.

First, achieving carbon neutrality relies on NCT. True carbon neutrality means net-zero emissions from human activities, where the anthropogenic atmospheric greenhouse gas increases are zero. In reality, it is challenging for other green technologies to reduce emissions from all human activities to zero. Therefore, significant advancement in NCT is essential for achieving ambitious carbon neutrality goals. Furthermore, NCT allows for a transition period in energy transformation. As climate change intensifies, the urgency of using negative carbon technologies to counteract the effects of high atmospheric carbon dioxide levels becomes increasingly critical. The global greenhouse effect is determined not by the incremental emissions of greenhouse gases, but by their existing concentrations. NCT provides a viable pathway for removing these gases from the atmosphere. One of the key benefits of developing NCT is that it provides a buffer for human activities, such as fossil fuel combustion. Through NCT, we can "offset" the greenhouse gases emitted by human activities. While the transition to clean energy and sustainable development is not yet complete, NCT can help mitigate climate change and "buy time" for this transition.

There are various types of negative carbon technologies, including Direct Air Capture (DAC) and Bioenergy with Carbon Capture and Storage (BECCS). Among these, Carbon Capture and Utilization (CCU) and Carbon Capture and Storage (CCS) are two primary technologies. CCU involves using captured carbon dioxide to produce goods such as plastics, batteries, and fuels, thereby promoting resource recycling. A rapidly growing application of CCU is Power-to-X (P2X), which utilizes electricity from renewable sources such as solar, hydro, or wind energy to convert water into green hydrogen (see Fig. 6.6). This hydrogen can be used directly for decarbonizing industrial processes or can react further with carbon dioxide or nitrogen from the air to produce various fuels and chemicals. CCS focuses on the safe underground storage of captured carbon dioxide to prevent its re-entry into the atmosphere.

Although CCUS is critical, its development faces various limitations. China and the United States are the countries with the most CCUS patents. Figure 6.7 illustrates the number of CCS patents filed in China, the United States, and globally since 2000. Since 2000, especially after 2004, there has been rapid advancement in CCS and related technologies, with the number of related patents increasing significantly. By 2012, the annual number of CCS patents filed peaked at over 2000. However, since

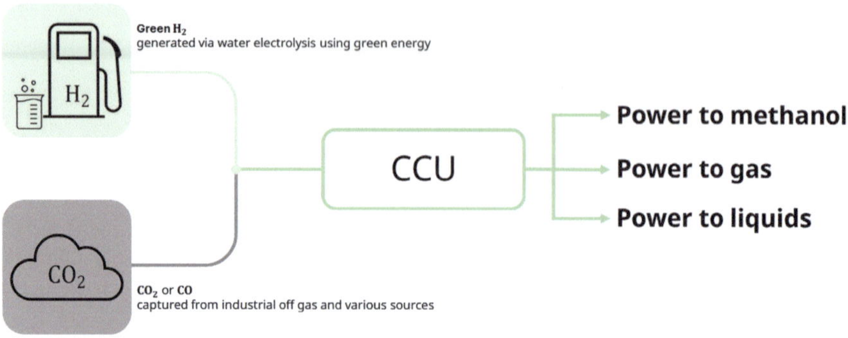

Fig. 6.6 Power-to-X

the COVID-19 pandemic, there has been a noticeable decline in CCS patent applications. Indeed, the impact of COVID-19 may have significantly hindered research and development activities. We must also consider other factors that restrict the advancement of CCUS technologies.

In addition to government support and subsidies, the incentives for investment in CCUS technologies depend on their revenues from the carbon market. The carbon market is a platform that allows for the trading of carbon emissions permits or carbon credits. The "negative emissions" generated by CCUS can be traded as carbon credits once verified. Consequently, a well-structured carbon market and sufficiently high carbon prices significantly impact the market returns and, thus, incentives for CCUS

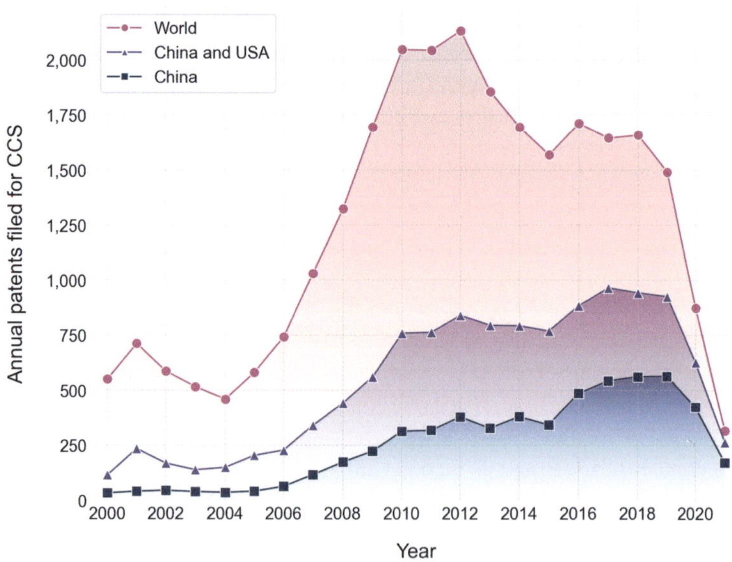

Fig. 6.7 Trend of the number of CCS patents filed. *Note* Adapted from Our World in Data (2024)

investments. We will discuss carbon market-related topics in subsequent chapters. Additionally, CCU can generate revenue by selling its products. However, the key challenge is that the global carbon market and its operational mechanisms are still in the exploratory stage. Therefore, effective policy design and comprehensive support programs are particularly important for incentivizing investments in CCUS at this stage.

6.3 Rebound Effect and Technological Lock-In

6.3.1 Understanding the Rebound Effect

Is green technology the silver bullet for solving energy issues? The answer is likely much more complex than we imagine. Technological advancements and improvements in energy efficiency do not necessarily lead to a reduction in energy consumption or fossil fuel use. Entrepreneurs who control fossil fuel assets or have heavily invested in non-renewable energy sectors are unlikely to simply wave the white flag and shift to renewable energy. In fact, substantial investments and research and development occur within these fossil fuel-related companies.

In economics, the rebound effect (also referred to as the Jevon Paradox) is a counterintuitive phenomenon (Jevons, 1865). It suggests that improvements in resource efficiency can lead to lower usage costs, which in turn may increase demand, resulting in greater resource consumption rather than a decrease. Additionally, efficiency gains from technological advancements can boost real income, accelerate economic growth, and increase population growth, further heightening resource demand. When the demand increase driven by technological innovation dominates, even outweighing the resource savings from efficiency improvements, the Jevons Paradox occurs.

At the micro level, examples of the rebound effect can be found throughout our daily lives. For instance, after purchasing a more efficient car, one might drive more. The macro rebound effect is more complex, involving adjustments at the entire economic level. Improvements in macro energy efficiency can influence energy demand through channels such as price changes and market reallocation. Figure 6.8 depicts a possible scenario of the rebound effect. Initially, the supply and demand curves for a product are in equilibrium at point $A(p_1, c_1)$. Assuming a technological shock occurs that enhances production efficiency, the production costs of the product significantly decrease. In a competitive market, the supply curve shifts to the right, establishing a new market equilibrium at point B. At this point, the market price drops (p_2), and the consumption of the product increases (c_2).

Observing and measuring the rebound effect is challenging at both micro and macro levels, requiring comprehensive data support and a proficient application of input–output or econometric models.

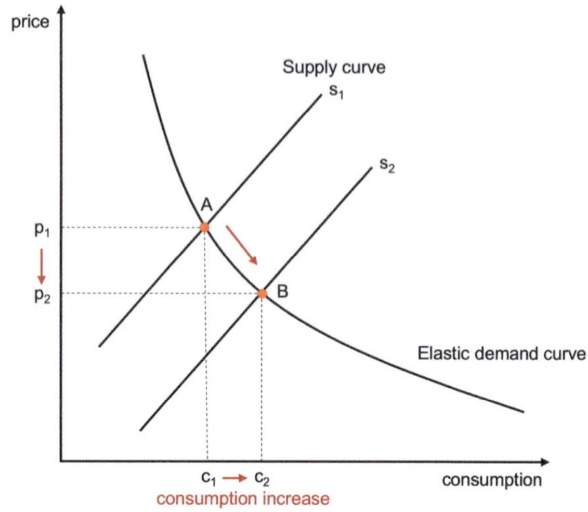

Fig. 6.8 Rebound effect

The rebound effect can vary significantly across different consumption sectors or products. For example, rebound effects for heating and electricity tend to be relatively small, while those for transportation and food consumption can be quite large (Hoolohan et al., 2013). A study on ten widely endorsed measures to reduce GHG emissions from UK households—seven of which involve using more efficient appliances—found that all these measures can collectively reduce household emissions under cost–benefit considerations. However, in terms of the magnitude of the rebound effect, measures affecting household energy use exhibit a moderate rebound effect (0–32%), while those impacting vehicle fuel use show a larger rebound effect (25–65%). Measures aimed at reducing food waste demonstrate a very large rebound effect (66–106%) (Chitnis et al., 2014). It shows that the rebound effect is also correlated with household income; lower incomes may lead to a larger rebound effect. Additionally, social norms, environmental values, and consumption habits can influence the rebound effect. Policies and market conditions in various countries and regions also play a role in shaping the rebound effect.

Overall, in most cases, the savings effects of technological advancements and efficiency improvements are not completely outweighed by the rebound effect. Additionally, technological progress can reduce the environmental impact of resource and energy consumption. In fact, scholars and policymakers are generally more inclined to believe that the Jevon Paradox does not occur, although some researchers express skepticism about it (Alcott, 2005).

Reading

William Stanley Jevons

William Stanley Jevons (Fig. 6.9) was a British economist and philosopher renowned for his significant contributions to economic theory. He was a key proponent of marginal utility theory, arguing that consumers consider the marginal utility of goods when making decisions. His work, *The Theory of Political Economy*, marked a transition in economics from classical theory to a more modern microeconomic analysis, particularly in the areas of price, value, and utility.

Fig. 6.9 William Stanley Jevons (1835–1882)

Jevons introduced the famous Jevons Paradox. In his 1865 book, *The Coal Question*, he posited that when the efficiency of a resource increases, the overall demand may rise even if the consumption of individual units decreases. This conclusion stemmed from his observations and studies of coal-burning factories during the Industrial Revolution. For instance, improvements in steam engine efficiency led to an increase in coal consumption. This observation challenges the simplistic assumption that technological advancements result in resource conservation and provides an important theoretical foundation for studying energy policy and its potential impacts.

6.3.2 Policy Implications of Rebound Effect

The rebound effect has important implications for energy efficiency policies as it highlights the complex relationship between improved energy efficiency and actual energy consumption. When consumers adopt more energy-efficient technologies, such as fuel-efficient vehicles or energy-saving appliances, the expected reduction in energy use may not materialize as anticipated. For policymakers, this suggests that merely promoting energy efficiency may not be sufficient to achieve desired energy savings or environmental goals. Moreover, the impacts of the rebound effect extend beyond energy savings to broader economic and welfare considerations. While increased energy efficiency can lower energy costs and stimulate economic activity, it may also lead to increased energy consumption, potentially offsetting environmental benefits. Therefore, a more nuanced approach that considers the potential of the rebound effect is crucial for effective policy design.

Economic instruments such as energy taxes and carbon taxes can mitigate the rebound effect by increasing energy prices, thereby reducing the economic incentives for energy consumption. For instance, if a carbon tax is levied on fossil fuels, the savings from using more efficient appliances or vehicles may be offset by higher costs associated with increased consumption. This creates economic incentives for consumers to not only adopt energy-saving technologies but also to favor renewable energy options. Policymakers must tailor and design these pricing mechanisms carefully to avoid negative impacts on income and equity. For example, using a portion of carbon tax revenues to fund further energy-saving initiatives or provide rebates to low-income households can ensure that policies remain fair while still promoting energy efficiency.

Another approach to reducing the rebound effect involves public awareness and educational campaigns that emphasize the importance of energy-saving behaviors alongside the adoption of energy-efficient technologies. Public awareness and behavior are key factors in the existence of the micro rebound effect. Educational campaigns can inform consumers about the potential for rebound effects and encourage them to adopt practices that reduce energy use, such as driving less or adjusting thermostat settings. For example, a campaign could highlight the benefits of combining energy-efficient appliances with behavioral changes, such as turning off lights when not in use or using public transportation. Using economic incentives to shape resident behavior is considered more effective than solely relying on educational campaigns. For example, Time-of-Use (TOU) pricing policies can significantly influence residents' electricity consumption habits.

Single policy instruments often struggle to effectively address the rebound effect, necessitating a combination of various policy tools. An effective policy mix should include measures such as technical standards, economic incentives, and behavioral guidance. For instance, technical standards can ensure the energy efficiency of new technologies, while economic incentives (such as subsidies and tax breaks) can promote the adoption of green technologies. Although the rebound effect is a concern, some scholars argue that excessive focus on it may detract from more significant issues, such as the welfare impacts of energy efficiency policies.

6.3.3 Understanding the Technological Lock-In

Technological lock-in refers to the phenomenon where a technology or technological system becomes widely adopted, making it difficult for alternative technologies to enter the market or gain acceptance due to various economic, social, and technical factors. This concept was first introduced by economist Paul David (1935–2023) in 1985. The core idea of technological lock-in is that once a technology or system is established, it tends to evolve along a specific path, making it challenging to shift away, even in the presence of potentially superior alternatives. In his paper, David highlighted that the historical context related to the mechanical limitations of early typewriters led to the adoption of the QWERTY keyboard, despite the development of more efficient options, such as the Dvorak Simplified Keyboard (David, 1985). To this day, the QWERTY layout remains commonly used.

Technological lock-in is closely related to path dependence, which refers to the profound influence of historical events and early choices on future development trajectories. Lock-in is a specific manifestation of path dependence, emphasizing the randomness and irreversibility of early technology choices. Early technology selections create cumulative effects that guide subsequent technological development along established paths, resulting in lock-in (Jin et al., 2024).

The existence of technological lock-in is associated with technological paradigms, which are cognitive frameworks and rules shared by a community of technology stakeholders (such as engineers, businesses, and technical institutions). These paradigms define the nature of technological problems and potential solutions, reducing the costs of technology promotion and dissemination through standardization. However, the presence of technological paradigms can also lead to technological progress following established directions, making disruptive innovation less likely. This results in technological change being more incremental rather than radical.

Technological lock-in is also related to increasing returns to adoption. As production scales up, the unit costs of technology applications decrease. Through multiple iterations, technologies improve incrementally. In contrast, new technologies typically enter the market on a smaller scale and lack the same economic advantages. Additionally, users may face learning costs when adopting new technologies and products. Disruptive innovations often fail to gain traction because they cannot complement or integrate with existing technological networks.

The effects of technological lock-in have profound implications for the energy sector, particularly in the context of energy transition and the development of green technologies. Established fossil fuel technologies have created extensive infrastructure and market networks, resulting in significant economies of scale and learning effects that hinder the rapid adoption of emerging green technologies, such as renewable energy and electric vehicles. This lock-in effect slows down the energy transition process and impedes the widespread application of low-carbon technologies. Furthermore, existing policies and market mechanisms often favor established technological paths, reinforcing the lock-in effect. To break this cycle, policymakers need to take proactive measures, such as providing R&D funding, implementing carbon taxes, and subsidizing green technologies, to promote diversification and sustainable development in energy systems (Zhang & Zhang, 2023). Overcoming technological lock-in can accelerate the energy transition, leading to more environmentally friendly and efficient energy use.

Critical Thinking

Scale Effects and the Price of Computer Storage

From 1960 to 2020, the price of computer storage media underwent a remarkable transformation, dropping exponentially from $1 trillion per TB to just $100 per TB (see Fig. 6.10). This dramatic price change reflects not only rapid advancements in electronic technology but also the significant benefits of technological iterations and economies of scale, leading to a swift reduction in costs. Nevertheless, existing storage technology paths also create high barriers to the entry of emerging storage technologies. New technologies must achieve substantial performance breakthroughs and retain their costs within a reasonable range to compete for market share against established storage products.

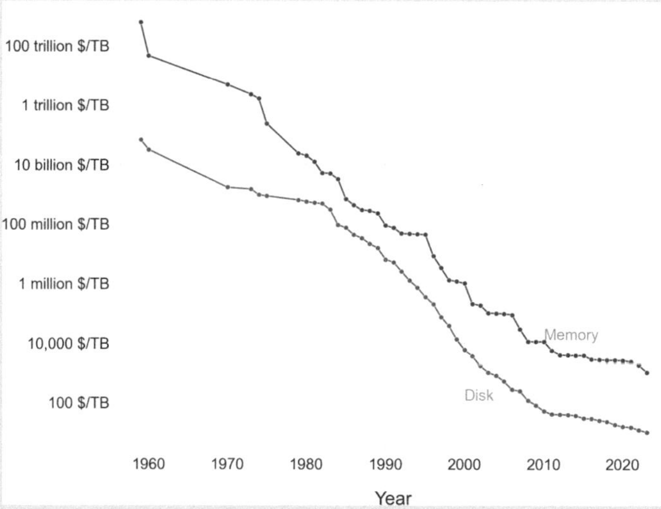

Fig. 6.10 Price of computer storage since 1960s. *Note* Adapted from Our World in Data (2024)

6.3.4 Energy Behavior Bias

The behavioral perspective is crucial for understanding the relationship between technological advancement and energy consumption. Energy behavior biases refer to systematic deviations in consumer decision-making during energy consumption, driven by cognitive biases and irrational tendencies. These biases include, but are not limited to, status quo bias, loss aversion, sunk cost effect, temporal and spatial discounting, and social norm influence (Frederiks et al., 2015). These biases often lead consumers to act differently from predictions made by rational choice models when faced with energy use and conservation decisions.

- *Status Quo Bias* indicates a tendency for individuals to prefer maintaining their current state rather than making changes. Even when changes could result in better outcomes, people often stick with the status quo due to the effort and risks associated with change.
- *Loss Aversion* refers to the phenomenon where the pain of losing is felt more acutely than the pleasure of gaining an equivalent amount. Consequently, individuals are more inclined to avoid losses than to pursue gains in their decision-making processes.
- *Sunk Cost Effect* occurs when individuals continue to invest time, money, or effort into a project or behavior despite unrecoverable investments. This tendency makes

it difficult for people to abandon ongoing projects, even when doing so would be more rational.
- *Temporal and Spatial Discounting.* Temporal discounting refers to the tendency of people to prioritize immediate benefits while overlooking future gains. Spatial discounting indicates that individuals are less concerned about distant events than immediate ones. Both biases can lead to a focus on short-term benefits when making long-term decisions.
- *Social Norm Bias* indicates that individuals' behaviors are affected by the actions of those around them and societal expectations. People tend to follow social norms, even when those norms are not always rational. This influence can manifest through descriptive norms (the prevalence of certain behaviors) and injunctive norms (the social approval of specific actions).

These energy behavior biases significantly impact the development of green technologies. Due to status quo bias and loss aversion, consumers may adopt a conservative stance toward new technologies, even when these technologies offer substantial long-term energy savings and environmental benefits. Such conservatism can delay the adoption and application of new technologies, affecting the pace and breadth of technological advancement. For instance, even if consumers recognize the benefits of energy-saving measures, the sunk cost effect may lead them to continue using outdated, energy-intensive devices instead of investing in more efficient alternatives. Additionally, temporal discounting may cause consumers to opt for lower short-term costs while overlooking long-term energy savings.

Behavioral biases can also amplify the rebound effect. For example, after adopting energy-saving technologies, consumers might increase their energy consumption in other areas (such as using air conditioning more frequently or purchasing higher powered appliances) due to perceived energy savings, thus negating some or all of the benefits of the energy-efficient technology. This behavior exacerbates the rebound effect, undermining the effectiveness of energy conservation policies.

Understanding energy behavior biases is essential for formulating effective energy policies and promoting green technologies. Policymakers should consider these biases when designing tools that can counteract consumers' irrational behaviors, such as interventions based on behavioral economics to encourage more sustainable energy choices. As an exciting emerging interdisciplinary research direction, integrating behavioral economics with energy economics, even intersecting with social psychology, neuroscience, and brain science, explores how interventions and corrections of behavioral biases can facilitate the energy transition. Indeed, to achieve green development, decision-makers and scholars must also overcome the "Status Quo Bias" and innovate new strategies to investigate pathways for addressing real-world issues, such as correcting energy behavior biases.

6.4 The U.S. Shale Gas Revolution

The U.S. shale gas revolution refers to the significant increase in shale gas production in the U.S. since 2007, driven by advancements in horizontal drilling and hydraulic fracturing technologies. By 2018, production had risen more than 12-fold. This shift allowed natural gas to rapidly replace coal as the primary fuel source for the U.S. power sector, leading to a gradual decline in carbon dioxide emissions since 2007. However, despite the relatively clean nature of shale gas, concerns have emerged regarding its negative impact on green innovation and potential long-term environmental consequences (Acemoglu et al., 2023).

The shale gas revolution has also triggered a rebound effect, where increased efficiency and lower costs of shale gas usage result in a rise in overall energy consumption. Although shale gas emits less carbon dioxide per unit of energy compared to coal, its broader use may lead to an increase in total emissions. This phenomenon exemplifies the Jevons Paradox, where improved energy efficiency can lead to increased energy consumption, negating the environmental benefits of energy savings.

Additionally, the shale gas revolution has directed innovation toward fossil fuel technologies rather than clean energy solutions. As shale gas gained traction, more investment and research resources have shifted to natural gas and coal-related technologies, resulting in a notable decline in patents for green technologies. This change in innovation focus impedes the development and application of clean energy, further delaying the transition to zero-carbon energy sources.

The implications of this directed innovation for energy transition are profound. While the use of shale gas may reduce carbon emissions in the short term, it can hinder the development of renewable energy in the long term, ultimately leading to a decrease in social welfare. Therefore, policymakers must implement effective measures, such as incentivizing research and development of green technologies, to ensure a smooth energy transition and avoid falling into the trap of dependency on fossil fuels.

References

Acemoglu, D., Aghion, P., Barrage, L., & Hémous, D. (2023). *Climate change, directed innovation, and energy transition: The long-run consequences of the shale gas revolution.* National Bureau of Economic Research, Paper No. 31657.

Alcott, B. (2005). Jevons' paradox. *Ecological Economics, 54*(1), 9–21.

Chitnis, M., Sorrell, S., Druckman, A., Firth, S. K., & Jackson, T. (2014). Who rebounds most? Estimating direct and indirect rebound effects for different UK socioeconomic groups. *Ecological Economics, 106*, 12–32.

David, P. A. (1985). Clio and the economics of QWERTY. *The American Economic Review, 75*(2), 332–337.

Du, K., & Li, J. (2019). Towards a green world: How do green technology innovations affect total-factor carbon productivity. *Energy Policy, 131*, 240–250.

Energy Information Administration (EIA). (2024). *Opendata—U.S. Energy Information Administration.* https://www.eia.gov/opendata/index.php#bulk-downloads
Filippini, M., & Zhang, L. (2016). Estimation of the energy efficiency in Chinese provinces. *Energy Efficiency, 9*(6), 1315–1328.
Frederiks, E. R., Stenner, K., & Hobman, E. V. (2015). Household energy use: Applying behavioural economics to understand consumer decision-making and behaviour. *Renewable and Sustainable Energy Reviews, 41*(1364–0321), 1385–1394.
Hoolohan, C., Berners-Lee, M., McKinstry-West, J., & Hewitt, C. N. (2013). Mitigating the greenhouse gas emissions embodied in food through realistic consumer choices. *Energy Policy, 63*, 1065–1074.
Jaunky, V. C., & Zhang, L. (2016). Convergence of operational efficiency in China's provincial power sectors. *The Energy Journal, 37*(01), 3–28.
Jevons, W. S. (1865). *The coal question.* Macmillan & Co.
Jin, W., Ma, M., & Zhang, L. (2024). Path dependence in environment-related innovation: Evidence from listed firms in China. *Applied Economics,* 1–18.
Moscona, J., & Sastry, K. A. (2022). Does directed innovation mitigate climate damage? Evidence from U.S. agriculture. *The Quarterly Journal of Economics, 138*(2).
Our World in Data. (2024). *Patents filed for renewable energy technologies.* https://ourworldindata.org/grapher/patents-filed-for-renewables
World Bank. (2024). *World development indicators.* https://databank.worldbank.org/source/world-development-indicators
Wu, Z., Fan, X., Zhu, B., Xia, J., Zhang, L., & Wang, P. (2022). Do government subsidies improve innovation investment for new energy firms: A quasi-natural experiment of China's listed companies. *Technological Forecasting and Social Change, 175*, 121418.
Zhang, H., & Zhang, L. (2023). Public support and energy innovation: Why do firms react differently? *Energy Economics, 119*, 106528.
Zhang, L. (2017). Correcting the uneven burden sharing of emission reduction across provinces in China. *Energy Economics, 64*, 335–345.
Zhang, L., & Kofi Adom, P. (2018). Energy efficiency transitions in China: How persistent are the movements to/from the frontier? *The Energy Journal, 39*(01).
Zhao, X., Ma, X., Chen, B., Shang, Y., & Song, M. (2022). Challenges toward carbon neutrality in China: Strategies and countermeasures. *Resources, Conservation and Recycling, 176*, 105959.
Zhou, N., Levine, M. D., & Price, L. (2010). Overview of current energy-efficiency policies in China. *Energy Policy, 38*(11), 6439–6452.

Chapter 7
Energy Market

The energy crisis has not yet overwhelmed us, but it will if we do not act quickly. It affects our economy, our environment, and our security. We need a new and effective energy policy now.
—Richard Nixon, Address to the Nation about National Energy Policy, 1973

7.1 Primary Energy Markets

7.1.1 Coal Market

Coal has long been a dominant energy source in industrial production since the British Industrial Revolution in the eighteenth century, and it still provides a third of the world's electricity nowadays and is directly involved in many chemical production processes as a raw material.

Its abundant and evenly distributed reserves make coal trading a nearly perfectly competitive market worldwide. Coal's reserves are far from exhausted, even though it is a non-renewable resource. Their static range is well over 100 years. Due to its abundance, the scarcity rent of coal is almost non-existent. In addition, coal is stored fairly evenly around the world, with abundant reserves in Australia, America, Europe and Asia. Trading centers, regions and unified markets organize coal transactions around the world on a daily or even more frequent basis.

70% of the international coal trade is thermal coal trade. Coal trade is still dominated by bilateral contracts between producers and wholesale customers, also known as over-the-counter (OTC) types. These contracts usually run for 10 years, with prices adjusted in the fourth quarter of each year based on the spot price of coal. Because these prices do not need to be published, the world coal market has always lacked transparency.

L. Zhang, *Sustainability Economics*, https://doi.org/10.1007/978-3-031-87829-9_7

According to data from the European Energy Exchange (EEX), coal prices spiked twice around 2010, when they were not competitive with oil and gas. But starting in 2012, coal prices started to become cheaper than nature gas. The following determinants can be cited to explain these developments. First, US coal exports decreased. Until the late 1990s, this country was the main exporter of coal, acting as a swing producer. Since then, this stabilizing force disappeared. Second, Chinese coal exports developed. In the past few decades, China's rapid economic growth led to a surge in demand for energy, mainly electricity and coal consumption. China's coal exports declined, forcing East Asian importers such as Japan and South Korea to obtain coal from more distant regions, resulting in higher global coal transportation costs. As China's demand for coal slowed, the surplus coal supply in the international market increased, and prices fell.

Rising coal supplies and lower natural gas prices caused coal prices to fall sharply in late 2022. The price drop attracted price-sensitive buyers such as China and India, although the depreciation of the Chinese Yuan and the Indian Rupee against the dollar partially offset the price decline. China and India increased imports in early 2023, with China even ending its unofficial ban on Australian coal. As of April 2023, China and India accounted for about 50% of global coal imports, as the two largest coal producers and consumers were also the largest importers.

Case

Russia-Ukraine Crisis and Coal Trade

In 2022, global coal flows shifted significantly due to geopolitical sanctions. Rail restrictions hindered Russian coal transport, tightening global supplies. Australia faced production disruptions from La Niña, and South Africa struggled with infrastructure issues. In response to high prices, China and India boosted domestic production, though India's imports still rose due to strong demand.

International thermal coal trade increased by 2% to 1,045 million tons, while metallurgical coal trade slightly declined to 307 million tons. Russian thermal and metallurgical coal exports fell by 10% and 14%, respectively. Australian thermal coal exports dropped by 7%, and metallurgical coal by 3%. U.S. thermal coal exports decreased by 2.8% due to rail issues and sector challenges, while metallurgical coal exports remained stable. Colombian thermal coal exports fell by 4%.

Indonesia's thermal coal exports rose by 9%, and Mongolia's metallurgical coal exports increased by 54%. Smaller suppliers like Mozambique, Botswana, and Tanzania also increased exports. Overall, thermal coal exports from other countries rose by 16% to 79 million tons, though this growth may be temporary due to logistical and cost challenges.

7.1.2 Crude Oil Market

Crude oil, being liquid at normal atmospheric pressure, is easy to store, transfer, and transport, making it cost-effective. It also has a high energy density. For example, a 50-L gasoline tank contains about 450 kWh of energy. Filling such a tank in two minutes equates to a gasoline pump's capacity of 13.5 MW. A gas station with eight pumps can sell the same energy as 54 wind turbines with a capacity of 2 MW each.

Crude oil reserves are unevenly distributed globally, with more than half concentrated in the Middle East and Central Asia. This region, along with Russia, forms the "Energy Oval", accounting for nearly 40% of the world's conventional crude oil supply. Extraction costs vary widely, with the Persian Gulf averaging less than $40 per barrel, while other regions can be twice as expensive.

Historically, the oil industry has seen significant developments. Figure 7.2 shows the timeline of the development of the oil industry. In the 1920s, the Aknacari Agreement among British Petroleum, Shell, and ExxonMobil, later joined by other companies, formed a cartel to control oil prices and market share. This "Seven Sisters" cartel dominated until the 1970s when OPEC emerged.

OPEC, formed in 1960 by Iran, Iraq, Kuwait, Saudi Arabia, and Venezuela, aimed to set oil prices above competitive market levels. From 1961 to 1971, in the 10 years, countries such as Qatar, Indonesia, Libya, the United Arab Emirates, Algeria, and Nigeria successively joined OPEC, and OPEC's market share continued to increase. However, internal challenges and external pressures, such as the 1973 oil embargo and the 1979 Iran-Iraq War, led to significant price fluctuations and economic impacts (Fig. 7.1).

The first oil crisis occurred in October 1973 when the Organization of Arab Petroleum Exporting Countries, led by Saudi Arabia, imposed an oil embargo on countries supporting Israel in the Fourth Middle East War. This embargo, lasting nearly a year, caused global oil prices to rise by nearly 300%, leading to significant economic impacts, including recessions in developed countries. The crisis highlighted the strategic importance of oil, prompting the U.S. to intervene in Middle Eastern affairs and establish the "oil-dollar system" with Saudi Arabia.

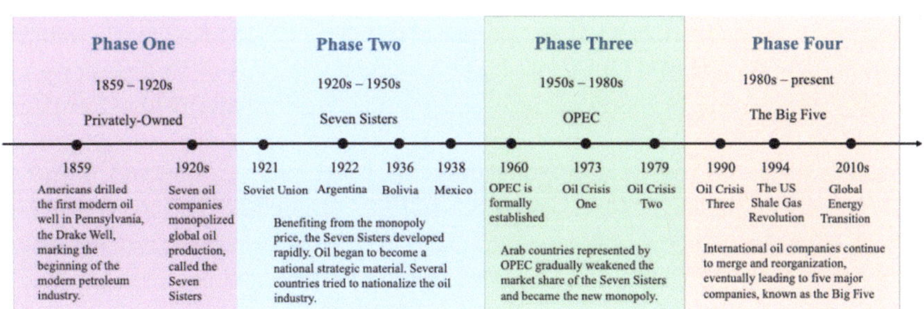

Fig. 7.1 Development timeline of oil industry

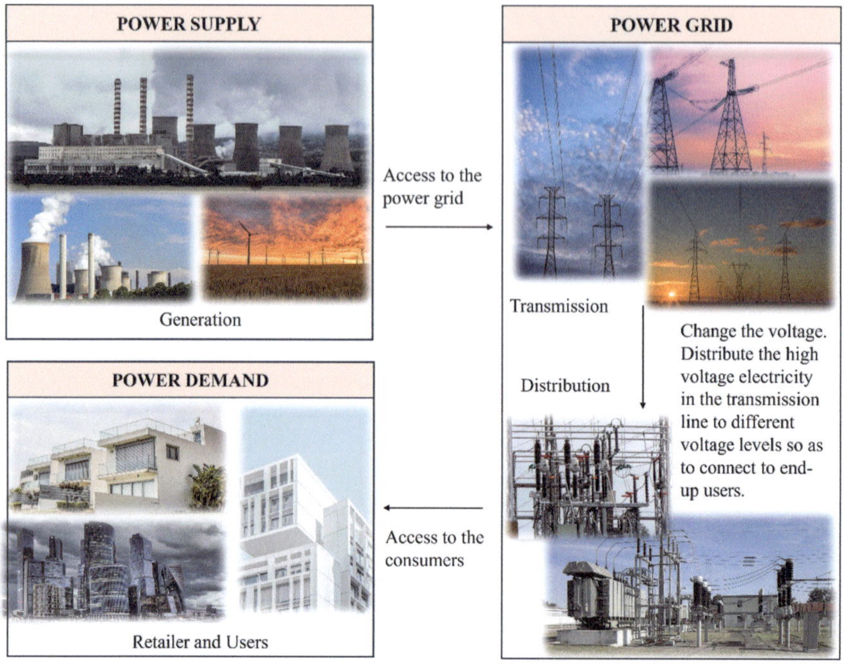

Fig. 7.2 Electricity industry segments

The second oil crisis (1979–1981) was triggered by the Iran-Iraq War, causing oil prices to rise from $15 to $39 per barrel. This time, oil companies chose to purchase oil from the North Sea instead of the Persian Gulf countries. OPEC's supply fell by more than 50%. In order to alleviate the decline in market share, in early 1986, Saudi Arabia decided to increase production, the cartel briefly broke down, and oil prices fell below $20 per barrel. This led to a 3% drop in US GDP. In response, Western countries adjusted their oil strategies, focusing on energy independence, developing non-OPEC oil sources, and establishing strategic oil reserves.

The third oil crisis in August 1990 was caused by Iraq's invasion of Kuwait, leading to a surge in oil prices from $21 to $46 per barrel. The US-led coalition's military intervention and increased OPEC production mitigated the crisis. This event underscored the importance of energy security and prompted further energy transformation efforts.

The three major oil crises highlighted the strategic importance of oil and prompted shifts in energy policies. These crises reduced OPEC's dominance and spurred investment in alternative energy sources. The 2021–2022 global energy crisis, driven by supply shortages and geopolitical tensions, further emphasized the need for secure energy supplies, reduced emissions, and stable prices.

The international oil trade has developed a well-structured wholesale market, including spot, forward, and financial derivative markets. Spot markets involve immediate delivery and payment, while forward markets involve future delivery with fixed prices. Derivatives markets, such as futures and options, allow financial institutions to participate without physical delivery.

Energy exchanges, like the New York Mercantile Exchange (NYMEX) and the Intercontinental Exchange (ICE), manage contracts and mitigate counterparty risk. These exchanges define benchmark crude oils, such as West Texas Intermediate (WTI) for the U.S. and North Sea Brent for Europe. Recently, Dubai Crude has become the benchmark for the Persian Gulf.

7.1.3 Nature Gas and LNG Market

Natural gas is another exhaustible resource. It is a combustible gas that is naturally generated and stored under certain geological conditions by ancient animals and plants buried deep underground. It is a gaseous mixture of hydrocarbons and non-hydrocarbons, with methane as its main component. It is a fossil fuel. Natural gas is produced in oil fields and gas fields, and a small amount is also produced in coal seams. In addition to methane, it usually contains ethane, propane and a small amount of hydrocarbons with a larger number of carbon atoms, as well as some non-flammable gases, such as carbon dioxide, hydrogen sulfide, nitrogen, helium, etc. Each cubic meter of commercial-quality natural gas produced produces 38 MJ (10.6 KW-h) of energy, which is equivalent to 1031 British thermal units (BTUs) per cubic foot of natural gas burned in the United States.

Liquefied natural gas (LNG) is purified, compressed and cooled liquefied natural gas. Its main component is methane, which is colorless, odorless, non-toxic and non-corrosive. Its volume is about 1/625 of the volume of the same amount of gaseous natural gas, and its mass is only 45% of the same volume of water. Under normal pressure, the calorific value of LNG is 52MMBtu (1MMBtu = 2.52 × 108 cal), and it has many excellent characteristics such as environmental optimization, easy transportation and storage, and high calorific value. LNG technology was developed for the import of natural gas from Japan, as the country still has no access to natural gas supply by pipeline. It is also used to import natural gas from North Africa and the Middle East to Europe. Compared with pipelines, LNG does not have technical and cost advantages. The capacity of a standard LNG chain is 3.5 to 4.8 billion t/y (4.8 × 6.6 billion m^3/y, respectively).

Historically, the natural gas industry is older than the oil industry, as the first natural gas deposits were developed in the New York City area as early as 1825. However, there were technical limitations to the exploration of natural gas until the 1950s when it was discovered as an associated gas with crude oil. Later, it was discovered that natural gas could be recovered as a by-product of coke production, and city gas networks were established. The construction of these regional networks laid the foundation for the construction of long-distance natural gas pipelines and made it possible to connect large natural gas deposits to each other.

High production and transportation costs are the key factors that restrict the development of the natural gas industry. Technological innovation has become the core of promoting development. The shale gas revolution that began in the late 1990s may change the energy market structure of the United States and even the world. Shale gas is an unconventional natural gas extracted from shale layers. Compared with conventional natural gas, shale gas reservoirs have the characteristics of self-generation and self-storage. Shale is both a source rock and a reservoir. It is not controlled by structure, has no closure, and no clear gas–water interface. Most gas-producing shales are widely distributed, thick, and generally contain gas, which enables shale gas wells to produce gas at a stable rate for a long time, with the advantages of long mining life and long production cycle. The world's shale gas resources are 456 trillion cubic meters, mainly distributed in North America, Central Asia and China, the Middle East and North Africa, Latin America, Russia and other regions. The potential of shale gas may be greater than that of conventional natural gas.

The discovery of shale gas has had a profound impact on global geopolitical relations and energy landscape. Within a decade, the United States' shale gas dry gas production increased from 0.39 trillion cubic feet in 2000 to 850 trillion cubic feet in 2012. The proportion of shale gas in the United States' natural gas production has increased from 2 to 37%. In 2009, the United States surpassed Russia to become the world's largest natural gas producer.

The energy revolution led by shale gas development has gradually increased the energy independence of the United States, and its economic benefits are obvious: First, the US energy industry has regained vitality. Due to the cheap price of natural gas, billions of dollars of investment plans have been invested in the US manufacturing industry; second, companies and consumers have obtained cheaper energy; third, in terms of employment, the United States has added more than 3 million jobs, and the salaries of these jobs will be higher than the US average. This energy revolution will also have an increasingly important impact on global competition. The United States has abundant natural gas reserves and low prices. This has promoted the development of energy-intensive manufacturing in the United States, which will have an impact on competitors in Europe and Asia.

7.1.4 Sustainability Issues of Primary Energy Markets

An important characteristic of fossil energy is that it is non-renewable and therefore cannot support the development of human society and economy for a long time. The index reserve-production ratio describes the ratio of annual primary fossil energy production to remaining proven reserves. At the current rate of production, primary energy will be exhausted within a few decades. On the other hand, people realize that fossil energy has a high environmental cost. The sustainable development of fossil energy has triggered great discussion.

Scholars and institutions have predicted the future of primary energy. BP (2024) simulates the trends of fossil energy in the current scenario (current trajectory) and a stricter emission restriction scenario (net-zero) based on historical data. Models show that global coal consumption will begin to decline in a few years. The main drivers are slowing economic growth and the global transition to low-carbon energy. China's coal consumption will peak in 2030 and then begin to decline rapidly. By around 2050, China's coal consumption will have fallen by at least 90% and will contribute 60% to global carbon reduction efforts. However, coal consumption in some emerging economies, such as India, continues to increase, slowing the decline in coal consumption.

Oil demand will increase first and then decline. Oil will continue to play an important role in the global energy system in the coming decades. The growth in total energy demand has driven the growth in oil demand. However, as the electrification of the transportation sector increases and vehicles become more fuel-efficient, oil consumption will decline. Industrial oil use will also decline rapidly, reflecting the reduction of diesel generators and the gradual choice of alternative fuels, such as electricity and hydrogen, for non-transportation industrial vehicles.

Natural gas demand may increase or decrease in the future, depending on the speed of energy transition. According to the current trajectory of global energy transformation, natural gas will continue to grow in the coming decades and account for 28% of global primary energy consumption in 2050. If countries around the world take more proactive actions to accelerate energy transformation to achieve net zero emissions, natural gas demand will peak around 2030 and begin to decline thereafter. Natural gas demand will be reduced to half of its current level in 2050. The decline in natural gas use in the building sector is the main cause. However, natural gas use in the transportation and industrial sectors is likely to increase, especially in the process of using natural gas to produce hydrogen in industry. The downward trend in natural gas use in the power sector is not obvious, because gas units are still used as backup and fast-balancing units to support the development of renewable energy. The decline in natural gas consumption is mainly driven by technological upgrading and energy transformation in developed economies.

7.2 Electricity Market

7.2.1 Properties and Structure of Electricity Market

Electricity is central to many aspects of life in modern societies and will become even more important as technologies such as electric vehicles and heat pumps expand their role in transportation and heating. Power generation is currently the world's largest source of carbon dioxide (CO_2) emissions, but it is also the industry-leading the transition to net-zero emissions through the rapid increase of renewable energy such as solar and wind power. At the same time, the current global energy crisis has put electricity security and affordability at the top of the political agenda in many countries.

The electricity industry has traditionally implemented vertically integrated operations of generation, transmission, distribution and sales, and has been regulated by the government in terms of access and profitability as a natural monopoly industry. Since the early 1980s, under the trend of liberalization and deregulation, countries such as the United Kingdom and the United States have successively carried out market-oriented reforms in the electricity industry, separating the generation, transmission, distribution and sales links vertically, introducing competition in the electricity generation and sales links that do not have economies of scale, and implementing more effective supervision in the electricity transmission and distribution links that have natural monopoly characteristics. The direct motivation for the reform is to reduce costs and improve efficiency by relaxing regulations and introducing competition. The theoretical basis of the reform is rooted in the traditional neoclassical economics theory of competitive equilibrium and welfare maximization, and benefits from the application of innovative progress in industrial organization theory in the fields of asymmetric information and natural monopoly regulation, mechanism design, etc., in the electricity industry. Researchers have published a series of classic papers, providing a theoretical basis for the design and construction of effective electricity markets.

The electricity industry has four main segments: generation, transmission, distribution, and consumption (as shown in Fig. 7.2). The first three involve the physical hardware of producing and distributing electricity, while the fourth is a purchasing and accounting function. Historically, all of these segments were vertically integrated within regional utilities that operated monopoly franchises to serve customers in each region. While some vertically integrated utilities are owned by local governments or associations of governments, most electricity was and still is sold through investor-owned utilities that are regulated by state agencies. Still other utilities (primarily municipal or cooperative distribution utilities) purchase all electricity and engage only in distribution and retail. Decades of regulation of the electricity industry based on the cost-of-service principle have raised concerns about the incentives provided to regulated utilities and the resulting efficiencies.

By the end of the twentieth century, the United Kingdom, Spain, and some states in the United States began to restructure their electricity sectors. Generators began to earn market prices, and independent generators could enter this market. Furthermore, in many regions, incumbents (vertically integrated, investor-owned utilities) are required to sell their generation equipment or operate it in a separate entity at market prices. In the early twenty-first century, some developing countries also got involved in this action. For example, since 2003, the Chinese government has led the separation of power plants and grids, breaking the vertical integration model in which the State Grid is responsible for all power production, supply, and dispatching.

With deregulation, decisions on the type and amount of investment in new generation capacity have shifted from government regulation to a decentralized, market-based process. Electricity trading is not conducted within a company or between neighboring utilities, but through centralized wholesale electricity markets. Wholesale energy markets are cleared under the supervision of market operators, and generator production is adjusted minute by minute to continuously balance supply and demand at each location. In order to increase flexibility and maintain market stability, market tools such as ancillary service markets and demand-side response contracts have emerged. Owners of high-voltage transmission and distribution lines continue to operate as natural monopolies. Therefore, they are still subject to economic regulation at the state or federal level. Investments in transmission and distribution assets are reviewed by regulators, and costs are recovered based on the principle of cost-of-service regulation.

Finally, renewable energy brings new challenges to the electricity system. First, the variability, intermittency and randomness of renewable energy will cause considerable problems for the security and reliability of electricity supply (Borenstein et al., 2002; Newbery et al., 2018). Second, renewable energy is generally concentrated in terms of geography and electricity generation time. The traditional electricity market adopts a single price clearing, which makes it difficult to reflect the high variability of renewable energy supply costs in time and space, which may also lead to transmission congestion (Green, 2008). Finally, since the fuel cost of renewable energy is close to zero, participating in the electricity market will put downward pressure on prices, reduce the profits of fossil fuel electricity generation companies, and weaken their incentives to invest in new capacity. Before electricity storage is economical, traditional electricity generation technology is still crucial to system stability and flexibility, and therefore poses a threat to supply adequacy in the long run (Praktiknjo & Erdmann, 2016).

7.2.2 *Electricity Liberalization*

With more advanced technologies, modern energy systems allow and embrace private sector involvement in energy production, distribution, and trading. Energy economists believe that liberalization and marketization of the electricity market are beneficial for cost reduction and improving system operational efficiency. Currently, developed countries and regions such as the UK, Europe, and the US have introduced electricity liberalization. In contrast, many developing countries, including China and India, are either in the early stages of attempting electricity market liberalization or have not yet prioritized it. The electricity liberalization in the UK has set a standard model known as the "British Model". This model typically involves privatizing some state-owned assets and separating grid maintenance from generation to establish competitive markets for electricity trading (Thomas, 2005).

The electricity liberalization process in the United States involved a significant shift from traditional regulatory frameworks to market-based mechanisms aimed at enhancing efficiency and competition in the electricity sector. This transition began in the late 1990s, with states adopting restructuring legislation that allowed for the introduction of wholesale electricity markets. The liberalization process was characterized by the establishment of competitive markets where electricity generation was no longer solely managed by local monopolies. Instead, market forces determined which power plants would operate based on their costs, promoting a more efficient allocation of resources. The introduction of market dispatch aimed to reduce production costs and improve the overall efficiency of the electricity grid, allowing for greater trade between power control areas (PCAs). While many areas eventually embraced market dispatch, some retained their traditional regulatory structures, leading to a mixed landscape of electricity generation and distribution. Overall, the liberalization process has enhanced the efficiency of resource allocation, facilitated greater inter-regional electricity trading, and provided substantial economic benefits to consumers and the electricity market as a whole (Cicala, 2022).

The Nordic electricity market, recognized as the world's first regional electricity market and a model for the European "Target Model", is currently undergoing significant transformation in response to increasing renewable energy generation, technological advancements, and the growing need for flexibility. One of the primary advantages of the Nordic electricity market design is its flexibility. The presence of numerous smaller bidding zones allows for localized pricing signals that reflect supply and demand dynamics more accurately. This granularity enables market participants to respond swiftly to fluctuations in renewable energy generation, which is particularly crucial as the share of variable RES, such as wind and solar, continues to rise. The flexibility inherent in this design not only supports the integration of renewables but also enhances the overall resilience of the electricity system (Lindberg, 2022). Figure 7.3 shows the trading volumes of electricity in Germany and the Nordics, covering both long-term market and OTC transactions from 2013 to 2018. From 2013 to 2018, the German market emerged as the largest long-term electricity market in Europe. In 2018, its trading volume reached 2100 TWh, approximately eight times the total electricity consumption in Germany. In contrast, the Nordic market hit a record low of 900 TWh in the same year.

The Nordic electricity market model stands out for its flexibility, support for cross-border trade, and promotion of regional cooperation, making it a model for other regions seeking to advance their energy transitions while effectively integrating renewable energy sources. As the global energy landscape evolves, the lessons learned from the Nordic market will be invaluable for shaping future electricity market designs.

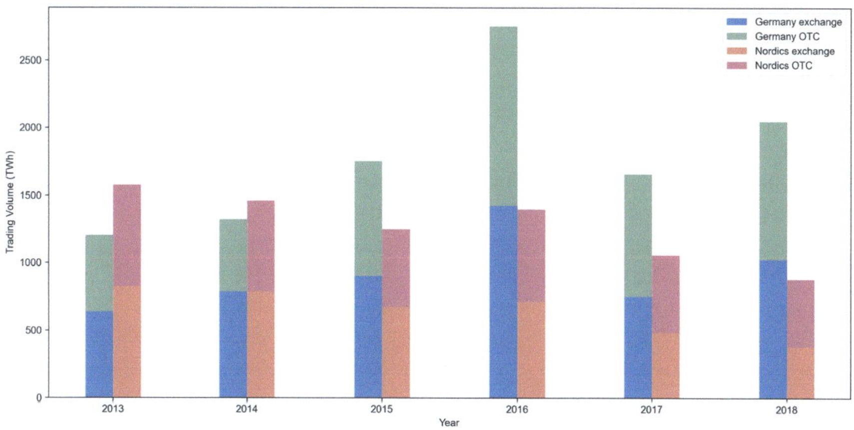

Fig. 7.3 Trading volumes in Germany and the Nordics, both long-term market and OTC, 2013–2018. *Data source* Lindberg (2022)

Reading

California Power Crisis

On the afternoon of September 8, 2011, the California power grid was operating at high load, and a 230 kV transformer tripped due to overload, causing a small-scale load blackout and partial load transfer. However, the load transfer caused a nearby 230 kV transmission line to overload and trip and local power grid frequency fluctuations. Due to the overly sensitive protective machine cutting action, the power grid suddenly lost a large power supply, resulting in more than 2.7 million people losing power supply in Southern California, Arizona and parts of Mexico, and a serious power grid safety accident.

The California power crisis that broke out in the early 21st century sounded the alarm for the market-oriented reforms that were in full swing. California electricity prices soared more than 500 times, causing the state to suffer multiple large-scale power outages. The state's largest power company filed for bankruptcy protection, and the economic recession severely damaged Gray Davis's position as governor. According to estimates, the crisis caused losses of US$40–45 billion.

7.3 Market Analysis Techniques

7.3.1 Energy Balance and Flow Diagrams

Energy balance provides a comprehensive overview of the energy flow of a region (nation), an entity or a process, and is an important quantitative tool for understanding the energy economy. Statistics bureaus and research institutions regularly publish energy balance information, which records the sources, transformations and final consumption of all energy during a certain observation period.

According to the content of energy, energy balance is divided into single energy balance and comprehensive energy balance. According to the size of the balance range, it is divided into regional energy balance, departmental energy balance, industrial and mining enterprise energy balance, and workshop (process) or key energy-consuming equipment energy balance. The energy balance table adopts a matrix form, with "columns" representing various primary and secondary energy sources and "rows" representing energy flows and various economic activities. The balanced relationship of the total matrix is:

$$\mathrm{TE} = \mathrm{TC} + \mathrm{TL} \tag{7.1}$$

where TE denotes the total amount of energy available for consumption in the region, TC denotes the total consumption of end-up users, and TL denotes the total losses during process, conversion, operation and transportation.

Energy flow diagrams are often used to illustrate energy balances. An energy flow diagram is a chart that shows the flow of energy (such as heat energy, electrical energy, etc.). It can be a regional, industrial enterprise, or a single device. After drawing the energy flow diagram, you can analyze the energy utilization efficiency and waste heat recovery based on the energy flow diagram or heat balance, further improve the possibility of energy utilization efficiency, and propose measures, methods and calculation plans to strengthen energy management, improve efficiency, reduce pollution, etc.

7.3.2 Perfect Competition and Monopoly

Economists emphasize that revealing the commodity attributes of energy and following market equilibrium pricing methods can improve resource allocation efficiency and reduce energy costs. However, based on the characteristics of different energy sources, there are multiple market structures in the current energy market. The analysis paradigm based on industrial structure is the key to analyzing the energy market.

Perfect competition

A perfectly competitive market, also known as a pure competition market or a free competition market, refers to a market in which there are many production and sales companies in an industry, and they all provide similar, standardized products to the market in the same way. Sellers and buyers are both recipients of the price of goods or services. Although a perfectly competitive market is a completely ideal model in economics, there are markets that are very close to perfectly competitive markets in the energy industry. For example, based on the homogeneity of coal products and the characteristics of easy overcapacity production, China's coal transactions are close to a perfectly competitive market.

In the short term, the price and quantity in the market are determined by the supply and demand in the market. The production capacity of the manufacturer is fixed, and there is no entry and exit in the market. The manufacturer can only adjust the quantity of variable factors. Under a given production scale, the manufacturer achieves marginal cost (MC) and marginal revenue (MR) equal to each other through the change of short-term variable cost (SVC) caused by the change of output and maximizes profits. Because the MR of the manufacturer in the perfectly competitive market will be equal to the market price, the short-term equilibrium of the perfectly competitive market satisfies the equation:

$$\mathrm{MC} = \mathrm{MR} = P \tag{7.2}$$

In the long term, the manufacturer can change its production capacity and adapt to the change of market price by entering or exiting the market. If the market price is higher than the average total cost (ATC) of the manufacturer, new manufacturers will enter the market; if the market price is lower than the average total cost of the manufacturer, some manufacturers will exit the market. This will cause the market supply curve to shift to the right and the price to fall until the excess profits of all manufacturers are consumed by competition. In the long-term equilibrium, all manufacturers in the market will obtain normal profits, which means that all manufacturers produce at a scale where ATC is equal to the market price:

$$\mathrm{ATC} = P \tag{7.3}$$

Monopoly

Monopoly means there is only one supplier in a market. A monopoly firm is a firm that provides all supplies for a market and can choose to produce at any point on the market demand curve. Typical monopoly cases mostly exist in industries protected or authorized by the government. For example, patents provide legal protection for production technology, resulting in some large pharmaceutical companies and computer chip companies having a monopoly position; companies obtain public utility franchises, such as power grid companies, oil and gas pipeline companies, or urban gas companies, heating companies, etc. There are many examples of monopoly in the energy market. For example, generally speaking, power grid companies and oil and gas pipeline companies are monopoly firms in the region. Sellers in the oil and gas market are concentrated in specific countries with oil resources, so the oil and gas market has monopoly characteristics in specific historical periods or within specific regions.

Natural monopoly is a special form of monopoly. The production technology of natural monopoly industries has the characteristics of increasing returns to scale, so the average cost and marginal cost have been showing a downward trend. As the output increases, the monopolist can continue to lower the price and maintain a certain profit. In this way, the most efficient enterprise will rely on its cost advantage to squeeze out other incumbent competitive enterprises from the market, monopolize the entire market, and use low costs to build barriers for potential competitors to enter the market, thus forming and maintaining a natural monopoly position.

Since a monopoly firm is the only supplier in the market, unlike firms in a perfectly competitive market, which are price takers, it can adjust its output and set the market price according to market demand. Given the market demand curve $P(Q)$, following the profit maximization condition of MC = MR, the first-order condition for the monopoly firm's profit maximization is

$$\mathrm{MR} = P(Q) + Q\frac{\mathrm{d}P(Q)}{\mathrm{d}Q} = \mathrm{MC} \tag{7.4}$$

where $\frac{\mathrm{d}P(Q)}{\mathrm{d}Q} < 0$, so $\mathrm{MR} = \mathrm{MC} < P$. Monopoly firms earn excess profits. Since marginal cost increases with output and market price decreases with output, the equilibrium output of a monopoly market is less than the equilibrium output of a perfectly competitive market, and the equilibrium price is higher than the equilibrium price of a perfectly competitive market. That is, in a monopoly market, prices are higher while output is lower. Manufacturers with monopoly characteristics, such as power grid companies and oil industry monopolists in the Gilded Age, can usually rely on their monopoly position to make huge profits.

7.3.3 Oligopoly and Cartel

An oligopoly market refers to a market with a few large manufacturers. OPEC and some non-OPEC oil-producing countries together constitute the global oil production oligopoly. Applying the oligopoly analysis model can help us better understand the global oil market and the strategic position of oil resources in the game between great powers.

The analysis of the oligopoly market involves the interaction of manufacturers in the market, so the framework of game theory is usually used to analyze the decision-making of manufacturers on price and output in the oligopoly market. According to the decision variables, oligopoly market competition can be divided into output competition and price competition; according to the decision order of manufacturers, oligopoly market competition can be divided into synchronous games and sequential games. According to these two classifications, the main oligopoly market competition models are divided into the Cournot model with the synchronous game and output competition, the Stackelberg model with the sequential game and output competition, the Bertrand model with the synchronous game and price competition, and the sequential game price competition model.

We use a simple example to illustrate the Cournot model. Assume that there are only two homogeneous firms in the market, and their marginal costs are the same, both constants, denoted by c, and the market inverse demand function is $P = a - bQ$. Using the first-order condition, we can obtain that

$$q_1 = q_2 = \frac{a - c}{3b} \tag{7.5}$$

The total equilibrium output on the market is

$$Q = q_1 + q_2 = \frac{2(a - c)}{3b} \tag{7.6}$$

The market equilibrium price is

$$P = \frac{a + 2c}{3} \tag{7.7}$$

The model results clearly show that the total output of an oligopoly market is smaller than that of a perfectly competitive market, but larger than that of a monopoly market; the price of an oligopoly market is higher than that of a perfectly competitive market, but lower than that of a monopoly market. Further calculations show that the social welfare of an oligopoly market is smaller than that of a perfectly competitive market, but larger than that of a monopoly market. Therefore, although an oligopoly market has certain oligopolistic characteristics, due to the increase in market competitors and the increase in output, social welfare increases compared to a monopoly market with only one producer.

When manufacturers face different marginal costs, c_1 and c_2, respectively, and assume that $c_1 < c_2$, the simultaneous equations are solved as follows:

$$q_1 = \frac{a - 2c_1 + c_2}{3b} \tag{7.8}$$

$$q_2 = \frac{a - 2c_2 + c_1}{3b} \tag{7.9}$$

Therefore, the lower the manufacturer's own cost, the greater its equilibrium output; the higher the cost of the rival manufacturer, the higher the manufacturer's equilibrium output. In other words, the manufacturer with lower cost in Cournot competition is more competitive. Taking the crude oil market as an example, the phenomenon we observed is roughly in line with the conclusion of the model: currently, countries with high reserves and low extraction costs, such as Saudi Arabia, Russia, Iraq, and Venezuela, are also the countries with the largest oil production. These countries rely on abundant production and low extraction costs to occupy a large market share in the crude oil market, which has a great impact on the price of the crude oil market.

When there is collusion among oligopolistic enterprises, they can unite through collusion to control market output and prices to increase profits. We usually call this behavior a cartel. In a cartel, when market manufacturers act with the purpose of maximizing the total profit of the cartel, they actually make decisions like monopoly manufacturers. In this case, manufacturers participating in the cartel will also obtain greater profits than under the Cournot competition. This is the motivation for the formation of a cartel. As stated, the purpose of OPEC is to coordinate and unify the oil policies of member countries, maintain the stability of international oil market prices, and ensure that oil-producing countries obtain stable income.

Reading

OPEC and Cartel

In the two oil crises in the 1970s, OPEC's collective actions significantly increased international oil prices, thereby obtaining huge benefits for member countries and exerting greater influence in international politics. During the two oil crises, OPEC established a price committee and acted in a cartel manner to set a unified fixed price. For the first time in history, it unilaterally raised the prices of Gulf crude oil, Arab light oil, North African crude oil, etc., despite the opposition of international oil companies and Western oil-consuming countries. Since then, the OPEC cartel has become one of the most influential international organizations in the crude oil market with its strong reserves, production cost advantages and independent pricing power for oil commodities.

The way OPEC exercises its cartel changed in the 1980s, from a price cartel to a production cartel. In the early 1980s, energy demand plummeted, oil demand fell, and finally the crude oil market collapsed in 1986. OPEC member countries drastically cut production to maintain the relative strength of international oil prices. From this period on, OPEC began to become a production cartel, using quotas to regulate the production of each member country to influence international oil prices. Although cartels have different ways of action, their purpose is to protect and maintain the benefits of their members.

Although cartels can earn high profits for their members, they are fragile and easy to collapse. In a cartel that aims to reduce market output and increase profits, since other manufacturers have cut production, individual manufacturers have the motivation to violate the agreement, increase production, and obtain higher profits. They also expect that if they cut production, other manufacturers will increase production, take away market share, and cause themselves to suffer losses. This is consistent with the phenomenon observed in OPEC in reality. Although OPEC controls crude oil production by setting production quotas for member countries, some member countries have violated the agreement and increased production to maximize their own interests. Studies have shown that the behavior of some member countries to increase production in violation of the agreement may even neutralize the predetermined price increase.

This phenomenon is more obvious in the "OPEC+" organization, which has less binding force and is looser. OPEC established the "OPEC+" mechanism in 2016 and signed a production reduction agreement with other non-OPEC oil producers to control the international crude oil supply and maintain international oil prices. The current members of OPEC+ are mainly some small and medium-sized oil-producing countries led by Russia, while other major oil-producing countries, including the United States, have not participated. So far, OPEC+ is still only an informal and temporary organization, and its production reduction agreement has very limited constraints on participating countries.

7.3.4 Market Failure

Market failure is common in the energy market due to imperfect market structures, public goods, externalities, and incomplete information. Competitive markets are typically efficient, but unregulated energy markets often fail to produce fair and efficient outcomes.

Imperfect market structure: The energy market includes monopolies, oligopolies, and cartels, such as the oil industry during the Gilded Age and power grids. These structures lead to higher prices and reduced output, necessitating government intervention through antitrust laws and price regulation. For example, OPEC's price collusion and mergers in the power and coal industries distort market dynamics, leading to higher consumer prices and reduced innovation. Regulatory bodies like the Federal Trade Commission (FTC) enforce antitrust laws to maintain competitive markets, benefiting consumers with lower prices and better products.

Externalities: The energy industry is a major source of greenhouse gases, causing significant negative externalities like environmental pollution and climate change. These externalities result in welfare losses that are not compensated by the responsible parties. Air pollution, for instance, is linked to 70% of diseases and 40% of deaths. Climate change impacts ecosystems, agriculture, and human health. Policies addressing these externalities are integrated with environmental and climate policies.

Incomplete information and public goods: Incomplete information in the energy market, especially in natural monopolies like electricity and oil pipelines, leads to information asymmetry between companies, consumers, and regulators. This affects price regulation. Energy services also have quasi-public goods attributes, known as Universal Service, ensuring all users access to reliable and affordable energy. For example, Chinese government policies mandate power grid companies to operate as public utilities, ensuring fair access to electricity. In the U.S., rural power cooperatives receive preferential loans and access to low-cost electricity, addressing rural electricity supply issues. Addressing negative externalities and ensuring universal energy service are crucial for sustainable development. These efforts mitigate environmental and health impacts, promote energy equity, and support the transition to a cleaner and more resilient energy system.

7.4 Energy Security

7.4.1 Energy Crisis and Security Concern

Since 2021, the energy crisis in Europe and other regions has intensified, making energy security a priority for many countries. The international energy landscape is evolving rapidly due to factors like the Ukrainian crisis, great power competition, and low-carbon transformation. The concept of energy security, initially focused on oil supply stability, now includes power security, energy network security, and key mineral security.

Power security has become increasingly important as the energy system shifts from oil to electricity. IEA (IEA, 2024b) estimates that 50% of global energy consumption will need to be met by electricity by 2050 to achieve net-zero carbon emissions. However, renewable energy sources like wind and solar have weaknesses such as volatility and unpredictability, increasing power system complexity and security risks. Aging infrastructure and poor monitoring have also led to international power outages, exacerbated by the European energy crisis since 2021.

Energy network security is a critical aspect of modern infrastructure, ensuring the reliable and safe delivery of energy to consumers. It involves protecting the physical and cyber components of energy systems from threats such as cyberattacks, natural disasters, and human errors. Cybersecurity is now a crucial part of energy security. Digitalization has made the power sector more connected and efficient but also more vulnerable to cyberattacks. Since 2017, cyberattacks on the energy sector have surged, with oil and gas infrastructure and power grids being primary targets (Table 7.1). For example, in May 2021, the Colonial Pipeline, the largest refined oil pipeline system in the United States, was attacked, forcing the company to shut down 5500 miles of pipelines, and U.S. gasoline futures rose by more than 4% at one point. In December 2023, a hacker attack caused 70% of gas stations in Iran to be disrupted, and this cyberattack had a major impact on the capital, Tehran. By prioritizing energy network security, we can safeguard our energy supply, support economic stability, and enhance national security.

Table 7.1 Summary of cyberattacks in the oil and gas industry in recent years

Year	Attacked company	Attacked system
2017	Petrobras (Brazil) Spain Natural Gas (Spain) Saudi Aramco	Gas station Office system SIS system
2020	Burisma Holdings (Ukraine) National Freight Company (Philippines) Energias de Portugal (Spain) Chevron Corporation (USA)	Email Cruise Sensitive data Unknown
2021	Shell (UK) Eversource (New England, USA) Colonial Pipeline (USA)	Network security equipment Cloud services Oil pipelines
2022	Amsterdam and Rotterdam (Netherlands) Antwerp (Belgium) Rosneft (Germany) Rompetrol (Romania) Fazprom (Russia) Ignitis Group (Lithuania) Creos (Luxembourg)	Portal Gas station Payment system Data IT services Refinery Port oil facilities
2023	Petro-Canada (Canada) BAZAN (Israel) Gas stations (Ukraine) China Petrochemical Industry Development Co., Ltd. (Taiwan, China) Gas stations (Iran)	Gas station payment system SCADS system Data
2024	Naftogaz (Ukraine)	Data Center

The security of critical minerals has become prominent as demand for resources like lithium, cobalt, nickel, and rare earths increases. The U.S., Europe, and other regions rely heavily on imports of these materials, raising security concerns. International competition for key minerals is intensifying, making their supply security a focal point of geopolitical struggles. Table 7.2 shows the top three countries with the largest supply of various key energy transition minerals. The supply of these minerals has a highly concentrated market structure, with the top three supplying 70% or more of the total supply.

Reading

Energy Security Transition and Power Grid Facilities

The energy transition has transformed the traditional energy supply structure, making a modern, smart, and reliable transmission network essential for energy security. For over a century, the electric grid has delivered electricity to homes, businesses, and industries. The clean energy transition is expanding electricity's role in economies, necessitating bigger, stronger, and smarter grids. However, grids currently receive insufficient attention.

Constructing and expanding electric grids is crucial to meet national energy and climate goals. Clean electricity and electrification reforms are vital as the world adopts more electric vehicles, heating and cooling systems, and hydrogen production. According to the IEA (2023), achieving national targets requires adding or renovating over 80 million km of grids by 2040. This is crucial for decarbonizing electricity and integrating renewable energy, with wind and solar PV expected to dominate new power generation capacity.

Modernizing and digitalizing the power grid is necessary for power security during the clean energy transition. As variable renewable energy sources like solar and wind increase, the power system must become more flexible. Traditional fossil energy units provide stable rotating power, unlike renewable energy, which requires comprehensive grid technology upgrades. Additionally, electronic components are more vulnerable to cyber-attacks, necessitating robust information security measures.

The power grid is a bottleneck in the transition to net-zero emissions. Despite rapid growth in renewable energy investment, global grid investment remains stagnant at around $300 billion per year. At least 3000 GW of renewable energy projects are waiting for grid connection, risking the grid becoming the weak link in the clean energy transition. Neglecting grid investment and reform will increase global CO_2 emissions and slow the energy transition. The IEA (2023) warns that limited investment and modernization will result in higher fossil fuel use and significantly higher CO_2 emissions, jeopardizing climate goals and increasing the likelihood of exceeding a 2 °C global temperature rise.

Table 7.2 Total supply for key energy transition minerals, 2023

Mineral	Top1		Top2		Top3		Top3 share (%)
	Country	Mining (kt)	Country	Mining (kt)	Country	Mining (kt)	
Copper	Chile	5311	Congo	2678	Peru	2644	47
Cobalt	Congo	157	Indonesia	20	Russia	8	77
Lithium	Australia	84	China	34	Chile	46	85

(continued)

Table 7.2 (continued)

Mineral	Top1		Top2		Top3		Top3 share (%)
	Country	Mining (kt)	Country	Mining (kt)	Country	Mining (kt)	
Nickel	Indonesia	1787	Philippines	391	New Caledonia	210	69
Graphite	China	1320	Mozambique	97	Madagascar	66	92
Magnet rare earth elements	China	44	Myanmar	11	Australia	5	85

Data source IEA (2024a)

7.4.2 Measures to Ensure Energy Security

In order to cope with energy shortages and price increases, governments have adopted price limits, subsidies, tax cuts and other measures to minimize the impact of rising energy prices on consumers and small and medium-sized enterprises. Although this move has caused countries to lose fiscal revenue, it can guarantee the basic energy consumption needs of their residents to a certain extent against the backdrop of skyrocketing energy prices.

Limiting energy prices and corporate excess profits

France, the United Kingdom, Spain and other countries have called for reforms to the EU's liberalized energy market, setting energy price caps and protecting consumers from excessive energy bills. The French government announced a price cap on natural gas prices, and through substantial tax cuts, it will cap the price increase of electricity at 4% between February 2022 and February 2023. The United Kingdom will increase the price cap on electricity, natural gas and other energy sources from April 1, 2022, by 54%. The UK energy price cap is adjusted twice a year to track wholesale energy and other costs, prevent energy companies from making excessive profits, and ensure that customers do not pay more than a fair price for their energy, but within the price cap, energy companies are allowed to pass on all reasonable costs to customers. The Spanish government announced an energy price cap plan in September 2021, promising to reduce residents' energy consumption expenditure to the 2018 level. Although the above move has caused some losses in fiscal revenue in various countries, it can guarantee the basic energy consumption needs of domestic residents to a certain extent against the backdrop of soaring energy prices. However, there are still voices of concern in the industry, believing that excessive intervention in energy prices will not only increase the government's fiscal burden, but also put public service departments at risk of bankruptcy.

In addition to measures such as energy price caps, some countries have imposed windfall taxes to limit the excess profits of energy companies. For example, Spain approved the imposition of windfall taxes on energy suppliers in September and is expected to obtain more than 2 billion euros in fiscal revenue; the UK also said it was considering imposing windfall taxes on energy companies that profit from rising natural gas prices.

Tax reductions and subsidies

The surge in energy prices has caused companies to face the challenge of rising production costs. Germany, the United Kingdom, France, Italy, Spain and other countries have introduced government subsidies and phased tax reduction policies to reduce energy tax rates on the power generation and user sides and reduce the burden on enterprises and users.

Subsidies to end consumers. The United Kingdom has introduced 4.2 billion pounds (about 5.8 billion U.S. dollars) of aid measures, including discounts on electricity and gas bills for low-income households; negotiated with energy-intensive enterprises and promised to provide loan support; and studied other measures to help consumers, such as providing more subsidies and discounts for retirees and low-income households. France provides low-income households with an "inflation subsidy" of 100 euros per household. Italy has allocated billions of euros in subsidies to relevant companies and approved the latest version of the 2022 budget bill at the end of 2021, planning to allocate 3.8 billion euros in fiscal subsidies to cope with the continued increase in electricity and gas bills in the first quarter of 2022. Germany has launched a one-time subsidy plan of 130 million euros for low-income households, which is planned to be issued when households receive winter electricity and gas bills in the summer of 2022. It also announced that the renewable energy tax will be reduced by more than 40% to reduce the burden on consumers from soaring energy prices. Greece has introduced a package of subsidies to protect Greek households from the impact of soaring oil, gas and electricity prices, including doubling the amount of electricity subsidies and providing discounts for residential electricity and natural gas.

Tax relief for small and medium-sized enterprises. Germany exempted about one-third of the renewable energy surcharge, which accounts for about one-fifth of household electricity bills. Spain suspended the 7% power generation tax and reduced the electricity value-added tax from 21 to 10%. The Czech Republic announced that it would temporarily cancel the value-added tax on electricity and natural gas consumption in November and December. Greece reduced the social rate of electricity bills. Brazil extended the time to end subsidies for coal-fired power plants from the original plan of 2027 to 2040.

Release of crude oil reserves

Faced with the tight supply and high oil price pressure in the crude oil market, several major consumer countries announced the release of crude oil reserves. On November 23, 2021, the United States announced that it would release 50 million barrels of crude oil from its strategic petroleum reserves. This batch of crude oil will be put on the market in mid-to-late December, of which 18 million barrels will be sold directly, and another 32 million barrels of high-sulfur crude oil will be short-term exchanges and will be returned to the strategic petroleum reserves from 2022 to 2024.

Example

China Reserve Crude Oil Auction

On November 24, the Chinese Ministry of Foreign Affairs stated that China would arrange the release of national reserve crude oil according to its own actual situation and needs, take other necessary measures to maintain market stability and publish relevant information in a timely manner. On September 14, China's National Grain and Material Reserves Administration issued the "Announcement of the National Petroleum Reserve Center on the Launch of the First Batch of National Reserve Crude Oil Bidding Transactions in 2021". On September 24, the first batch of 7,383,616.78 barrels of national reserve bonded crude oil was completed through bidding in the national reserve oil trading system. This is the first time since the construction of the first oil reserve base in China in 2006 that national reserve crude oil has been put on the market through bidding transactions to ease the pressure of rising raw material prices for production enterprises.

Following the United States, many countries have announced the release of crude oil reserves. On November 24, Japan announced the release of about 4.2 million barrels of national oil reserves. On the same day, South Korea's Ministry of Foreign Affairs and the Ministry of Trade, Industry and Energy also stated that they would release some crude oil from the national strategic crude oil reserves to ease inflationary pressure. India proposed that it plans to release 5 million barrels of crude oil from its strategic oil reserves. In addition, the British government stated that it would allow private oil reserves to release 1.5 million barrels of crude oil.

The energy crisis has led to increased politicization of energy commodities. The U.S. and Europe have escalated energy sanctions against Russia. Europe has accelerated efforts to reduce dependence on Russian energy, increasing imports from the Middle East, Australia, and the U.S. By the end of 2022, the U.S. became the largest crude oil supplier to the EU, and the EU became the largest export destination for U.S. liquefied natural gas (LNG). Russia's energy exports have shifted to the Asia–Pacific region, with China and India becoming major trading partners. In the clean energy sector, the U.S. has increased support for green industries through measures like the Inflation Reduction Act (IRA). The European Union has launched the Critical Raw Materials Act (CRMA) and the Carbon Border Adjustment Mechanism (CBAM), leading to green trade protectionism. In 2022, the U.S. established a mineral security partnership with several countries and further established a critical mineral buyers' club in 2023. The G7 Summit in May 2023 emphasized diversifying the supply chain of critical minerals and green products. These developments highlight the need for coordinated global energy governance to address new risks and promote sustainable development.

References

Borenstein, S., Bushnell, J. B., & Wolak, F. A. (2002). Measuring market inefficiencies in California's restructured wholesale electricity market. *American Economic Review, 92*, 1376–1405.

British Petroleum (BP). (2024). *Bp energy outlook 2024 edition.* BP. https://www.bp.com/content/dam/bp/business-sites/en/global/corporate/pdfs/energy-economics/energy-outlook/bp-energy-outlook-2024.pdf

Cicala, S. (2022). Imperfect markets versus imperfect regulation in US electricity generation. *American Economic Review, 112*(2), 409–441.

Green, R. J. (2008). Electricity wholesale markets: Designs now and in a low-carbon future. *The Energy Journal, 29.*

International Energy Agency (IEA). (2023). *Nuclear power and secure energy transitions.* IEA. https://www.iea.org/reports/nuclear-power-and-secure-energy-transitions

International Energy Agency (IEA). (2024a). *Critical minerals dataset.* IEA. https://www.iea.org/data-and-statistics/data-product/critical-minerals-dataset

International Energy Agency (IEA). (2024b). *World energy outlook 2024.* IEA. https://www.iea.org/reports/world-energy-outlook-2024

Lindberg, M. B. (2022). The power of power markets: Zonal market designs in advancing energy transitions. *Environmental Innovation and Societal Transitions, 45*, 132–153.

Newbery, D., Pollitt, M. G., Ritz, R. A., & Strielkowski, W. (2018). Market design for a high-renewables European electricity system. *Renewable and Sustainable Energy Reviews, 91*, 695–707.

Praktiknjo, A., & Erdmann, G. (2016). Renewable electricity and backup capacities: An (un-) resolvable problem? *The Energy Journal, 37*(SI2), 89–106.

Thomas, S. (2005). British experience of electricity liberalisation: A model for India? *Economic and Political Weekly*, 5260–5268.

Part III
Environment-Related Sustainability Issues

Chapter 8
The Origins of Environmental Problems

Getting and spending, we lay waste our powers; Little we see in Nature that is ours; We have given our hearts away, a sordid boon!
"The World is Too Much with Us", William Wordsworth (1807)

8.1 Understanding Environmental Problems

Environmental problems represent one of the most pressing global challenges of our time, affecting ecosystems, human societies, and economies. Understanding these problems requires exploring their complexity, origins, and impacts. This chapter aims to provide a foundational understanding of environmental problems by defining their scope, exploring their classification, and tracing their evolutions. By examining these aspects, we can better comprehend the systemic nature of environmental challenges and lay the groundwork for promoting effective solutions.

8.1.1 Definition of Environmental Problems

Environmental problems are significant disruptions to natural systems caused by human activities. These issues range from broad systemic challenges to specific localized impacts, requiring a comprehensive understanding from multiple perspectives.

At their core, environmental problems arise when human activities disrupt the delicate balance of Earth's ecosystems, affecting both living organisms (plants and animals) and non-living elements (air, water, soil). The environment is a complex web of interconnected components, from vast forests to microscopic soil organisms, all of which can be impacted by these disturbances. As shown in Fig. 8.1, there are three critical dimensions to consider:

L. Zhang, *Sustainability Economics*, https://doi.org/10.1007/978-3-031-87829-9_8

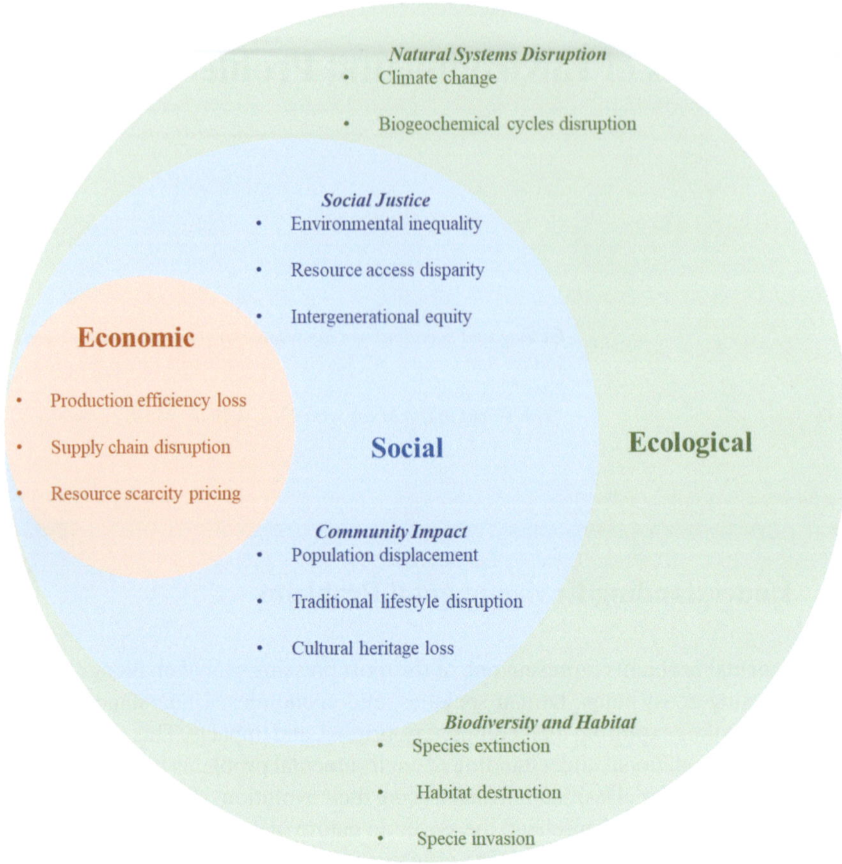

Fig. 8.1 Multidimensional framework of environmental problems

Ecological Dimension: Direct impacts on natural systems, including biodiversity loss and ecosystem degradation.
Social Dimension: Effects on human communities, such as public health concerns and displacement of vulnerable populations.
Economic Dimension: Financial implications of environmental degradation, including resource depletion and market inefficiencies.

The United Nations Environment Programme (UNEP) defines environmental problems as human-induced phenomena that overwhelm the natural environment's capacity for self-repair. This includes major challenges like climate change, pollution, and the depletion of natural resources. These problems reflect conflicts between natural systems and human activities, often exceeding ecosystems' self-repair capabilities.

Environmental problems can be classified by their geographic scale and type:

- **Local Scale**: Issues primarily in urban and industrial settings, such as air pollution from vehicles and industries, water contamination from effluents, and soil pollution in industrial and agricultural areas.
- **Regional Scale**: Problems that transcend local boundaries, like acid rain affecting distant ecosystems, regional biodiversity loss, and cross-boundary air pollution.
- **Global Scale**: Complex challenges like climate change, ocean acidification, plastic pollution, and ozone layer depletion.

Some typical environmental problems are summarized in Table 8.1, causing severe damage.

The historical evolution of environmental problems highlights distinct phases reflecting humanity's changing relationship with nature.

In the pre-industrial phase (before 1750), environmental challenges were localized, including deforestation for agriculture, limited water pollution from early manufacturing, and soil degradation from primitive farming. These impacts, though significant locally, remained geographically contained.

The Industrial Revolution phase (1750–1950) saw a dramatic intensification of environmental problems. Severe urban air pollution, exemplified by London's "Great Smog" of 1952, showcased the deadly potential of unchecked industrial emissions. Large-scale water pollution from manufacturing transformed major waterways, while intensive resource extraction permanently altered landscapes. Chemical contamination introduced new environmental hazards, including persistent organic pollutants and radioactive materials.

Table 8.1 Examples of severe environmental problems

Category	Description	Source
Local scale	Tokyo Groundwater Heavy Metal Contamination: 38% of samples exceed standards; some areas 12× above guidelines	Japan Ministry of Environment Annual Water Quality Monitoring Report in 2019
Regional scale	Southeast Asian Coral Bleaching: 76% of Sulawesi waters' coral reefs were affected (vs. 20–30% historical average)	Indonesia Ministry of Marine Affairs and Fisheries Coral Reef Monitoring Report in 2020
Resource depletion	Australian Groundwater Depletion: Murray-Darling Basin aquifers dropping by 1.4 m annually; 32% faster depletion rate than natural recharge	Australian Water Resources Annual Assessment in 2023
Environmental pollution	Singapore Electronic Waste: 60,000 tons of e-waste annually; only 6% properly recycled; growing at 5× the rate of municipal waste	Singapore National Environment Agency Report in 2022
Soil degradation	New Zealand Pastoral Soil Compaction: 68% of dairy farm soils exceed critical compaction thresholds, reducing productivity by 23%	New Zealand Ministry for Primary Industries Soil Quality Report in 2021

The Post-Industrial phase (1950–present) is marked by complex, interconnected global environmental challenges. Climate change, with its far-reaching impacts on weather patterns, ecosystems, and human societies, represents the quintessential modern environmental problem. Accelerating biodiversity loss and disruption of global nutrient cycles threaten Earth's ecosystem stability. Transboundary pollution issues have become more prevalent, necessitating unprecedented international cooperation for effective management.

Environmental problems are multidimensional and interconnected, requiring a comprehensive understanding of their definitions, classifications, and historical evolution. By exploring these aspects, we can better appreciate the systemic nature of these issues and the importance of integrated solutions. This foundational understanding is crucial for addressing the root causes of environmental problems and achieving sustainable development.

8.1.2 Understanding the Origins

Understanding the origins of environmental problems is crucial for developing long-term solutions, improving research, policymaking, and interdisciplinary collaboration. By examining the roots of these issues, we can better understand their causes and impacts, guiding us toward a sustainable future.

Exploring the origins of environmental issues reveals patterns and drivers often missed in single-focus studies. Historical and systemic analyses, such as the Industrial Revolution's contribution to carbon emissions or the socio-economic factors behind deforestation, highlight the interconnected nature of environmental challenges. These insights foster new approaches that integrate economics, sociology, and environmental science. The "Coupled Human and Natural Systems" (CHANS) framework is a good example, which emphasizes the dynamic interactions between human activities and natural processes, enriches academic discourse and supports evidence-based solutions.

Practically, identifying the root causes of environmental problems is essential for designing effective policies, fostering technological innovation, and advancing public education. Policymakers who understand systemic drivers like unsustainable consumption, inefficient resource management, or inadequate regulations can develop strategies addressing core issues. For example, reducing urban air pollution might involve emission controls and investments in public transportation and renewable energy. Technological advancements, such as biodegradable plastics, carbon capture technology, and circular economy models, demonstrate how addressing root causes can lead to impactful solutions. Public education campaigns also empower individuals to make informed decisions and advocate for systemic change, as seen in efforts linking deforestation to climate change.

The complexity of environmental problems necessitates interdisciplinary collaboration, drawing on insights from economics, sociology, and environmental science. Economists analyze the financial mechanisms behind resource exploitation and the economic costs of environmental degradation. Sociologists examine social dimensions, including unequal distribution of environmental risks and cultural factors influencing sustainable practices. Environmental scientists study ecosystem processes and human impacts, identifying tipping points like deforestation and coral reef degradation.

A compelling example of interdisciplinary research is the global food system. Agriculture, a leading cause of deforestation, biodiversity loss, and greenhouse gas emissions, is deeply intertwined with economic, social, and environmental systems. Economists analyze inefficiencies in food supply chains and economic incentives driving unsustainable farming. Sociologists explore cultural and social barriers to plant-based diets and inequalities in access to nutritious food. Environmental scientists study intensive agriculture's impacts on soil health, water resources, and ecosystems. Together, these disciplines develop strategies like sustainable agriculture practices, agroecology, and policies to reduce food waste and encourage dietary shifts, highlighting the value of collaboration in addressing environmental challenges.

In conclusion, tracing the origins of environmental problems is foundational for addressing today's pressing challenges. By examining these issues through academic, practical, multiscale, and interdisciplinary lenses, we can develop comprehensive strategies essential for achieving long-term sustainability, fostering equity, and ensuring a livable planet for future generations.

8.1.3 Overview of Environmental Problems

Environmental problems threaten ecosystems, human well-being, and socioeconomic development. These challenges are interconnected and systemic, reflecting the complex interplay between natural and human systems. This overview covers four core areas: resource allocation, air pollution, water resources, and land use.

Resource allocation: the heart of sustainability

How we manage resources like oil, water, and soil profoundly impacts our planet. Overfishing, deforestation, and inefficient consumption disrupt the balance that supports life. Economic pressures, population growth, and unsustainable practices exacerbate these issues. For example, fossil fuel extraction depletes reserves and contributes to pollution, while unsustainable agriculture degrades soil and depletes water resources. Sustainable resource management, including concepts like the circular economy, aims to balance human needs with the planet's capacity to regenerate.

Air pollution: a breath of trouble

Air pollution, visible as smog and haze, is a health crisis. Pollutants from vehicles and industries cause respiratory and cardiovascular diseases, with over 7 million premature deaths annually (WHO, 2021). Beyond local impacts, air pollution has regional and global consequences, such as transboundary haze and climate change. Mitigating air pollution requires systemic changes, including renewable energy, public transportation, and stricter emission standards.

Water resources: a thirsty world

Water scarcity and pollution affect billions globally. Freshwater resources are unevenly distributed and threatened by climate change, over-extraction, and pollution. Agriculture, using 70% of global freshwater (FAO, 2020), often conflicts with industrial and domestic needs. Pollution from industrial discharges, agricultural runoff, and untreated wastewater contaminates water supplies, harming ecosystems and human health. Integrated water management strategies are essential to balance demands, protect ecosystems, and enhance resilience to climate change.

Land use: the ground beneath our feet

Land use changes, driven by urbanization, agriculture, and deforestation, significantly contribute to environmental degradation. Deforestation for agriculture and infrastructure leads to habitat loss, decline in biodiversity, and increased greenhouse gas emissions. Urban sprawl consumes fertile land and disrupts ecosystems. Land degradation, including desertification and soil erosion, threatens food security and livelihoods. Sustainable land management practices, such as agroforestry and reforestation, are crucial for mitigating these impacts and ensuring long-term stability.

8.2 Environmental Problems Over Time

The evolution of environmental challenges has paralleled humanity's technological and societal development. Historical analysis shows how human–environment interactions have transformed from localized impacts to global systemic challenges, progressing through distinct phases: pre-industrial local issues, industrialization's systematic impacts, and today's complex global challenges requiring international cooperation.

8.2.1 Pre-industrial Environmental Challenges

Before industrialization, human societies maintained a relatively balanced relationship with their environment, though significant regional impacts occurred. Early civilizations often faced environmental limitations influencing their development. For example, in medieval England, King Edward I's 1272 proclamation against sea coal use marked an early attempt at air pollution control, leading to the Environmental Pollution Management Act of 1388, the world's first environmental legislation.

Ancient Mesopotamia provides a compelling case study. The Sumerians developed sophisticated irrigation systems that transformed arid landscapes into productive agricultural lands. However, repeated flooding and evaporation increased soil salinity, leading to a dramatic shift from wheat to barley and a decline in agricultural yields by 1700 BCE, contributing to societal disruption. Hunter-gatherer societies also influenced their environments, with early human hunting contributing to megafauna extinctions in isolated ecosystems (Martin & Klein, 2021). However, at that stage, these impacts were often reversible due to the smaller scale of human activity compared to natural regenerative capacity (Ponting, 2007).

8.2.2 *Industrial Revolution and the Emerging Crisis*

The Industrial Revolution, beginning in late 18th-century Britain, marked a watershed moment. Manufacturing output surged, with British coal production increasing from 5.23 million tons in 1750 to 81.32 million tons by 1850, and iron production rising from 12,000 tons to over 2 million tons. This expansion fundamentally transformed humanity's relationship with the environment, establishing patterns of resource exploitation that shape contemporary challenges.

The scale of industrial production created intensive demands for natural resources, particularly fossil fuels and minerals. The transition to fossil fuels enabled extraordinary industrial output but introduced new environmental stresses. By 1900, global coal consumption reached 800 million tons annually, leading to significant air pollution and landscape degradation.

Urban industrial activity generated unprecedented environmental pressures. The 1952 London Great Smog, which claimed thousands of lives, highlighted the severe consequences of unchecked industrial pollution. This and similar disasters prompted early environmental regulation and pollution control efforts. Colonial resource extraction systems also created new forms of environmental degradation, initiating lasting patterns of inequality and global challenges.

Visible environmental degradation in the late nineteenth century sparked initial organized responses, including environmental legislation and conservation movements (Ponting, 2007). However, these efforts were limited compared to the scale of industrial impacts. Addressing global environmental problems requires coordinated international responses and systematic sustainability approaches (Rockström et al., 2009).

8.2.3 *Post-industrial Era and Globalization*

The post-World War II period saw a dramatic intensification of environmental challenges, transforming localized issues into complex global phenomena.

The environmental crisis escalation (1950s-1970s)

The post-war economic boom triggered unprecedented environmental pressures. According to BP's Statistical Review historical data, global petroleum consumption saw dramatic increases during the post-war period, fundamentally reshaping global emissions patterns. Industrial production and energy consumption accelerated, creating multifaceted environmental challenges. The transportation and industrial sectors became major contributors to increasing environmental pressures during this period.

Industrial expansion introduced new forms of pollution, with petrochemical industries and synthetic materials production creating persistent organic pollutants. Chemical contamination became increasingly complex, as documented by Carson's "Silent Spring" in 1962. High-profile environmental disasters, such as the Minamata disease crisis and the Cuyahoga River fires, catalyzed environmental consciousness and legislation in industrialized societies.

However, early environmental regulations were limited, focusing on end-of-pipe pollution control rather than underlying production processes. They often addressed individual pollutants or specific media separately, failing to account for interconnected environmental systems. The geographic scope of protection was also limited, with little consideration for transboundary pollution or global commons.

Global environmental impact and international cooperation (1970s-1990s)

The late twentieth century saw a dramatic shift in environmental degradation patterns as global trade expanded. Between 1970 and 1990, global trade volume quadrupled, altering the geographic distribution of industrial production and its environmental impacts. This period marked the rise of complex global supply chains, with raw material extraction, manufacturing, and consumption increasingly separated across national boundaries.

Heavy industries and resource extraction relocated to developing regions. For example, Chile's copper production increased from 686,000 tons in 1970 to 2.2 million tons by 1990. Similar shifts occurred in steel production, with East Asian output growing by three times. Environmental impact displacement also grew by about three times, with carbon leakage through trade accounting for 25% of global emissions by 1990. This industrial migration often occurred in regions with less stringent environmental regulations, leading to significant environmental challenges, including water pollution and hazardous waste accumulation.

The emergence of electronic waste and accelerated resource exploitation had far-reaching consequences for global ecosystems. Since the mid-1990s, Guiyu, China, has received over 1 million tons of e-waste annually, leading to severe soil and water contamination.

The growing recognition of global environmental challenges spurred unprecedented international cooperation. The 1972 Stockholm Conference on the Human Environment established the UNEP and produced the Stockholm Declaration, articulating key principles of international environmental governance. The discovery of the Antarctic ozone hole in 1985 led to the Vienna Convention and the Montreal

Protocol, successfully phasing out ozone-depleting substances. The 1992 Rio Earth Summit produced landmark agreements, including the UN Framework Convention on Climate Change (UNFCCC) and the Convention on Biological Diversity (CBD), though implementation often fell short, particularly in developing nations.

Global environmental crisis in the twenty-first century (2000–present)

The twenty-first century presents unprecedented challenges in resource consumption and environmental sustainability. As shown in Fig. 8.2, global material extraction surged from 30 billion tons in 1970 to 106.6 billion tons in 2024 (UNEP & IRP, 2024). This period brought a more sophisticated understanding of Earth's environmental systems and their limits, exemplified by the planetary boundaries framework developed in 2009. This framework identifies nine critical environmental thresholds, including climate change, biodiversity loss, and ocean acidification.

The Arctic case in 2007, where sea ice reached a historic minimum, demonstrated how crossing one planetary boundary could trigger cascading effects across others. Modern environmental problems, such as microplastics and endocrine-disrupting chemicals, often involve "invisible threats" that cross multiple planetary boundaries simultaneously.

Technology has played a dual role in creating and monitoring environmental problems. Digital infrastructure introduces new challenges, but also enhances our ability to detect and understand environmental limits. The global financial crisis of

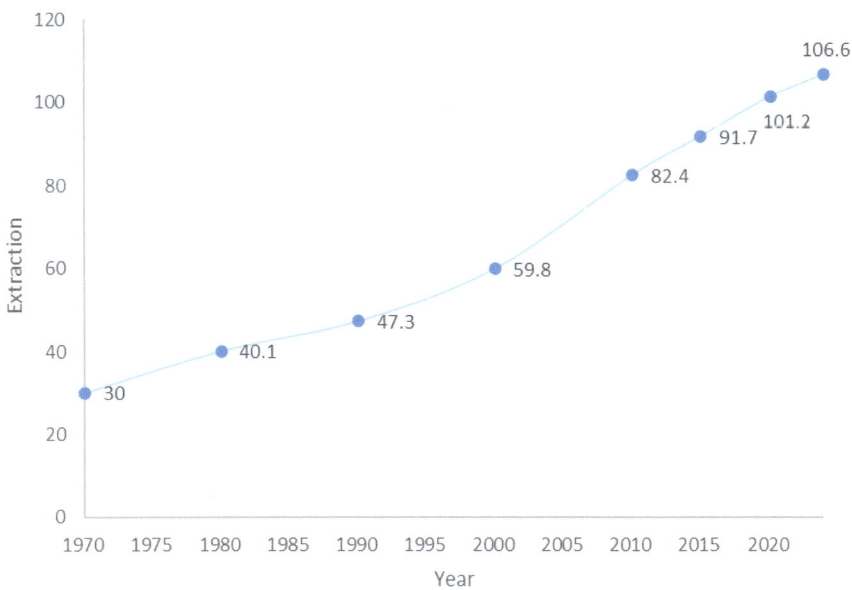

Fig. 8.2 Global material extraction (1970–2024). *Data source* Global Material Flows Database, UNEP & IRP (2024)

2008 highlighted the deep integration of environmental problems with economic systems, driving transformative developments in global environmental governance.

The 2015 Paris Agreement marked a significant evolution in international climate action, establishing a framework for limiting global temperature rise. Market-based mechanisms, such as the European Union's Emissions Trading System and China's national carbon trading market, have emerged as crucial tools in this new governance landscape. China's commitment to peak carbon emissions before 2030 and achieve carbon neutrality before 2060 represents a significant development in global environmental governance.

These developments exemplify the evolution of 21st-century environmental governance, where national actions, market mechanisms, and international frameworks increasingly interact to address global challenges. The integration of economic instruments with environmental policy creates new opportunities for balancing environmental protection with development priorities.

Table 8.2 shows the major environmental issues and policies in history.

8.3 Key Drivers of Environmental Problems

Understanding the drivers of environmental problems requires a systematic framework and detailed analysis. As Herman Daly noted in "Beyond Growth: The Economics of Sustainable Development" in 1996, these challenges arise from allocating finite natural resources among competing uses while navigating the interconnections between climate change, pollution, overpopulation, and energy consumption. The IPAT equation introduced in Chap. 5 provides a foundational framework for examining how human activities influence environmental outcomes.

Herendeen's (1998) logarithmic version of the IPAT formula allows for more sophisticated analysis. Studies using this framework reveal that while population growth increases environmental pressure, its impact is significantly modulated by consumption patterns and the broader set of technological, institutional, and cultural factors in the T component. The WRI (WRI, 2023) shows that environmental impacts vary significantly across regions with similar population and affluence levels, highlighting the crucial role of these additional factors.

Modern environmental challenges can be understood through multiple, interconnected pathways. As societies advance and economies grow, they strain natural systems through complex interactions between various drivers. These problems stem from conflicts between human development and natural systems, manifesting through five key categories: social, economic, technological, institutional, and resource-related factors.

Understanding these drivers requires examining both their individual effects and their interactions. Some drivers, like population growth and technological change, directly influence environmental outcomes, while others operate through indirect pathways, creating feedback loops and cascading effects that amplify challenges. The

Table 8.2 Major environmental problems and policies in history

Year	Event/Policy	Significance
Pre-Industrial Era		
1272	King Edward I's Sea-Coal Ban	First documented air pollution control measure
1388	Environmental Management Act	First comprehensive environmental legislation
Industrial Revolution		
1863	Alkali Act	Pioneer industrial pollution control legislation
1876	Rivers Pollution Prevention Act	Water pollution control frame establishment
1947	Nature and Heritage Protection Act	Modern ecological protection law foundation
1948	Donora Smog Disaster	Catalyzed air quality legislation
1952	Great London Smog	Led to the Clean Air Act implementation
Post-Industrial Era: The Environmental Crisis Escalation		
1956	Minimata Disease Crisis	Reveal severe health impacts of Hg pollution
1962	Publication of "Silent Spring"	Launched modern environmental movement
1963	Clean Air Act	First comprehensive air quality legislation
1967	Organic Pollution Control Act	First all-around environmental frame in Asia
1969	Environmental Protection Act	Pioneering European environmental legislation
1969	Cuyahoga River Fire	Catalyzed water pollution control efforts
Post-Industrial Era: Global Environmental Impact and International Cooperation		
1972	Clean Water Act	Reliable water quality protection establishment
1972	Stockholm Conference	Initiated global environmental governance
1985	Discovery of Antarctic Ozone Hole	Triggered international ozone protection efforts
1985	Vienna Convention	First global frame for ozone protection
1987	Montreal Protocol	First binding global environmental treaty
1992	Rio Earth Summit	Established sustainable development principles
1992	UN Convention on Climate Change	Created global climate action framework
1992	Convention on Biological Diversity	Framework for global biodiversity protection
1997	Kyoto Protocol	Framework for addressing climate change
Post-Industrial Era: Global Environmental Crisis in the 21st Century		
2005	Emissions Trading System	World's first major carbon market

(continued)

Table 8.2 (continued)

Year	Event/Policy	Significance
2006	Global Warming Solutions Act	First all-around U.S. state climate legislation
2007	Arctic Sea Ice Minimum	Demonstrated critical climate tipping point
2009	Planetary Boundaries Framework	New paradigm for environmental limits
2011	China's First Carbon Trading Pilots	Largest DC Carbon Market Initiative
2015	Paris Agreement	New era of global climate governance
2017	National ETS Announcement	World's largest carbon market formation
2020	Global Net-Zero Commitments	Major economies pledge carbon neutrality

following sections explore each category of drivers in detail, providing a comprehensive analysis of how these factors shape environmental outcomes and interact to create persistent problems.

8.3.1 Social Drivers

Social factors are foundational drivers shaping the intensity and nature of environmental challenges. IPCC identifies social dynamics as key drivers influencing climate action and environmental protection across sectors. Population dynamics, inequality, consumption patterns, and cultural norms consistently modify the impacts of other environmental drivers, including technological innovation and institutional responses.

Population

Population dynamics drive environmental challenges through growth, demographic transitions, and urbanization. Population growth increases resource demands and environmental pressures through consumption and waste generation. Urban areas, housing 56% of the world's population, consume 60–80% of global energy and produce about 70% of carbon emissions (UN Habitat, 2022). Demographic transitions and development stages shape environmental impacts, with aging populations in developed nations driving increased energy consumption and youth-heavy populations in developing regions correlating with rapid urbanization. For example, China's urban population expanded by 500 million between 1980 and 2020, triggering a sixfold increase in energy consumption.

Social Inequality

Social inequality shapes environmental outcomes through differential resource access, varying environmental burden distribution, and disparate adaptive capacities. Inequality and environmental degradation form a self-reinforcing cycle, where disparities in resource access lead to uneven environmental impacts, exacerbating social inequalities. The top 1% of global emitters contribute 17% of lifestyle consumption emissions, while the bottom 50% contribute less than 10% (UNEP, 2024). In developing nations, poverty often leads to immediate survival imperatives that override long-term environmental considerations, resulting in ecological damage.

Lifestyle and consumption patterns

Lifestyle and consumption patterns influence environmental outcomes through resource intensity, consumption scale, and diffusion of consumption norms. The population of the middle class might reach 4.9 billion by 2030, with Asia accounting for 65% of this growth. Each 1% increase in the middle-class population correlates with a 1.3% increase in energy consumption and a 1.5% increase in carbon emissions in emerging economies (IEA, 2023). Changing consumption patterns, such as increased meat consumption in China, drive deforestation and increased methane emissions.

Culture

Cultural and psychological factors influence environmental outcomes through value systems, social norms, and behavioral patterns. Cultural values shape environmental attitudes, which manifest in social norms governing acceptable behaviors. South Korea's successful transformation in waste management and recycling practices demonstrates how cultural shifts can drive behavioral change. The emergence of "flygskam" (flight shame) in Scandinavia, leading to decreased air travel and increased rail usage, further illustrates this dynamic.

Intergenerational dynamics

Intergenerational dynamics shape environmental outcomes through value transmission, institutional change pressure, and innovation dynamics. According to the Climate Inequality Report from the World Inequality Lab (WIL, 2023), older generations have historically contributed disproportionately to global carbon emissions, while younger generations bear the brunt of climate adaptation costs. This growing awareness has fueled youth-led climate movements, such as the Global Climate Strike, which exemplify the role of younger generations in pressuring governments to adopt equitable climate policies. These dynamics underscore how intergenerational inequality drives institutional change and innovation in climate governance.

Education

Educational factors influence environmental outcomes through knowledge transmission, skill development, value formation, and behavioral change facilitation. Environmental education develops environmental literacy and action competence. Finland's comprehensive environmental education program led to a 35% reduction in energy consumption and a 45% increase in recycling rates by 2023. Obviously, countries with comprehensive environmental education programs show higher public support for environmental policies and participation in conservation initiatives.

The interplay between these social drivers creates complex feedback loops and emergent patterns of environmental impact. Understanding these drivers and their interactions is crucial for developing effective environmental solutions.

8.3.2 Economic Drivers

Environmental problems stem from the complex interactions between economic activities and ecological systems. Three critical economic drivers influence environmental outcomes: industrialization, globalization, and market failures. Industrialization generates environmental stresses through resource intensity, infrastructure transformation, and consumption pattern changes. Globalization accelerates transnational resource flows and lifestyle diffusion, creating challenges around supply chain complexity and economic standardization. Market failures, characterized by externalities, misaligned incentives, and information asymmetries, obscure the true ecological costs of economic activities, leading to overconsumption and insufficient environmental protection.

Industrialization

Industrialization generates environmental stresses through three pathways: resource intensity, infrastructure changes, and altered consumption behaviors. Industrial production often leads to habitat destruction, biodiversity loss, and increased resource extraction. Infrastructure changes, such as urban expansion and industrial complex development, reshape ecological landscapes, disrupting natural systems. For instance, in Ethiopia, the cities of Galan, Dukem, and Bishoftu have collectively allocated approximately 3060 ha (30.6 km^2) of land for industrial development between 2006 and 2021, primarily converting agricultural land into industrial use. Additionally, industrialization influences societal consumption patterns, embedding resource-intensive practices within broader societal structures. This shift often leads to increased demand for energy and raw materials, further exacerbating environmental pressures.

Globalization

Globalization exerts environmental pressures through complex supply chains, lifestyle diffusion, and production standardization. Global supply chains distribute environmental impacts across geographical boundaries, making ecological consequences increasingly opaque. For instance, approximately 22% of global CO_2 emissions are embedded in international trade (Our World in Data, 2024a). The rapid transmission of resource-intensive lifestyle models from developed to developing economies normalizes high-consumption patterns, creating accelerated environmental pressures. Global standardization processes create uniform environmental pressures across diverse contexts, as multinational corporations establish standardized production protocols.

Market failure

Environmental resources often exhibit characteristics of public goods or common resources, leading to market failures. The "tragedy of the commons" occurs when individuals pursuing their own interests deplete shared resources, creating long-term environmental issues. Market failures arise from external costs and poorly defined property rights, preventing markets from accurately valuing environmental impacts. According to the International Monetary Fund, global fossil fuel subsidies, including unpriced environmental damages, amount to $5.9 trillion in 2020. The Coase theorem suggests that well-defined property rights and minimal transaction costs could resolve environmental externalities efficiently, but real-world factors often prevent this.

Beyond externalities, market failures stem from misaligned incentives and information asymmetries. Economic systems often prioritize short-term financial performance over long-term ecological sustainability, creating barriers to environmental protection. Information asymmetry complicates decision-making, as market participants frequently lack a comprehensive understanding of environmental impacts. According to the Climate Inequality Report 2023, in the United States, while the national average is about 21 tons CO_2e per capita, the richest 10% emit approximately 70 tons CO_2e per capita. This stark contrast in emissions patterns demonstrates how wealth inequality translates into environmental impact inequality, with the richest groups needing to make the most substantial reductions to meet climate targets (WIL, 2023).

8.3.3 *Technological Drivers*

Over the past few decades, technological advancements have driven unprecedented economic growth and improved living standards. However, this progress has also come with significant environmental costs. According to World Bank (2023), global GDP grew from $1.367 trillion in 1960 to over $100 trillion in 2022, while global carbon dioxide emissions increased by nearly 300% during the same period. This growth pattern underscores the intricate relationship between technological development and environmental impact.

Energy technologies

The evolution of energy technologies reflects human progress but also exacerbates environmental challenges. Despite the rapid development of renewable energy, fossil fuels remain dominant, accounting for approximately 80% of global energy consumption in 2022 (IEA, 2023). For example, China's coal power capacity exceeded 1.3 billion kilowatts by mid-2023, causing severe air pollution. Additionally, inefficiencies in energy systems, such as India's high transmission and distribution losses, further compound environmental burdens (World Bank, 2023). Emerging technologies like hydraulic fracturing have improved extraction efficiency, but they also introduce significant environmental risks, such as groundwater contamination and methane leaks.

Industrial technology

Advances in industrial technology have increased productive capacity but also environmental burdens. The cement industry, for example, accounts for approximately 8% of global carbon emissions (IEA, 2023). Rapid urbanization in India and Southeast Asia has intensified environmental pressures, with India's cement production doubling from 2010 to 2023. Mining technologies, while increasing production, have led to severe environmental degradation, such as the 2019 Brumadinho dam collapse in Brazil. Industrial waste management remains a challenge, with significant amounts of toxic waste lacking effective treatment.

Agricultural technologies

Modern agricultural technologies have revolutionized food production but imposed severe ecological pressures. Precision agriculture, while improving resource efficiency, has expanded monoculture practices, reducing biodiversity. The Green Revolution's technological package increased food production but caused groundwater contamination and eutrophication. Livestock production technologies contribute significantly to greenhouse gas emissions and deforestation. Chemical agricultural technologies, such as pesticides, have disrupted ecosystems, while genetic modification introduces ecological uncertainties.

Electronic technology

The digital revolution has transformed modern life but carries significant environmental implications. Electronic waste reached 53.6 million tons in 2019 and is projected to grow to 74.7 million tons by 2030 (Global E-waste Monitor, 2020). Planned obsolescence in consumer electronics fosters frequent upgrades, increasing waste. Raw material extraction for electronics, particularly rare earth elements, causes extensive environmental damage. E-waste management remains a challenge, with only 17.4% of e-waste properly recycled, leading to severe environmental impacts in processing centers like Ghana's Agbogbloshie.

8.3.4 Institutional and Governance Drivers

In global environmental governance, institutions and governance drivers are intermediaries through which economic and social factors affect the environment. They not only reflect environmental challenges but also shape them. The effectiveness of these frameworks, which coordinate social, economic, and ecological systems, determines our ability to respond to environmental issues. Institutional designs and governance models profoundly affect resource allocation, environmental policy, and ecosystem health. From international climate negotiations to local environmental protection, the quality of institutions and governance determines whether environmental problems are perpetuated or alleviated.

According to UNEP's 2023 Global Governance Report (UNEP, 2024), about 67% of major environmental challenges stem from institutional deficiencies rather than technical or economic constraints. This highlights that the root causes of environmental problems often lie in complex institutional structures and governance mechanisms. These drivers reshape human interactions with the environment through legal frameworks, policy incentives, accountability mechanisms, and resource allocation methods. Understanding these drivers requires a systemic, dynamic, and holistic perspective on environmental governance.

Institutional failure

Institutional failures are real-world obstacles that undermine our ability to address environmental challenges. Weak environmental regulations are a critical issue. For example, Brazil's environmental laws are comprehensive on paper but weak in enforcement, allowing large-scale deforestation in the Amazon to continue unchecked. Governance fragmentation is another problem. Environmental issues do not respect administrative boundaries, yet institutional frameworks remain territorial. In the Mekong River Basin, a lack of coordinated governance leads to catastrophic outcomes as each nation pursues its narrow interests. Corruption further weakens regulation, transforming environmental laws into negotiable recommendations. In Indonesia, environmental permits for mining and plantations are often obtained through opaque processes involving substantial bribes. Accountability deficits also hinder regulation. Many regulatory bodies lack mechanisms to track, assess, and penalize violations, making penalties minimal and ineffective. Political cycles add complexity, as short-term electoral rhythms incentivize immediate economic gains over long-term environmental planning.

Governance drivers

Governance challenges are systemic obstacles that undermine effective environmental decision-making. Short-term political dynamics create a sustainability dilemma, as democratic systems incentivize short-term thinking that undermines long-term planning. For example, U.S. presidential transitions can lead to substantial policy reversals, creating uncertainty for long-term environmental planning. Poor global governance also exacerbates environmental issues. International environmental governance is fragmented, with weak coordination and misaligned national interests. The Paris Agreement, while symbolically significant, lacks robust enforcement mechanisms. Policy incoherence further drives systemic contradictions. Many governments proclaim ambitious carbon reduction targets while maintaining substantial fossil fuel subsidies, effectively nullifying their commitments.

8.3.5 *Resource Drivers*

The global landscape of resource distribution is increasingly unequal, creating systemic pressures. As populations grow and technology advances, the demand for natural resources escalates, posing challenges for ecological management and sustainable development. Resource drivers include the distribution of natural resources, climate change impacts, and ecological system vulnerabilities, influencing economic development, social stability, and environmental sustainability.

Resource distribution and ecological complexity are significant drivers of environmental issues. Their extraction and utilization trigger and accelerate environmental challenges, creating complex causation chains beyond immediate geographical boundaries. The geopolitical dimension of resource distribution is starkly illustrated by rare earth metals, with 70% concentrated in China, transforming them into strategic drivers of global environmental dynamics. China's reliance on coal has led to higher air pollution levels compared to regions using alternative energy sources. Similarly, the demand for minerals for electric vehicle batteries drives environmental degradation and social disruption in the Democratic Republic of Congo, causing deforestation, ecological disruption, and social conflicts.

Resources drive environmental problems through continuous transformation mechanisms. The Amazon rainforest, once a robust carbon sink, now risks becoming a net carbon emitter due to deforestation and agricultural expansion. In Brazil's Atlantic Forest, over 90% of the original forest cover has been decimated, driving species like the golden lion tamarin toward endangerment. Resources fundamentally restructure ecosystems.

Analyzing environmental problem drivers reveals an interconnected system where social, economic, technological, institutional, and resource factors continuously interact. No single driver operates in isolation; they form a complex network shaping environmental challenges. Understanding these relationships is crucial for developing comprehensive strategies to address environmental issues, setting the stage for exploring the broader complexity of environmental problems in the next section.

8.4 The Complexity of Environmental Problems

In today's globalized and interconnected socio-economic systems, environmental issues have evolved into complex challenges driven by a web of interconnected factors. These drivers—spanning economic, technological, social, policy, and resource dimensions—not only influence one another but also interact dynamically across domains, regions, and timescales. This complexity makes simple, linear analyses and isolated solutions inadequate for addressing the root causes of environmental challenges. Recognizing the interdependencies, non-linear dynamics, and systemic nature of these problems is crucial.

8.4.1 Interdependency Between Pollution and Growth

The Environmental Kuznets Curve (EKC) discussed in Chap. 5 offers a framework to understand how these multiple drivers interact through different stages of economic development. It suggests that environmental degradation follows an inverted U-shaped pattern as the economy develops. This relationship can be understood through an economic process. Initially, environmental impact was thought to increase linearly with economic growth, expressed simply as:

$$e = \alpha y \tag{8.1}$$

where e represents environmental degradation and y represents income per capita. This understanding is deepened by recognizing that the environmental impact of economic activity α tends to change with development level y, as depicted in Eq. 8.2:

$$\alpha = \beta^0 - \beta^1 y, \tag{8.2}$$

where β_0 represents the initial environmental intensity and β_1 reflects improvements in environmental efficiency. Combining these relationships yields the characteristic EKC equation:

$$e = \beta^0 y - \beta^1 y^2. \tag{8.3}$$

As shown in Fig. 8.3, this relationship reflects three distinct phases of development. This relationship can be understood through an economic thinking process. In the early phase, environmental degradation is limited primarily to subsistence activities and biodegradable wastes. As economies enter rapid industrialization, environmental pressures intensify through accelerated resource extraction, increased waste generation, and limited environmental controls. However, at the highest development levels, several factors emerge to reduce environmental impact: structural shifts toward information-intensive industries, enhanced environmental awareness, stronger regulations, and improved technologies.

Contemporary scholarly discourse suggests that although the perceived trade-off between economic growth and environmental quality may be less pronounced than previously postulated, favorable environmental outcomes necessitate proactive interventions rather than passive dependence on economic development alone. Such interventions require the synchronous advancement of technological innovation, policy frameworks, and societal behavioral modifications. The synergistic relationship between economic and technological drivers serves as an illustrative example. When firms invest in green innovation, environmental improvements can catalyze broader technological advances. For instance, Tesla's focus on electric vehicle batteries led to innovations in energy storage, AI, and manufacturing processes. Other drivers, such as social and policy factors, also interact with economic drivers in complex ways.

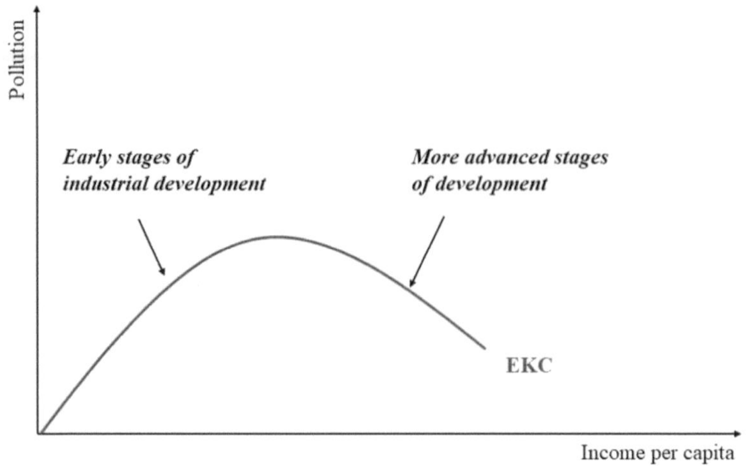

Fig. 8.3 Environmental Kuznets curve

For example, China's aggressive electric vehicle policies have reduced emissions, restructured global automotive markets, and influenced consumer preferences.

8.4.2 Nonlinear Dynamics of Environmental System

Feedback mechanisms

Environmental systems react to external drivers through positive and negative feedback loops. Positive feedback loops are self-reinforcing cycles that can accelerate environmental problems. For example, global warming causes Arctic ice to melt, reducing the earth's reflectivity and increasing heat absorption, which further accelerates ice melting. Negative feedback loops, on the other hand, stabilize environmental systems. For instance, increased atmospheric carbon dioxide can stimulate plant growth, enhancing carbon absorption and partially offsetting emissions. These feedback mechanisms highlight the complexity of environmental systems, which can quickly escalate or self-regulate based on specific interactions.

Critical thresholds of environmental problems

Environmental systems often have critical thresholds or tipping points, beyond which rapid and potentially irreversible changes occur. For example, overfishing can lead to the sudden collapse of marine ecosystems, causing severe economic repercussions. These thresholds can also involve time lags and cumulative effects. Greenhouse gas emissions, for instance, may not show their full impact for decades, complicating scientific understanding and policy responses.

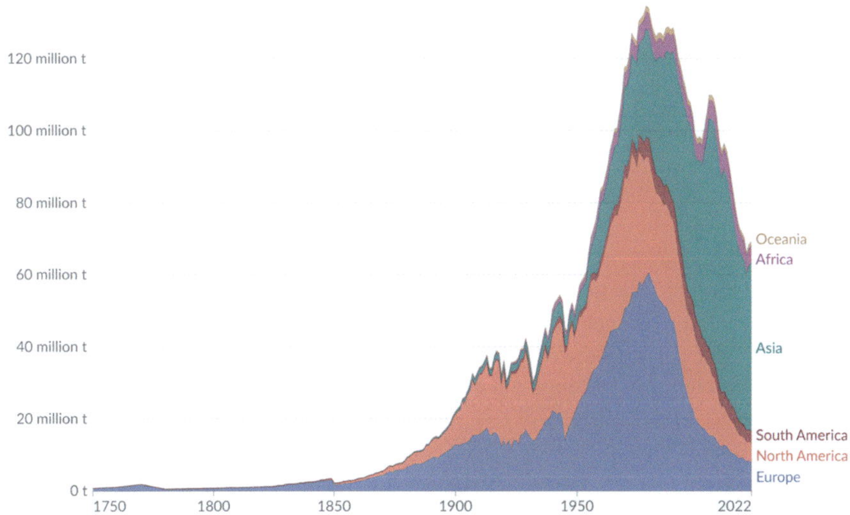

Fig. 8.4 Tipping points of global sulfur dioxide emission by world region. *Note* This figure is plotted by Our World in Data (2024b)

Understanding these critical thresholds is crucial for environmental policy and management. The case of sulfur dioxide emissions, as illustrated in Fig. 8.4, provides a clear example of how different regions approach and navigate such tipping points. The data reveals distinct regional patterns: Europe led the initial rise in sulfur pollution, followed by North America in the mid-nineteenth century, both reaching their peaks in the 1970s–1980s before declining. The early declining trends in Europe and North America's emissions may suggest that reaching critical pollution levels could have prompted environmental responses like policies and technologies. Asia's emissions, on the other hand, show a dramatic increase since the late twentieth century, reflecting its later industrialization phase, while Africa and Oceania have maintained relatively low emission levels throughout.

As demonstrated by regional variations in sulfur dioxide emissions, these tipping points are not merely environmental phenomena but are deeply intertwined with socio-economic development patterns. This understanding has profound implications for developing regions, suggesting the potential for "leapfrogging" past the high-emission development stages experienced by earlier industrialized nations through early adoption of cleaner technologies and pre-emptive policy measures.

8.4.3 *Systemic Challenges*

Systemic challenges in environmental management involve complexity, uncertainty, and interdependence. These challenges require a holistic understanding that transcends disciplinary boundaries and recognizes the intricate interactions across economic, social, technological, and ecological domains.

At the core of these challenges are interconnected drivers and their nonlinear interactions. Resource management, for example, involves dynamic relationships between economic incentives, technological capabilities, institutional frameworks, and ecological constraints. Changes in one domain, such as agricultural irrigation techniques, can cascade through multiple systems, affecting urban water supply, industrial productivity, and regional ecological balance.

Effective pollution mitigation requires a synchronized approach that integrates energy transition strategies, industrial restructuring, technological innovation, and social behavioral change. China's electric vehicle policy exemplifies this systemic approach, addressing environmental, economic, and technological challenges simultaneously. Land use management also reveals the tension between immediate economic returns and long-term ecological sustainability.

Navigating environmental challenges demands a shift in governance and problem-solving approaches. The systemic nature of these challenges, marked by interdependencies, feedback mechanisms, time-lagged effects, and critical thresholds, renders traditional linear methods inadequate. Instead, adaptive strategies that embrace uncertainty, prioritize continuous learning, and recognize the interconnectedness of environmental, economic, and social systems are needed.

The path forward requires a multifaceted approach, emphasizing transdisciplinary collaboration, adaptive policy mechanisms, robust monitoring systems, technological and social innovations, and holistic evaluations of environmental-economic interactions. Developing a clear understanding of environmental challenges and implementing targeted solutions is essential.

Reading

The American Dust Bowl

The Dust Bowl (seen in Fig. 8.5) was a major ecological disaster in American history, driven by economic ambition, environmental vulnerability, and mismanagement. In the early 20th century, rapid economic expansion and agricultural mechanization led to extensive prairie land cultivation, prioritizing short-term gains over sustainability. The semi-arid climate made the ecosystem fragile, and the Great Depression further pressured farmers to maximize land use.

Fig. 8.5 Farmer and sons walking in the face of a dust storm. *Source* Library of Congress (1936)

Between 1931 and 1934, drought and aggressive land cultivation set the stage for disaster. In May 1934, a massive dust storm swept across the western prairie, creating a 2400-kilometer-long and 3400-meter-high dust band. This event destroyed crops, dried up streams, cracked the land, and killed livestock, forcing hundreds of thousands into "ecological migration."

Beyond immediate destruction, the Dust Bowl caused social and economic upheaval, collapsing agricultural economies and displacing populations. However, it also led to significant changes in environmental policy. The U.S. government introduced the Soil Conservation and Domestic Allotment Act and established the Soil Conservation Service, marking a shift to proactive environmental management. The Dust Bowl highlighted the complex relationships between economic development, technology, and sustainability, showing how short-term pursuits can undermine long-term stability.

References

Food and Agriculture Organization (FAO). (2020). *The state of food and agriculture: Overcoming water challenges in agriculture.* https://www.unwater.org/news/state-food-and-agriculture-2020-overcoming-water-challenges-agriculture

Global E-waste Monitor. (2020). *The global e-waste monitor 2020: Quantities, flows and the circular economy potential.* United Nations University.
Herendeen, R. A. (1998). *Ecological numeracy: Quantitative analysis of environmental issues.* Wiley.
International Energy Agency (IEA). (2023). *World Energy Outlook 2023.* https://www.iea.org/reports/world-energy-outlook-2023?wpappninja_v=o5bj1zhuo
Library of Congress. (1936). *Farmer and sons walking in the face of a dust storm.* https://www.loc.gov/pictures/item/2017760335
Martin, P. S., & Klein, R. G. (2021). *Quaternary extinctions: A prehistoric revolution.* University of Arizona Press.
Our World in Data. (2024a). *Share of CO_2 emissions embedded in trade.* https://ourworldindata.org/grapher/share-co2-embedded-in-trade
Our World in Data. (2024b). *Global sulphur dioxide (SO2) emissions by world region.* https://ourworldindata.org/grapher/so-emissions-by-world-region-in-million-tonnes
Ponting, C. (2007). *A new green history of the world: The environment and the collapse of great civilizations.* Penguin Books.
Rockström, J., Steffen, W., Noone, K., Persson, Å., Chapin, F. S., Lambin, E. F., Foley, J. A., et al. (2009). A safe operating space for humanity. *Nature, 461*(7263), 472–475.
UN Habitat. (2022). *World cities report 2022: Envisaging the future of cities.* https://unhabitat.org/world-cities-report-2022-envisaging-the-future-of-cities
United Nations Environment Programme, & International Resource Panel (UNEP & IRP). (2024). *Global resources outlook 2024—Bend the trend: Pathways to a liveable planet as resource use spikes.* https://wedocs.unep.org/handle/20.500.11822/44901
United Nations Environment Programme (UNEP). (2024). *Keeping the promise—Annual report 2023.* https://www.unep.org/annualreport
World Inequality Lab (WIL). (2023). *Climate inequality report 2023: Fair taxes for a sustainable future in the global South.* https://wid.world/news-article/climate-inequality-report-2023-fair-taxes-for-a-sustainable-future-in-the-global-south/
World Bank. (2023). *Annual report 2023: A new era in development.* http://documents.worldbank.org/curated/en/099092823161580577/BOSIB055c2cb6c006090a90150e512e6beb
World Resources Institute (WRI). (2023). *Setting enterprise targets: Modeling downstream water use and consumption.* https://www.wri.org/research/setting-enterprise-targets-modeling-downstream-water-use-consumption
World Health Organization (WHO). (2021). *Ambient air pollution: Health impacts.* https://www.wri.org/research/setting-enterprise-targets-modeling-downstream-water-use-consumption

Chapter 9
Dynamic Resource Allocation

> *Our knowledge of dynamic processes is necessarily inferior to our ability to describe stationary conditions.*
> *"On the Accuracy of Economic Observations", Oskar Morgenstern* (1974)

9.1 Resource Allocation

9.1.1 Resource Allocation Theory

In real life, people are always faced with limited resources. The limited nature of resources refers to the fact that a specific type of resource has a clearly limited value in the traditional sense, such as the proven coal reserves in a certain area. It can also be an invisible constraint faced by an individual, such as the sum of a day's work or leisure time that an individual can allocate. A fundamental problem facing economists is how to allocate limited resources to maximize social or individual welfare. It is the unremitting pursuit of optimizing resource allocation that has driven the dynamic development of modern economic theory. Among many economic theories, resource allocation theory directly answers this question. The study of resource allocation theory is based on the scarcity of resources. Resource scarcity means that available resources are always limited, while human needs are unlimited. Therefore, effective allocation must be carried out to meet different needs and goals. The core issue of resource allocation theory is how to achieve the optimal allocation of resources and maximize social welfare through market mechanisms, government intervention, and other means when resources are limited.

The history of resource allocation theory can be traced back to the end of the eighteenth century, when Adam Smith proposed the importance of market mechanisms in resource allocation in his book "The Wealth of Nations." Limited resources are one of the core issues of resource allocation theory because the scarcity of resources determines that they must be allocated effectively to meet different needs. Since then, economists such as David Ricardo, Alfred Marshall, and John Maynard Keynes have

L. Zhang, *Sustainability Economics*, https://doi.org/10.1007/978-3-031-87829-9_9

further developed this theory, emphasizing the different roles of the market and the government in resource allocation. In the mid-twentieth century, Herbert Simon proposed the theory of limited rationality and decision-making, which provided a new perspective for resource allocation theory.

From different research fields or perspectives, the definition of resource allocation is diverse. The optimal resource allocation theory is regarded as "given technology and consumer preferences, how to allocate limited economic resources to the production of various products according to certain rules in order to maximize individual needs" (Koopmans & Beckmann, 1957). Later, resource allocation refers to the allocation and adjustment of the resources needed by firms so that resources can match the specific needs of production practice, thereby forming a competitive advantage (Powers & McDougall, 2005). Then, resource allocation is defined as "the allocation of productive materials between different uses" (Britannica, 2010). Although there are different definitions of resource allocation, one consistent idea holds: the essence of resource allocation is to design optimal mechanisms to maintain the best proportional relationship and value orientation under limited resources, to improve resource utilization efficiency, and to meet individuals' growing needs.

As resource allocation theory developed, scholars turned to the study of its practical application. Early research on its practical application mainly focused on natural resources, such as water resources, land resources, mineral resources, and forest resources. In recent years, the practical application of resource allocation has become more extensive, from national defense and national education to specific departments of enterprises. From a national perspective, optimal resource allocation is an important way to achieve sustainable development. Reasonable resource allocation plays a vital and even decisive role in the regional economy, society, and ecological environment. At the firm level, resource allocation can reduce transaction costs, improve capital utilization efficiency, foster enterprise growth, and help firms form unique competitive capabilities. At the same time, it plays a key role in maintaining unique competitive advantages for enterprises and helps achieve good performance.

In the modern economy, social production is characterized by specialization and scale. Accordingly, resources should be allocated effectively to adapt to social mass production. From the perspective of the whole society, there are three types of resource allocation methods: First, resources are adjusted according to plan. This resource allocation method reflects the use and prediction of social planners on the planned proportion of social production and enterprise operation. It is a subjective adjustment process, known as the "visible hand." The second is market regulation. The essence is to give play to the power of the price of resources in the market. Through the guiding role of prices, resources are allocated to the most efficient production, which is called the "invisible hand." Third, the market and plan are combined to achieve the optimal allocation of resources. Nowadays, the above-mentioned social resource allocation methods are used widely. However, there is no consensus on which method is more effective.

Resource allocation theory also studies the mechanism of resource allocation. Based on different resource allocation methods, the mechanism for achieving optimal resource allocation is different under different situations (thus different underlying assumptions). Table 9.1 summarizes different resource allocation mechanisms that are commonly seen in the economic literature. Each mechanism describes the optimal allocation of resources under specific assumptions. The optimal allocation of resources depends not only on the way resources are allocated but also on the goals of social planners. By setting specific social planning utility functions and allocation mechanisms, resources are rationally allocated to individuals to achieve optimization.

Table 9.1 Different resource allocation mechanisms

Mechanisms	Key points	References
Walras equilibrium	• Market clearing means that supply and demand in all markets are completely equal under equilibrium prices • Price mechanism guides resources to where they are most needed through price signals, thus achieving optimal allocation of resources	Walras, 1874
Lindahl equilibrium	• Honestly share costs based on individuals' willingness to pay for public goods to ensure fairness • The optimal allocation of public goods is achieved when the supply quantity ensures that the aggregate willingness to pay across all individuals equals the marginal cost	Lindahl, 1919
Nash equilibrium	• Given the strategies of other participants, participants choose their own optimal strategies, that is, no one has the motivation to change strategies unilaterally • The strategy combination of participants constitutes a stable state, and the strategy change of a single participant will not increase its benefits	Nash, 1950
Fair allocation	• The resource allocation results are fair among all participants, and each participant believes that he or she has received a reasonable share • The resource allocation process is transparent and fair, ensuring that all participants have equal opportunities and rights in the allocation process	Rawls, 1971
Core allocation	• Stability ensures that resource allocation is stable among all participants and no subset of participants can gain greater benefits by reallocating resources • Fairness ensures that each participant gets a reasonable share and achieves optimal resource allocation	Anderson, 1978
Pareto optimality	• Pareto improvement means that it is impossible to make at least one person better off without making others worse off by reallocating resources • Pareto efficiency ensures that resources are used most effectively and maximizes the overall welfare of society	Pareto, 1964

9.1.2 Dynamic Resource Allocation

The development of dynamic resource allocation theory began in the mid-twentieth century with advancements in computer technology and operations research, leading to the formation of dynamic programming and optimal control theory. Richard Bellman's dynamic programming method and the game theory research of John von Neumann and Oscar Morgenstern provided important tools for this field. Dynamic resource allocation focuses on the flexible adjustment of resources over time and space to meet changing needs, ensuring optimal utilization under varying conditions (Harrigan, 1985). This theory is crucial for sustainable development as it balances short-term and long-term goals.

Resource allocation theory and dynamic resource allocation theory are key areas in economics and management, aiming to optimize resource use for maximum social welfare and economic benefits. Both theories rely on principles like supply and demand, marginal utility, and optimal control, using mathematical models and optimization algorithms. They are applied to various practical problems, including production scheduling, inventory management, traffic management, energy allocation, and environmental protection. Static resource allocation theory deals with fixed-point resource allocation, suitable for short-term decisions, while dynamic resource allocation theory considers changing resources and demand over time, making it ideal for long-term planning and complex problems (Hurwicz, 1973). Dynamic models are more complex but offer higher adaptability to changes in resources and demand.

Dynamic resource allocation theory has a wide range of applications in the field of energy and environmental economics because it can effectively cope with dynamic changes in resources and demand.

First, the supply and demand relationship of energy and environmental resources is often affected by factors such as seasonality, climate change, and policy adjustments (Ebinger, 2011). Dynamic resource allocation theory can accurately predict these changes and adjust resource allocation strategies through time series models and prediction algorithms.

Secondly, dynamic resource allocation theory is highly adaptable and flexible, and can respond quickly to resource shortages or sudden increases in demand, thereby optimizing resource utilization efficiency. In addition, dynamic resource allocation theory can fit the characteristics of energy and environmental economics, such as the intermittency and instability of renewable energy (Adefarati & Bansal, 2017), and ensure the stability of energy supply and the sustainability of environmental protection by dynamically adjusting resource allocation.

In general, static resource allocation theory and dynamic resource allocation theory have many similarities in goals and theoretical basis, but there are significant differences in time dimension, model complexity, and adaptability. Dynamic resource allocation theory has higher flexibility and adaptability in dealing with changes in resources and demand and is suitable for a wider range of practical application scenarios.

With the development of computer technology, big data, large models, and hardware facilities, the theory of dynamic resource allocation has surpassed a single model (Gupta et al., 2019). Today's resource allocation relies more on a system that integrates various optimization models and has become increasingly complex in terms of data interaction, interface optimization, and human–computer interaction. This progress has made dynamic resource allocation an important tool for solving complex resource allocation problems. By integrating multiple optimization models, the system can more flexibly respond to different types of uncertainties and changes and improve resource utilization efficiency and decision-making accuracy. At the same time, the progress of data interaction and interface optimization enables decision-makers to understand and manage the resource allocation process more intuitively, and the enhancement of human–computer interaction further improves the adaptability and responsiveness of the system.

In addition, the development of modern hardware facilities, such as high-performance computers and cloud computing platforms, makes it more efficient to process massive data and complex calculations. The application of large models, such as deep learning and artificial intelligence technology, can extract valuable information from massive data and provide more accurate prediction and optimization solutions. The combination of these technologies enables the dynamic resource allocation system to not only respond to market changes in real time but also foresee future trends and needs, adjust in advance, and ensure the optimal allocation of resources. In short, with the continuous advancement of technology, dynamic resource allocation has become a complex system that integrates multiple optimization models and advanced technologies. It not only improves the efficiency and accuracy of resource allocation but also enhances the flexibility and adaptability of the system, enabling decision-makers to make more informed decisions under uncertain conditions and achieve efficient use of resources and sustainable development.

Example

China's Time-of-use Electricity Price Policy in 2021

With the rapid development of China's economy and the acceleration of urbanization, electricity demand has continued to increase. Thus, electricity supply is facing tremendous pressure. In the summer of 2021, many parts of China experienced extremely high temperatures, which led to a rapid increase in electricity demand and tight electricity supply. High temperatures not only increase residential and commercial electricity demand, but also put higher requirements on the stable operation of the power system. In response to this challenge, the Chinese government has implemented a time-of-use electricity price policy. The policy aims to optimize the allocation of power resources and ensure the stability and reliability of power supply by increasing electricity prices during peak electricity demand periods. It has successfully alleviated the electricity shortage caused by unstable supply and increased demand due to high temperatures by encouraging users to change their electricity usage behavior and reduce electricity load during peak hours. Time-of-use electricity pricing is a typical example of dynamic resource allocation. Through the time-of-use electricity price policy, the allocation strategy of electricity resources can be adjusted in real time, which significantly reduces electricity costs, and thus ensures a stable power supply.

9.2 A Simple Economic Model

9.2.1 A Two-Period Model

The fundamental principle of dynamic resource allocation is to maximize the present value of net benefits by reasonably distributing limited resources between the present and the future. This involves determining an optimal temporal distribution of the resource.

To illustrate the basic idea of dynamic resource allocation, we provide a simple numerical example. The basic model for determining dynamic resource allocation across two time periods relies on a set of straightforward assumptions frequently encountered in such analyses. First, assume a fixed supply of resources to be distributed over two periods. Also, assume that the demand function is equivalent for the two periods, with the marginal willingness to pay given by the formula $P = 8 - 0.4Q$, and the marginal cost of supplying that resource set to be constant at 2 per unit.

Figure 9.1 depicts this two-period model. At present, the assumption is not sufficient to determine the best distribution for Period 1 and Period 2 as we are not aware of the exact number of depletable resources available for distribution. Obviously, the abundance of resources will affect the result of the optimal efficiency allocation. For instance, in extreme cases where the allocation amount is sufficiently large, an increase in consumption during Period 1 would have no impact on the consumption level in Period 2. In this case, it is only necessary to choose the amount of consumption in the first and second periods, respectively, a result that consumers in both periods would prefer. If the total number of resources is limited, the amount used in the first period will affect the amount used in the second period. In this case, resources have to be allocated according to certain criteria to achieve optimal efficiency.

If the total supply is larger than 30 units, an optimal allocation should distribute the resources equally to each period, irrespective of the discount rate. When the total supply equals 30, demand in both periods can be fully met, ensuring that the quantity of resources used in Period 1 does not affect Period 2. This scenario is optimal because the net present value cannot be increased any further. However, if the supply falls below 30 units, for instance, to 20 units, the optimal allocation should be different. To satisfy the dynamic efficiency criterion, the allocation should ensure that the total

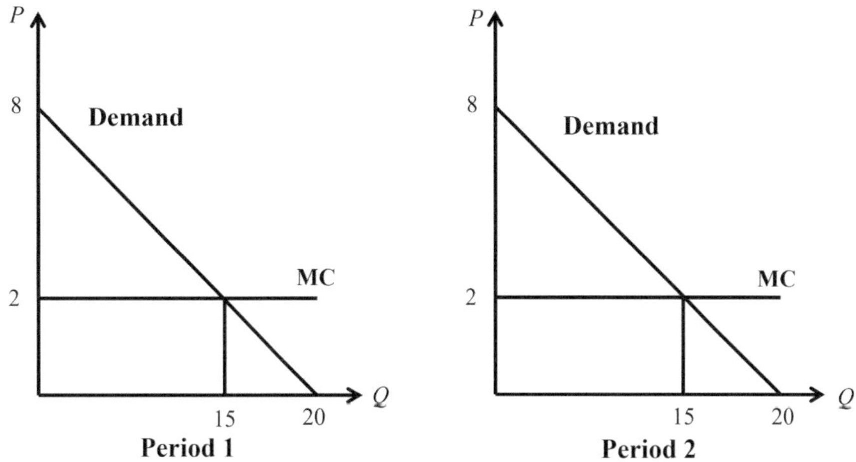

Fig. 9.1 Two-period dynamic resource allocation

present value of net benefits is maximized, achieved by summing the discounted net benefits for each individual period. For instance, let's examine the present value of a specific allocation: 15 units allocated to the first period and 5 units to the second. The present value for the first period corresponds to the geometric area between the demand and supply curves, which amounts to 45. For the second period, the present value is calculated as the area between the demand and supply curves from zero to five units, adjusted by the discount factor $1/(1 + r)$. Using a discount rate of 10%, the net present value in the second period is 22.73, resulting in an aggregate net present value of 67.73 when combined with the first period.

To find the optimal state, one approach is to examine all possible ways to divide the total available amount (20 units) between Q_1 and Q_2 for the two periods. The division that results in the maximization of the combined net present value is the optimal solution. To ensure dynamic efficiency, the allocation must meet the condition where the present value of the marginal net benefit for the last unit used in Period 1 equals the present value of the marginal net benefit for Period 2. The two-period allocation problem can be visualized graphically, as presented in Fig. 9.2. In the diagram, the net benefit curve for the first period is read from left to right, while the corresponding curve for the second period is read in the opposite direction. The efficient allocation is determined by the intersection of these curves, where the above conditions are met. To calculate the total net present value, the two areas below the marginal net benefit curve for the two periods are combined (up to the optimal allocation). By doing so, the best allocation and net present value are obtained.

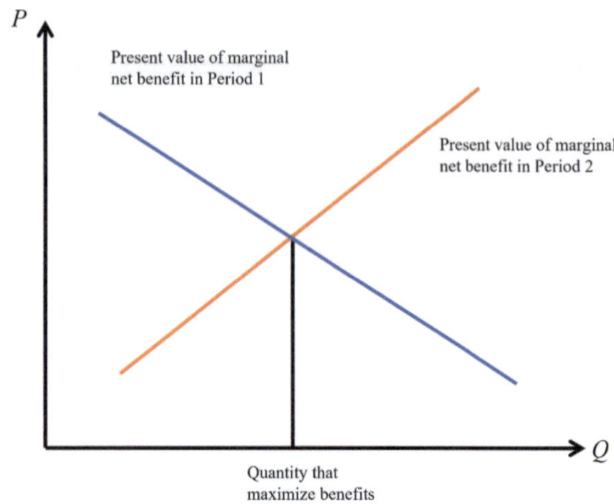

Fig. 9.2 Dynamic efficient allocation

9.2.2 Mathematics of Dynamic Resource Allocation

The two-period model mentioned above reveals the basic ideas of dynamic resource allocation theory. However, resource allocation in the real world far exceeds two periods. The two-period model is a simple explanation of resource allocation. Generally, a resource allocation model spanning multiple periods does not add much complexity. To better understand the allocation and efficiency issues related to a depletable resource in multiple periods, the following introduces simple mathematics of a multiperiod resource allocation model.

To understand how to achieve optimal resource allocation across different periods, assume that each period t has a similar demand curve as follows:

$$P_t = a - bq_t, \tag{9.1}$$

where P refers to the price of each unit of resource and q denotes to the number of resources consumed in period t.

Then, the benefits of consuming this resource in each period t are easy to obtain. Simply, the benefit for each period t is given as the integral of the demand curve:

$$\text{Benefit}_t = \int_0^{q_t} (a - bq)dq = aq_t - \frac{b}{2q_t^2}. \tag{9.2}$$

Besides, the marginal cost of one-unit resource in each period is assumed to be constant at c. The total cost of each period t is given as follows:

$$\text{Total cost}_t = cq_t. \tag{9.3}$$

Finally, the total quantity of resources is set to be constant at a certain number $\hat{Q}$. And the time discount rate is set to r. Then the problem turns into a maximization problem under certain resource constraints. To find out the optimal resource distribution across different periods t, consider the following problem:

$$\max_q \sum_1^n \frac{aq_i - \frac{bq_i^2}{2} - cq_i}{(1+r)^{i-1}} + \lambda\left(\hat{Q} - \sum_1^n q_i\right). \tag{9.4}$$

When the quantity of the resource $\hat{Q}$ is less than the number that is normally demanded, the quantity of each period t must satisfy the following conditions to maximize benefit:

$$\frac{a - bq_i - c}{(1+r)^{i-1}} - \lambda = 0, i = 1, 2, 3 \ldots n; \tag{9.5}$$

$$\hat{Q} - \sum_1^n q_i = 0. \tag{9.6}$$

By solving this problem, one could figure out how to achieve optimal distribution of resources across different periods. The basic idea of the condition is equivalent to the core idea of the two-period model: The net marginal present value of each period should be equal, and the sum of the resources distributed is equivalent to the quantity of the fixed resource $\hat{Q}$. A key implication is that the difference between P and MC (P-MC) grows at a rate of r over time. This difference is critical in determining the optimal consumption across time. When n is set to 2, the problem reduces to the two-period model discussed earlier.

9.2.3 Efficiency and Fairness in Resource Allocation

Efficiency and fairness are two key concepts in resource allocation theory. They are interrelated and need to be balanced. Since resources are limited, the core of resource allocation theory is how to maximize social welfare under constraints. To achieve this goal, efficiency should be pursued in resource allocation. Efficiency refers to the optimal use of resources. Given resource constraints, the allocation mechanism should maximize output and social welfare. Through a reasonable allocation plan, resources flow to where they are most needed, thereby achieving the optimal allocation of resources. The core of efficiency is to reduce waste and improve the effectiveness of production and services.

Efficiency involves two perspectives: production efficiency and allocation efficiency (see Fig. 9.3). Production efficiency refers to how to obtain the most output with the least input in the production process. Allocation efficiency focuses on the

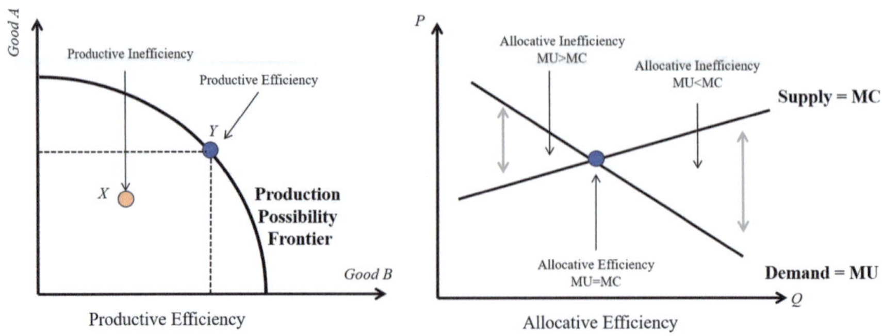

Fig. 9.3 Productive efficiency and allocative efficiency

optimal allocation of resources between different uses so that each resource unit can produce the greatest utility. However, the market mechanism does not always achieve efficiency automatically, especially in the case of market failure. Government intervention helps to correct market failure and ensure the effective use of resources.

Fairness focuses on the fairness and rationality of resource allocation. Fair resource allocation means that each individual or group can obtain a share of resources that matches their contribution and needs. Fairness not only involves the distribution of income and wealth, but also includes equal opportunities and the protection of social welfare. Fairness also emphasizes intergenerational fairness. The goal of efficient resource allocation must prioritize a balance between meeting current demands and safeguarding the ability of future generations to meet their needs. In addition, fairness is also reflected in the utility function of resource allocation, which will have a profound impact on the results of resource allocation.

In resource allocation theory, efficiency and fairness often need to be weighed. The pursuit of efficiency may lead to unequal resource allocation, while the pursuit of fairness may reduce resource utilization efficiency. Therefore, economists and policymakers need to find a balance between efficiency and fairness to achieve sustainable social development and maximize overall welfare.

Reading

Fairness in the Theory of Justice

John Rawls proposed a concept of fairness in his Theory of Justice. Fairness is mainly reflected in the following two aspects:

Minimum maximization principle: The minimum maximization principle proposed by Rawls emphasizes that the welfare of the most disadvantaged groups should be given priority. The role of the social welfare utility function here is to ensure that the result of resource allocation can maximize the utility of the most disadvantaged groups, thereby achieving overall social fairness. This means that, in the process of resource allocation, policymakers should pay attention to those at the bottom of society or the most vulnerable groups and ensure that their basic needs are met and their quality of life is improved. In this way, the allocation of social resources is not only to improve overall welfare, but also to narrow social inequality and promote social harmony and stability.

Difference principle: The difference principle states that social and economic inequality should maximize the benefits of the most disadvantaged groups. The role of the utility function in the difference principle is to ensure that resource allocation can maximize efficiency and utility on the basis of fairness by measuring the marginal utility of different groups. This means that in the process of resource allocation, policymakers need to consider the marginal utility of each group, that is, the utility gain of each additional unit of resources to different groups. In this way, resources can be allocated to those who need them most, thus achieving the optimal allocation of resources and maximizing social welfare.

Through these two principles, Rawls's Theory of Justice closely combines fairness with the social welfare function. This ensures that resource allocation can promote overall welfare while taking care of the interests of the most vulnerable groups. This view of justice not only emphasizes the efficiency of resource allocation, but also emphasizes the fairness of resource allocation. It ensures that every member of society can get a reasonable share in resource allocation, thereby achieving overall social justice.

9.3 Dynamic Resource Allocation: Future Perspectives

9.3.1 Data and Technological Challenges

Currently, the realization of dynamic resource allocation is no longer based on simple mathematical models but integrates large-scale and complex models into the system. Various general or specific systems are used to perform dynamic resource allocation to achieve specific goals. In recent years, dynamic resource allocation technology has made significant progress, especially in the field of artificial intelligence (AI). Modern dynamic resource allocation systems, such as deep reinforcement learning (DRL) and game theory-based systems, can achieve optimal resource allocation in complex and dynamic environments. DRL models continuously optimize resource allocation strategies by simulating and learning from feedback in the environment, making them highly adaptive and flexible. Game theory-based models, on the other hand, find the equilibrium point of resource allocation by analyzing the interactions between different participants to ensure the maximization of the interests of all parties. These systems are characterized by their ability to process large amounts of real-time data, respond quickly to environmental changes, and make optimal decisions under uncertain conditions.

However, modern dynamic resource allocation systems require a large amount of high-quality data, especially real-time data. Basic data is needed to build the models, while responding to changes in demand and environment requires analysis of real-time data. The acquisition and processing of real-time data are core to dynamic resource allocation systems because they reflect current resource demand and environmental changes. Through the analysis of real-time data, the system can quickly adjust the resource allocation strategy to ensure efficient use of resources and stable operation. Currently, dynamic resource allocation systems face major challenges in managing and processing real-time data, which requires powerful data storage and processing capabilities, as well as efficient data integration and analysis methods.

In modern dynamic resource allocation systems, data management is always a challenge (Grabowski et al., 2016). These systems need to process large amounts of basic and real-time data from various sources, such as sensors, user input, and external databases. Basic data provides information about system operation, such as the initial state and historical usage of resources, while real-time data reflects current resource requirements and environmental changes. Managing data requires not only powerful storage and processing capabilities but also ensuring the accuracy and reliability of data. Any errors in basic data or delays in real-time data may lead to suboptimal decisions in resource allocation, affecting the overall efficiency of the system. Additionally, the diversity and complexity of data increase the difficulty of management, requiring effective data integration and analysis methods to support optimization decisions in dynamic resource allocation.

Integrating advanced technologies, such as deep learning models, large-scale models, and remote sensing, into dynamic resource allocation systems is also a major challenge. Deep learning models require large amounts of external data for training, and the data acquisition and processing process can be very complex. The complexity and computing requirements of machine learning algorithms also place higher demands on the system's hardware and software. Remote sensing technology can obtain real-time data at low cost, with advantages of large scale and high precision. However, connecting remote sensing data to the system and achieving rapid response is challenging. Ensuring that these technologies can be seamlessly integrated and run efficiently is one of the main technical challenges facing dynamic resource allocation systems. The application of these technologies requires powerful computing power and efficient data processing algorithms to ensure that the system can respond and adjust in real time.

In dynamic resource allocation systems, data security and privacy protection are also important challenges. Since the system needs to process large amounts of sensitive data, such as user behavior data and enterprise operation data, it is crucial to ensure the security of this data during transmission and storage. Additionally, compliance with relevant privacy protection regulations and standards, such as General Data Protection Regulation (GDPR), is a challenge that dynamic resource allocation systems must face. Any data leakage or privacy violation incidents may seriously impact the credibility of the system and user trust and, in severe cases, may even lead to legal consequences. Therefore, dynamic resource allocation systems must adopt advanced encryption technology and security protocols to ensure the security of data during transmission and storage.

Reading

Energy Management Systems Are Facing Challenges

Energy management system (EMS) is a platform for monitoring, controlling, and optimizing power systems. It ensures efficient operation by collecting and analyzing real-time data, such as power generation, transmission, distribution, and consumption. EMS enables dynamic resource allocation, improves energy efficiency, reduces waste, and supports renewable energy integration. A notable example is China Southern Power Grid's smart grid project, launched in 2019, which uses advanced technologies for optimal power resource management. The project includes infrastructure like smart meters, sensors, and communication networks for comprehensive monitoring.

However, EMS implementation faces challenges, including data diversity and complexity, real-time processing demands, and data accuracy and reliability. EMS must integrate data from various sources, such as power plants, substations, and user-side systems, which vary in format and complexity. Real-time monitoring requires efficient data processing to ensure timely responses. Additionally, data accuracy is critical, as errors or delays can lead to suboptimal decisions, impacting grid stability and efficiency.

9.3.2 Policy Challenges

Dynamic resource allocation faces a series of institutional and policy challenges in its implementation. First, policy uncertainty and changes may affect the stability and predictability of resource allocation (Bai et al., 2023). For example, the government may adjust its energy subsidy policy or introduce new environmental regulations, and these changes may significantly impact resource allocation strategies. To meet these challenges, the dynamic resource allocation system needs to be flexible and adaptable, and able to adjust quickly to respond to policy changes. Specifically, the system needs to establish a real-time monitoring and feedback mechanism to obtain policy change information in a timely manner and adjust the resource allocation strategy through algorithms to ensure efficient operation despite policy changes.

Institutional coordination and integration are also important challenges (Friebel & Raith, 2010). Dynamic resource allocation involves multiple stakeholders, including government departments, enterprises, and the public. Coordination and cooperation among these stakeholders are essential for effective resource allocation. However, there may be conflicts of interest and communication barriers between different departments and institutions, resulting in inefficient resource allocation. To overcome these barriers, it is necessary to establish an effective communication mechanism and cooperation framework to ensure that the interests of all parties are balanced and coordinated. For example, a cross-departmental coordination committee can be established to hold regular meetings to discuss and solve problems in resource allocation, ensuring that the needs and interests of all parties are fully considered and met.

Regulatory and compliance issues are also major challenges facing dynamic resource allocation. The resource allocation system needs to comply with relevant laws, regulations, and standards, such as environmental protection regulations and data privacy protection regulations. Any violation of regulations may lead to legal risks and reputational losses. Therefore, the dynamic resource allocation system needs to establish a strict compliance management mechanism to ensure that all operations and decisions comply with relevant regulations and standards. Specific measures include establishing a compliance review process, conducting regular internal audits and assessments to ensure that the system's operation complies with legal requirements, and timely adjusting and improving operations and processes that do not comply with regulations. Through these measures, the dynamic resource allocation system can effectively respond to institutional and policy challenges, achieve optimal resource allocation, and improve the overall efficiency and reliability of the system.

Reading

Policy Uncertainty and Stranded Asset

Policy uncertainty has a significant impact on the stranding of coal-fired power plant assets. In recent years, with the continuous changes in global environmental and energy policies, especially the establishment of carbon emission limits and carbon trading markets, the operation and investment of traditional coal-fired power plants have faced great challenges. In China, the launch of the national carbon emission trading market has significantly increased the operating costs of coal-fired power plants. The price fluctuations of carbon emission rights and the uncertainty of quota allocation methods have put many coal-fired power plants under financial pressure.

For example, Mo et al. (2021) have shown that although China's coal-fired power units have a long technical lifespan, their operations are already close to financial breakeven. The future trend of carbon prices, the method of carbon quota allocation, and the price transmission relationship between the carbon market and the electricity market have the most significant impact on the operation of coal-fired power units. Under the scenario of carbon quota auctions, even at the current carbon price scenario of 50 yuan/ton CO_2 (4% annual growth rate), the average remaining lifespan of all coal-fired units will be reduced by 5.43 years, and the cumulative carbon dioxide emissions from 2020 to 2050 will be reduced by 22.73 billion tons. Under the carbon price scenario of 100 yuan/ton CO_2 (4% annual growth rate), the average lifespan of coal-fired units will be shortened by 11.73 years, and existing coal-fired units will complete their exit six years earlier.

9.3.3 Dynamic Renewable Resource Allocation

Renewable energy is characterized by cleanliness, reproducibility, and environmental friendliness. It mainly includes solar energy, wind energy, hydropower, biomass energy, and geothermal energy. These energy resources are abundant and can reduce dependence on fossil fuels and greenhouse gas emissions. However, the intermittent and unstable nature of new energy, as well as its distributed characteristics, requires the dynamic resource allocation system to have real-time, predictive, and flexible adjustment capabilities to ensure the stability and reliability of the energy supply.

Intermittent and unstable energy: Renewable energy such as solar and wind energy is characterized by intermittent and unstable nature, meaning their power generation is affected by weather and time (Mlilo et al., 2021). Solar power generation cannot generate electricity at night and on cloudy days, while wind power generation depends on changes in wind speed and direction. This instability requires the dynamic resource allocation system to have real-time monitoring and prediction capabilities, to predict energy production based on weather forecasts and historical data, and to dynamically adjust energy allocation strategies. When solar power generation is insufficient, the system can increase the supply of other energy sources to meet electricity demand. Additionally, the application of energy storage technologies such as battery energy storage systems can store electricity when there is excess power generation from new energy sources and release electricity when there is insufficient power generation, balancing supply and demand.

Distributed energy: Renewable energy usually exists in a distributed form, such as rooftop solar panels and small wind turbines. This distributed nature requires a dynamic resource allocation system that can flexibly manage and coordinate multiple distributed energy resources, optimize energy transmission and distribution, reduce energy losses, and improve overall energy efficiency (Khan et al., 2019). The system needs to have efficient communication and control capabilities, be able to monitor the operating status of each distributed energy source in real time, and dynamically adjust according to demand and supply. When solar power generation in a certain area increases, the system can reduce the supply of other energy sources in the area and optimize overall energy utilization. Additionally, the application of microgrid technology can realize the independent operation and interconnection of distributed energy sources, improving the flexibility and reliability of the energy system.

Environmentally friendly energy: The use of new energy sources has less impact on the environment and helps reduce greenhouse gas emissions and environmental pollution. The dynamic resource allocation system needs to consider environmental factors while optimizing energy distribution, maximizing the use of clean energy, reducing dependence on fossil fuels, and promoting sustainable development. The system can prioritize the allocation of clean energy such as wind and solar energy and reduce the use of traditional energy such as coal and oil, thereby reducing carbon emissions and environmental pollution. Additionally, the system needs to have environmental monitoring capabilities, be able to monitor the impact of energy use on the environment in real time, and adjust the energy allocation strategy based on the monitoring results to ensure a balance between environmental protection and energy utilization. During periods of high wind and solar power generation, the system can reduce the operating time of thermal power generation and reduce pollutant emissions.

References

Adefarati, T., & Bansal, R. (2017). Reliability and economic assessment of a microgrid power system with the integration of renewable energy resources. *Applied Energy, 206*, 911933.

Bai, Y., Tang, S., Lu, C., Chen, B., & Yan, M. (2023). Economic policy uncertainty and natural resource policy in the United States. *Resources Policy, 83*, 103598.

Britannica, E. (2010). *Britannica book of the year 2010*. Encyclopaedia Britannica, Inc.

Ebinger, J., & Vergara, W. (2011). *Climate impacts on energy systems: Key issues for energy sector adaptation*. World Bank Group.

Friebel, G., & Raith, M. (2010). Resource allocation and organizational form. *American Economic Journal: Microeconomics, 2*(2), 1–33.

Grabowski, M., Rizzo, C., & Graig, T. (2016). Data challenges in dynamic, large-scale resource allocation in remote regions. *Safety Science, 87*, 7686.

Gupta, S., Qian, X., Bhushan, B., & Luo, Z. (2019). Role of cloud ERP and big data on firm performance: A dynamic capability view theory perspective. *Management Decision, 57*(8), 1857–1882.

Harrigan, K. R. (1985). *Strategic flexibility: A management guide for changing times*. Simon and Schuster.

Hurwicz, L. (1973). The design of mechanisms for resource allocation. *The American Economic Review, 63*(2), 1–30.

Khan, M. W., Wang, J., Ma, M., Xiong, L., Li, P., & Wu, F. (2019). Optimal energy management and control aspects of distributed microgrid using multi-agent systems. *Sustainable Cities and Society, 44*, 855–870.

Koopmans, T. C., & Beckmann, M. (1957). Assignment problems and the location of economic activities. *Econometrica: Journal of the Econometric Society, 25*(1), 53–76.

Mlilo, N., Brown, J., & Ahfock, T. (2021). Impact of intermittent renewable energy generation penetration on the power system networks—A review. *Technology and Economics of Smart Grids and Sustainable Energy, 6*, 25.

Mo, J., Zhang, W., Tu, Q., Yuan, J., Duan, H., Fan, Y., Pan, J., Zhang, J., & Meng, Z. (2021). The role of national carbon pricing in phasing out China's coal power. *iScience, 24*(6), 102655.

Morgenstern, O. (1974). On the accuracy of economic observations: Foreign trade statistics. In *Illegal transactions in international trade*. North-Holland

Powers, J. B., & McDougall, P. P. (2005). University start-up formation and technology licensing with firms that go public: A resource-based view of academic entrepreneurship. *Journal of Business Venturing, 20*(3), 291–311.

Chapter 10
Air Pollution

This glorious and ancient city should wrap her stately head in clouds of smoke and Sulphur, so full of stink and darkness "The Inconvenience of the Aer and Smoak of London", John Evelyn (1772)

10.1 Air Pollution and Sustainable Development

10.1.1 History of Air Pollution

Almost everyone is exposed to air pollution daily, whether they realize it or not. The term "air pollution" is frequently mentioned in academic discussions, news coverage, and government reports. However, it took a long time to develop a clear understanding of air pollution and its broad social and economic consequences.

Historically, the recognition of air pollution as a health threat dates back to Hippocrates around 400 BC. Roman writers also noted its negative impacts. Frontinus (ca AD 96) highlighted the benefits of improving air quality through better aqueducts. Over the centuries, various cultures documented air pollution, such as Shen Kuo in ancient China.

The industrial revolution significantly increased the emission of harmful substances into the air, leading to major environmental events. The Great London Smog of 1952, which caused approximately 12,000 premature deaths, marked a turning point in public awareness. This event led to the Clean Air Act of 1956 in the UK, aimed at controlling emissions and improving air quality. Similarly, the US introduced the Clean Air Act in 1967 to regulate air emissions and protect public health.

L. Zhang, *Sustainability Economics*, https://doi.org/10.1007/978-3-031-87829-9_10

Early evidence of air pollution's impact was primarily local. However, scientists in the nineteenth century discovered the exchange of air pollution between countries. In the late 1960s, one idea that links sulfur emissions from major European emitters to acid rain in Scandinavia attracted the public's attention. This led to the 1972 Stockholm Conference, which emphasized the need for international cooperation to address transboundary pollution.

The Convention on *Long-range Transboundary Air Pollution* (LRTAP) was established to reduce sulfur dioxide and other pollutants. The first Sulphur Protocol, signed in 1985, aimed for a 30% reduction in sulfur emissions. This international framework has proven effective in combining science, monitoring, and policy development to address air pollution. In North America, the US established the *National Atmospheric Deposition Program* (NADP) in 1977 to monitor and address the long-range transport of pollutants. Similar to Europe, the US recognized the need for international cooperation to tackle air pollution and its ecological impacts.

Since 2000, advancements in atmospheric science, global meteorological monitoring, and space remote sensing have made observing cross-regional pollutant transmission more accessible. Atmospheric models simulate pollutant transport, providing theoretical explanations, while ground and air sensors continuously transmit real-time data, enhancing our understanding.

In the 1990s, the deterioration of forest health in polluted European regions drew public and political attention. Pollutants like sulfur dioxides (SO_2), nitrogen oxides (NO_x), and particulate matters (PM) significantly impacted forests, crucial for carbon sequestration and ecological balance. Acid rain from SO_2 and NO_x emissions weakened trees, while ground-level ozone and PM damaged leaf tissues. International actions, such as the 1992 Earth Summit and the Kyoto Protocol, addressed forest degradation and air pollution. At COP26 in 2021, over 100 world leaders committed to ending deforestation by 2030, emphasizing forests' role in mitigating climate change.

Since the early twenty-first century, air pollution has remained a significant global challenge. Rapid industrialization, urbanization, and vehicular emissions have deteriorated air quality. Key pollutants like $PM_{2.5}$, PM10, NO_x, SO_2, and VOCs are linked to severe respiratory and cardiovascular diseases. The WHO estimates air pollution causes approximately 8 million premature deaths annually. Air pollution also exacerbates climate change by increasing greenhouse gases like CO_2 and CH_4.

Technological advancements have been crucial in mitigating air pollution. Innovations like electrostatic precipitators, fabric filters, and wet scrubbers reduce industrial emissions. Catalytic converters in the transportation sector lower NO_x and CO emissions. Renewable energy technologies, such as solar, wind, and hydroelectric power, offer cleaner alternatives to fossil fuels. Emerging technologies like carbon capture and storage (CCS) and hydrogen fuel cells show promise, exemplified by the successful CCS implementation at the Boundary Dam Power Station in Canada.

Despite technological advancements, air pollution remains a severe issue. The widespread adoption of these technologies faces financial and policy challenges. Governments have implemented regulations and international cooperation, such as the Paris Agreement, to control pollution. However, economic and political factors often compromise these strategies. Public awareness and participation have grown, but sustained behavioral changes require ongoing effort. The rapid pace of urbanization and industrialization adds complexity to pollution sources, making air pollution a persistent challenge. Future efforts must involve integrated measures, continuous technological innovation, reinforced policy frameworks, and active public engagement to effectively address this global issue.

10.1.2 Definition of Air Pollution

Today, both academia and the public have developed a comprehensive understanding of the causes, dissemination, and impacts of air pollution. According to WHO (2024), air pollution is defined as the presence of harmful substances in the atmosphere, which can be gases, particulates, or biological molecules. These pollutants originate from various sources, including industrial emissions, vehicle exhaust, chemical solvents, and natural events such as wildfires and volcanic eruptions. Nowadays, it is widely recognized that air pollution has severe adverse effects on human health, leading to respiratory and cardiovascular diseases. Additionally, air pollution negatively impacts the environment by contributing to climate change, acid rain, and the depletion of the ozone layer.

The atmosphere contains numerous invisible and subtle pollutants. Air pollutants are substances present in the atmosphere that can have various adverse impacts on human health and ecosystems. These substances can exist as solid particles, liquid droplets, or gases, often in the form of aerosols (solid particles or liquid droplets dispersed and carried by gases). Pollutants can originate from natural sources or human activities and are generally classified into primary and secondary pollutants. Primary pollutants are directly emitted from sources such as volcanic ash and fossil fuels. Secondary pollutants, on the other hand, form in the atmosphere through chemical reactions involving primary pollutants. Below are some details of primary and secondary pollutants, as summarized in Fig. 10.1.

Primary pollutants: Primary pollutants are substances directly emitted from a specific source into the environment. These pollutants are harmful at the point of emission. Common primary pollutants include:

- Sulfur dioxide (SO_2): Mainly from burning fossil fuels like coal and oil.
- Carbon monoxide (CO): Primarily from vehicle exhaust and industrial emissions.
- Nitrogen oxide (NO_x): Mainly from transportation and industrial activities.
- Particulate matter (PM): Includes PM_{10} and $PM_{2.5}$, originating from industrial emissions, vehicle exhaust, and construction site dust.
- Volatile organic compounds (VOCs): Emitted from solvent use, paints, and chemical production.

Secondary pollutants: Secondary pollutants are not emitted directly. Instead, they form in the atmosphere through chemical reactions involving primary pollutants. Common secondary pollutants include:

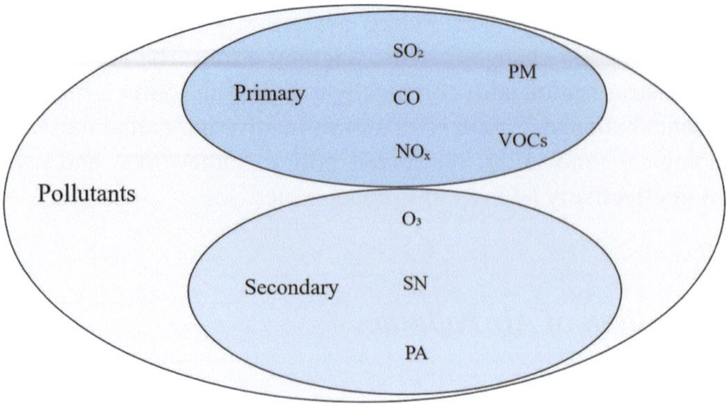

Fig. 10.1 Pollutants in the air

- Ozone (O_3): Formed by photochemical reactions between nitrogen oxides and volatile organic compounds in the presence of sunlight.
- Sulfates and nitrate (SN): Formed from sulfur dioxide and nitrogen oxide in the atmosphere.
- Photochemical smog (PA): A complex mixture of pollutants formed through photochemical reactions, including ozone, peroxyacetyl nitrate (PAN), and other oxidants.

10.1.3 Trade-Off Between Development and Clean Air

The intricate balance between economic development and environmental sustainability, particularly air quality, remains a pivotal concern in contemporary economic discussions. Economic development often necessitates industrialization, urbanization, and increased energy consumption, which invariably leads to elevated levels of air pollution. This air pollution, primarily from fossil fuel combustion, industrial emissions, and vehicular exhaust, poses significant health risks and environmental degradation. Emissions from factories, power plants, and transportation systems release pollutants such as particulate matters ($PM_{2.5}$ and PM_{10}), nitrogen oxides (NO_x), sulfur dioxides (SO_2), and volatile organic compounds (VOCs), which deteriorate air quality.

While economic development drives improvements in living standards, infrastructure, and technological advancements, it simultaneously fosters job creation, enhances income levels, and contributes to overall societal well-being. Nevertheless, these benefits often come at the expense of environmental health. To mitigate this trade-off, policymakers and researchers advocate for sustainable development practices. This includes the adoption of cleaner technologies, stringent environmental regulations, and the promotion of renewable energy sources. For instance, transitioning to electric vehicles, implementing carbon capture and storage (CCS) technologies, and enhancing energy efficiency in industrial processes can significantly reduce air pollution without stifling economic growth.

Moreover, it is crucial to recognize the economic costs associated with poor air quality. Health-related expenses, loss of productivity, and environmental cleanup efforts can impose substantial financial burdens on economies. Therefore, investing in clean air initiatives not only preserves environmental integrity but also ensures long-term economic resilience. The economic costs of pollution are often underestimated, yet they can significantly impact public health systems and labor markets, leading to a decrease in overall economic productivity. While there is an inherent trade-off between development and clean air, it is not an insurmountable challenge. Through innovative policies, technological advancements, and a commitment to sustainable practices, it is possible to achieve a harmonious balance that promotes both economic prosperity and environmental health. The path forward necessitates a collaborative effort from governments, industries, and communities to prioritize both economic growth and environmental sustainability.

10.2 Sources of Air Pollution

10.2.1 Local Pollutants

Local pollutants refer to contaminants generated within a specific area, significantly impacting the local environment and public health. In urban areas, the primary sources of these pollutants are industrial emissions, transportation, and construction activities. Factories and power plants release particulate matters ($PM_{2.5}$ and PM_{10}), nitrogen oxides (NO_x), sulfur dioxides (SO_2), and volatile organic compounds (VOCs). Additionally, vehicle exhaust contributes significantly to carbon monoxide (CO) and other harmful pollutants. The high population density and frequent industrial activities in cities exacerbate air pollution, leading to severe health issues such as respiratory and cardiovascular diseases.

Conversely, rural areas contend with local pollutants mainly from agricultural activities, household combustion, and natural sources. Agricultural practices, such as fertilization and livestock farming, release ammonia (NH_3) and methane (CH_4). The use of pesticides also contributes to local pollution. Furthermore, the burning of biomass fuels (like wood and crop residues) for cooking and heating in rural households releases significant amounts of smoke and harmful gases, affecting air quality and residents' health. Although rural air pollution is generally less severe than in urban areas, the lack of effective pollution control measures means the impact on the local environment and public health remains substantial.

Pollutants do not remain confined to their source areas and can be transported over long distances by wind and weather patterns. For instance, urban pollutants can travel to rural areas, exacerbating air quality issues there. The geographical location of pollution sources plays a crucial role in this process. Industrial zones located upwind of residential areas can significantly impact air quality downwind. Similarly, rural pollutants can be carried into urban areas, particularly during certain weather conditions, such as temperature inversions, which trap pollutants close to the ground. This cross-regional transmission of pollutants complicates efforts to manage air quality and necessitates coordinated policies and measures across different regions.

The sources of local pollutants differ markedly between urban and rural areas. Urban pollutants predominantly stem from industrial and transportation sources, whereas rural pollutants are primarily derived from agricultural and household combustion activities. Developing targeted pollution control measures and policies for different areas, considering the dynamics of pollutant transport and geographical factors, is crucial for improving air quality and protecting public health.

Reading

Controlling Straw Burning in China

The introduction of straw-burning control measures in China stems from the severe environmental challenges posed by the burning of agricultural waste. As one of the largest agricultural producers in the world, China generates a significant amount of crop straw annually, a considerable portion of which is directly burned, leading to the emission of particulate matter and harmful gases. This has become a major contributor to regional smog, posing serious threats to public health and ecosystems. To address this issue, the Chinese government incorporated straw-burning management into its environmental policy framework in the late 1990s and designated it as a key task in the *Air Pollution Prevention and Control Action Plan*, aiming to achieve dual goals of environmental benefits and sustainable agricultural development.

Key measures include banning straw burning through laws like the Air Pollution Prevention and Control Law, promoting straw recycling for energy and soil improvement, supporting farmers with subsidies and training, and using satellite monitoring for enforcement. These efforts have reduced burning, improved air quality, and advanced sustainable agricultural practices.

10.2.2 Global Pollutants

Ozone plays distinctly different roles in various layers of the atmosphere: the depletion of the stratospheric ozone layer increases harmful ultraviolet radiation, while ground-level ozone pollution directly threatens human health and ecosystems. Stratospheric ozone, primarily found in the Earth's stratosphere, forms the ozone layer, which absorbs ultraviolet radiation from the sun, thereby protecting life on the Earth. The presence of the ozone layer reduces the amount of harmful ultraviolet radiation reaching the surface, thus lowering the risks of skin cancer, cataracts, and other health issues, while also safeguarding ecosystems. However, in the 1970s, scientists first discovered that chlorofluorocarbons (CFCs) and other chemicals could deplete the ozone layer, leading to its thinning and the formation of ozone holes. A notable historical event was the discovery of the Antarctic ozone hole in 1985, where scientists observed a 40% reduction in the ozone layer over Antarctica (NASA, 2009). To address this global pollutant, the international community adopted the Montreal Protocol in 1987 (UNEP, 2018), which successfully reduced the emissions of ozone-depleting substances. These measures are expected to restore the ozone layer to its 1980 levels within the next few decades.

Ground-level ozone is formed by photochemical reactions involving NO_x and VOCs under sunlight. This type of ozone pollution is prevalent worldwide, especially in highly industrialized and urbanized areas. As depicted in Fig. 10.2, Asia is generally suffering from ozone pollution, with the problem being more prominent in Central Asia and South Asia. It can cause respiratory diseases, exacerbate asthma, and other health problems, and damage plants, reducing crop yields and affecting ecosystems. A notable historical event of ground-level ozone pollution was the 1943 photochemical smog incident in Los Angeles, USA (California Air Resources Board, 2012). During this event, the city experienced severe smog formed from vehicle emissions and industrial discharges under sunlight, leading to widespread respiratory issues among residents. To combat this global pollutant, international organizations such as the WHO and UNEP have set stringent air quality targets. For instance, the WHO's global air quality guidelines recommend that the 8-h mean concentration of ozone should not exceed 100 μg per cubic meter (WHO, 2021). These international goals and agreements aim to reduce ozone pollution and protect human health and the environment.

10.2.3 Climate Change-Induced Air Pollution

Climate change significantly exacerbates air pollution through a variety of mechanisms. The rise in global temperatures leads to more frequent and intense heatwaves, which in turn increase the concentration of ground-level ozone. Higher temperatures can also accelerate the chemical reactions that produce secondary pollutants, such as $PM_{2.5}$, further degrading air quality. Climate change contributes to the increased frequency and severity of wildfires, which release large quantities of pollutants, including carbon monoxide, nitrogen oxides, and fine particulate matter, into the atmosphere. These pollutants pose serious health risks, including respiratory and

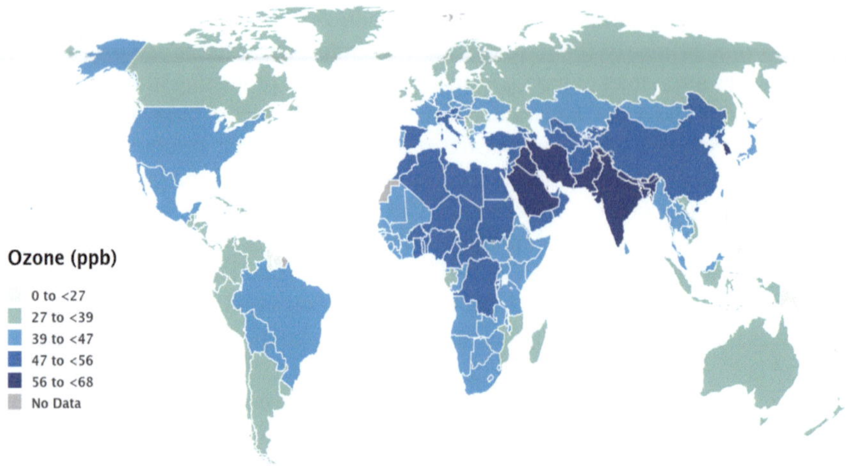

Fig. 10.2 Average seasonal population-weighted ozone concentrations in 2020. *Note* This map is plotted by Health Effects Institute (2024a), original data from the Institute for Health Metrics and Evaluation (2024)

cardiovascular diseases. Overall, climate change not only directly affects temperatures but also indirectly exacerbates air pollution through complex chemical reactions and environmental changes.

The frequency and severity of wildfires have surged due to climate change, making wildfires a major source of air pollution. Wildfires emit vast amounts of pollutants, including carbon monoxide, nitrogen oxides, and fine particulate matter, which can travel long distances and affect air quality far from the fire's origin. For instance, the 1871 Peshtigo Fire, the deadliest wildfire on record, devastated large areas of Wisconsin, Michigan, and Illinois, resulting in approximately 2,500 deaths (Hultquist, 2015). More recently, Canada experienced its worst wildfire season on record in 2023, with fires raging from March to November, causing widespread destruction (Natural Resources Canada, 2023). Additionally, the 2023 Lahaina Town Fire in Hawaii was the deadliest wildfire in the United States in a century, claiming at least 114 lives (Oatis, 2023). Because pollutants can be carried by the wind, wildfires can sometimes have impacts far beyond the area where the fire is burning. As Fig. 10.3 shows, the problem of annual $PM_{2.5}$ emissions from wildfires is more pronounced in developed areas, such as North America and Australia. The smoke from these wildfires can cause acute health problems, such as asthma attacks and other respiratory issues, and has been an unneglectable problem affecting people's health.

Climate change-induced extreme weather events, such as hurricanes and floods, can disrupt industrial activities and infrastructure, leading to accidental releases of pollutants. Flooding can cause chemical spills from industrial sites, contaminating both air and water. A notable example is Hurricane Katrina in 2005, which not only caused widespread destruction but also led to the release of large quantities of toxic

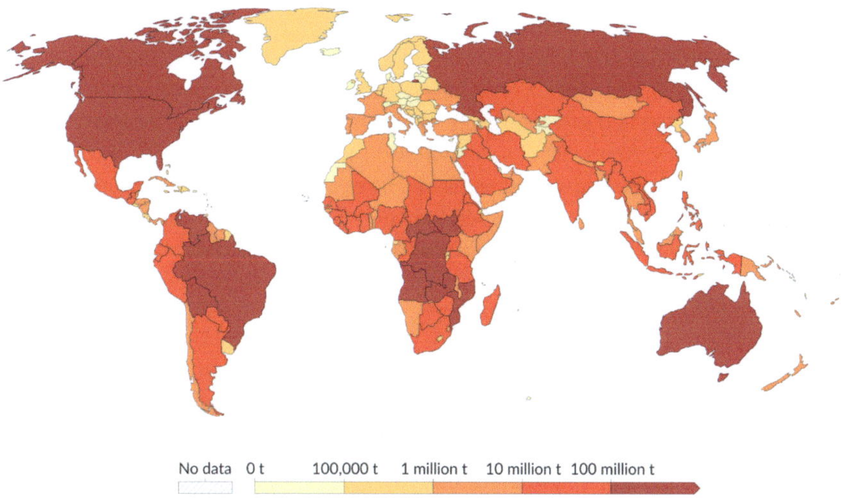

Fig. 10.3 Annual $PM_{2.5}$ emissions from wildfires in 2023. *Note* Wildfire, characterized by its uncontrolled and rapid spread, can occur in various types of vegetation and wildlands, including forests savannahs, grasslands, and various other vegetation types. These incidents are identified using satellite imagery, which detects thermal anomalies indicators of active burning areas. This map is plotted by Our World in Data (2024). Original data includes Global Wildfire Information System (2025)

chemicals in the New Orleans area, severely polluting the air and water (Santella et al., 2010). The increased use of diesel generators during power outages caused by extreme weather can elevate levels of nitrogen oxides and particulate matter in the air. These events highlight the interconnectedness of climate change and air pollution, underscoring the need for integrated policies to address both issues simultaneously.

10.3 Effects of Air Pollution

10.3.1 Direct Versus Indirect Effects

The direct effects of air pollution primarily involve the costs associated with reducing emissions. These costs can be substantial and include investments in cleaner technologies, upgrading industrial processes, and implementing pollution control measures. For instance, industries may need to install scrubbers and filters or adopt alternative energy sources to comply with environmental regulations. These measures, while necessary for reducing harmful emissions, often require significant financial outlays. Additionally, businesses might face increased operational costs due to the need for regular maintenance and monitoring of pollution control equipment. Governments also incur significant costs in enforcing regulations, conducting inspections, and monitoring air quality to ensure compliance. This includes funding for environmental agencies, developing and maintaining air quality monitoring networks, and investing in research and development for new pollution control technologies. Furthermore, governments often provide subsidies or incentives to encourage industries to adopt cleaner technologies, which add to public expenditure. On a personal level, individuals often bear costs to protect themselves from air pollution, such as purchasing air purifiers, wearing masks, and seeking medical treatments for pollution-related health issues. These personal defense measures, while essential for health, add another layer of financial burden on households.

Air pollution has numerous indirect effects that extend beyond the immediate area of emission. Externalities refer to the unintended consequences of pollution that affect third parties who are not directly involved in the polluting activity. For instance, air pollution can lead to health problems such as asthma, cardiovascular diseases, and premature deaths, imposing healthcare costs on society. These health impacts can reduce workforce productivity and increase public health expenditures. Spillovers occur when the effects of air pollution spread to neighboring regions. Pollutants can travel long distances through the atmosphere, affecting air quality in areas far from the original source. This can lead to regional environmental degradation, affecting agriculture, water quality, and natural ecosystems. For instance, acid rain, caused by sulfur dioxide and nitrogen oxides, can damage forests, lakes, and soil in regions downwind of industrial areas.

The global nature of air pollution means that pollutants emitted in one country can cross national borders, affecting neighboring countries and leading to international environmental and health issues. This necessitates cooperation and agreements to effectively manage and reduce air pollution. Without coordinated efforts to address the transmission of air pollution, people could risk significant environmental disasters due to the inherent negative externalities of unchecked pollution. A notable example is the transboundary air pollution in Southeast Asia, where forest fires in one country can cause severe problems in neighboring countries, impacting millions of people. These cross-border impacts underscore the need for coordinated global efforts to address air pollution, as actions taken in one region can have far-reaching consequences.

Reading

Transboundary Haze Pollution in Southeast Asia

Southeast Asia has long been plagued by transboundary haze pollution, primarily originating from peatland fires in Sumatra and Kalimantan, Indonesia. The smoke spreads to neighboring countries such as Malaysia and Singapore, significantly affecting air quality. To address this issue, ASEAN countries signed the ASEAN Agreement on Transboundary Haze Pollution in 2002, promoting regional cooperation in governance. Indonesia has taken measures to reduce pollution sources by prohibiting peatland development and enhancing fire prevention and control, while Malaysia and Singapore have provided technical and financial support. Although some progress has been made in mitigation, the haze problem remains unresolved, highlighting the conflict between economic development and ecological protection in transboundary environmental governance, as well as the lack of enforcement in regional cooperation mechanisms. This offers valuable lessons for Southeast Asia to strengthen collaborative governance and achieve sustainable development, while also demonstrating that resolving transboundary pollution requires long-term commitment and concerted efforts from the international community.

10.3.2 Perceived Versus Real Pollution

Understanding the direct and indirect impacts of air pollution requires accurate and timely measurement of its severity. Today, scientists have achieved real-time and accurate monitoring of air pollution through the use of sophisticated technologies and data from multiple sources, such as terminal sensors, high-resolution remote sensing, and pollutant model inversion. However, even when air pollution is accurately monitored, effectively communicating this information to the public is equally important. Since people's beliefs about air pollution depend largely on their own knowledge, past experiences, intuitive feelings about visible or invisible pollutants, and access to real pollution information, there could be a huge disparity between subjective perceptions and objective measurements of air pollution, creating an information asymmetry problem that could greatly bias individual decisions.

Perceived air pollution refers to the subjective assessment of air quality by individuals or communities. This perception can be influenced by visible factors such as smog, odors, and media reports. People's awareness and concern about air pollution often depend on their personal experiences and the air pollution information they receive. Even if the actual levels of pollutants are not critically high, perceived air pollution can lead to behavioral changes, such as avoiding outdoor activities or advocating for policy changes. Typically, the public's emotions and behaviors regarding air pollution are based on their own perceptions, which to some extent deviate from the actual pollution, thus leading to bias. Real air pollution, on the other hand, is the objective measurement of pollutants in the air. These measurements are typically obtained through monitoring stations and scientific instruments. Governments and researchers are more able to make decisions based on real air pollution information because they not only have easy access to accurate data but also have a more comprehensive understanding of the information behind it.

The relationship between perceived and real air pollution is complex. While real air pollution provides the factual basis for understanding environmental and

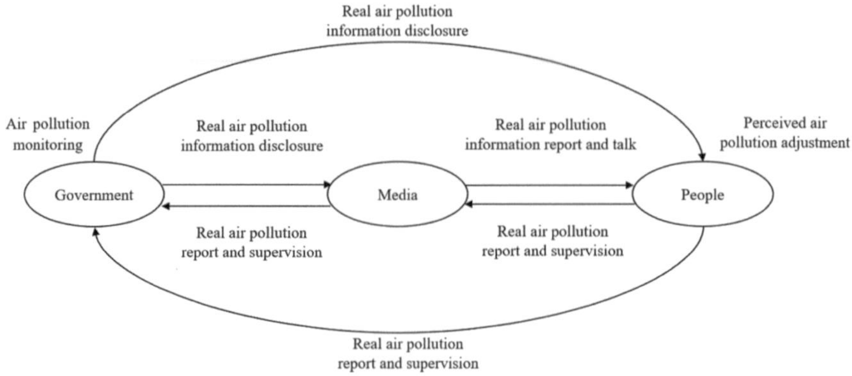

Fig. 10.4 Air pollution information interactions and adjustment

health impacts, perceived air pollution shapes public awareness and policy responses. Discrepancies between the two can occur in many situations. People might perceive air quality as poor due to visible smog, even if pollutant levels are within safe limits. Conversely, harmful pollutants that are not visible or odorous might be underestimated by the public. Moreover, the ease of access to air quality information plays a more important role. In daily life, people rarely monitor air pollution themselves but rely more on public data provided by the government or non-governmental organizations. Bridging the gap between perceived and real information requires authoritative, accurate, timely, understandable, and accessible air quality data. Figure 10.4 depicts how people interact with government and public media to correct their information bias over air pollution. People are able to learn from the disclosed authentic information from the government and reports from various media sources, adjusting their perceived air pollution information. Meanwhile, they could also report and take active efforts on supervision to promote air quality.

10.3.3 Effects on People

Air pollution profoundly affects various facets of an individual's life. Primarily, it deteriorates health by contributing to respiratory and cardiovascular diseases, thereby escalating mortality rates and healthcare expenditures. Beyond health implications, air pollution influences daily behaviors, curtailing outdoor activities and impairing cognitive functions, which subsequently diminishes productivity and elevates accident rates. In the labor market, it manifests as reduced worker efficiency and increased workplace injuries, further burdening economic resources. Moreover, air pollution exacerbates social inequalities, as vulnerable individuals often endure higher exposure and greater health risks, underscoring the imperative for comprehensive policies to mitigate its pervasive effects.

Air pollution has significant negative impacts on health, with different pollutants causing various health issues. Particulate matter ($PM_{2.5}$ and PM_{10}), among the most common pollutants, can penetrate deep into the lungs, leading to respiratory diseases such as asthma and chronic obstructive pulmonary disease (COPD), and increasing the risk of cardiovascular diseases like heart attacks and strokes. Nitrogen dioxide (NO_2), primarily from traffic and industrial emissions, can cause decreased lung function and respiratory inflammation with long-term exposure. Ozone (O_3), formed by chemical reactions between other pollutants in sunlight, can exacerbate asthma symptoms and cause lung diseases. According to UNEP (2023), air pollution is estimated to cause more than 8 million premature deaths each year, making air pollution one of the greatest environmental threats to global public health. Figure 10.5 shows the general health effects of air pollution. According to estimates from different research, indoor and outdoor air pollution has caused millions of deaths. Although the number of deaths varies by year and type of pollution, the health effects of air pollution are apparent.

Beyond health impacts, air pollution also affects people's daily decisions in many aspects. Air pollution impacts various aspects of individuals' lives, including consumption behavior, travel decisions, time allocation, and driving behavior. Financial resources are increasingly directed toward healthcare and preventive measures, such as purchasing air purifiers and masks, which reduce disposable income for other goods and services. This shift in spending can lead to cognitive biases, where individuals prioritize immediate protective measures over long-term investments, resulting in suboptimal decision-making.

Elevated pollution levels discourage outdoor activities, compelling individuals to spend more time indoors, which negatively impacts mental well-being due to reduced physical activity and limited social interactions. Additionally, air pollution influences

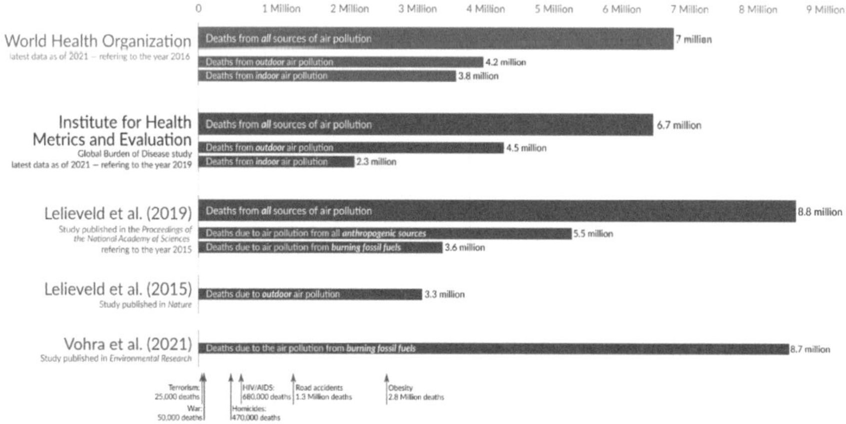

Fig. 10.5 How many people die from air pollution each year? *Note* This figure is plotted by Roser (2021), with original data coming from other sources including WHO (2021), Vohra et al. (2021) and Lelieveld et al. (2015, 2019)

travel decisions, with individuals avoiding highly polluted areas for tourism in the short term and considering relocation to regions with better air quality in the long term, affecting local economies and migration patterns. Furthermore, air pollution can impair drivers' vision and reaction times, increasing the likelihood of traffic accidents. These examples underscore the pervasive impact of air quality on individual lives and highlight the necessity for effective pollution control measures.

Air pollution significantly impacts labor market performance, affecting labor productivity, workplace injuries, and labor supply. Exposure to pollutants like particulate matter (PM2.5) and ozone reduces cognitive and physical functions, leading to fatigue and decreased concentration, which lowers productivity. Pollutants also increase the risk of workplace injuries by causing respiratory issues and eye irritation, which can distract workers and slow their reaction times. Additionally, high pollution levels deter individuals from working in polluted environments, reducing the available labor supply. Chronic health issues from long-term exposure further decrease labor force participation. Moreover, the varying degrees of impact on different demographic groups can exacerbate inequalities in labor market performance, such as wages and working hours, as lower-income individuals often face higher exposure and greater health risks, leading to increased economic disparities. These effects underscore the need for effective pollution control measures to protect workers and maintain economic stability.

10.3.4 Effects on Firms

Air pollution, as a source of negative externality, imposes significant economic and environmental impacts on both firms and society. Firms face increased costs and heightened policy risks when addressing emissions related to air pollution. However, if managed effectively, tackling air pollution can also present opportunities for technological advancement, market expansion, and brand enhancement. The impact of air pollution on firms can be seen from multiple perspectives.

Firstly, air pollution necessitates increased investment in green technology. Firms are compelled to develop and implement advanced technologies that reduce emissions and minimize their environmental footprint. As Fig. 10.6 shows, China's public firms are increasing their green patent applications. This reflects the fact that as the pollution problem becomes salient, large firms in China are taking active action to address issues related to pollution. Their green technology includes innovations such as renewable energy systems, energy-efficient machinery, and pollution control devices. By adopting these green technologies, firms not only help mitigate pollution but also unlock new market opportunities. For example, firms that lead in producing eco-friendly products or implementing sustainable practices can attract environmentally conscious consumers and gain a competitive advantage. Moreover, investing in green technology enhances a firm's corporate reputation, showcasing its commitment to sustainability and social responsibility. This improved reputation can lead to

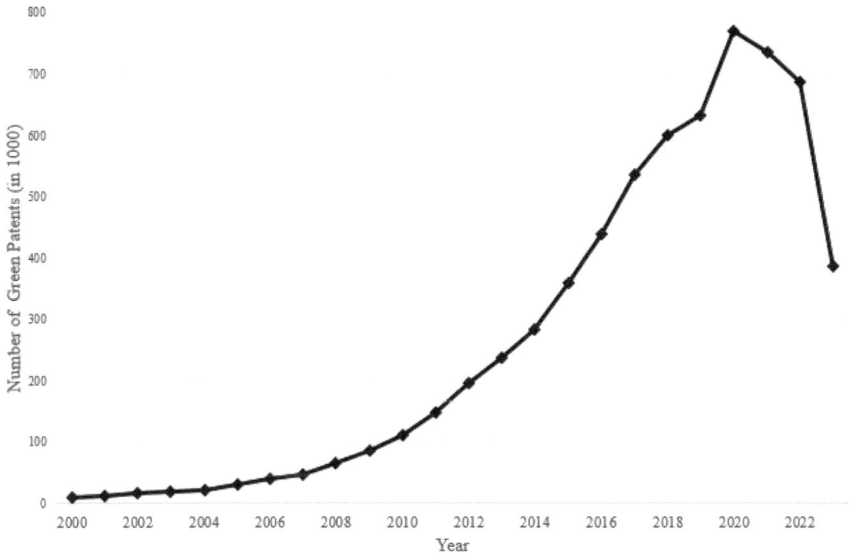

Fig. 10.6 Public firm's green technology in China, 2000 to 2022. *Note* The numbers of green patents are sum of the patent application of listed firms in China during a given year. Data source: CNRDS (2024)

increased customer loyalty, stronger investor relations, and potentially higher market valuation.

Sometimes, firms must adopt effective emissions strategies to comply with environmental regulations and avoid penalties. This often involves investing in advanced pollution control technologies, optimizing production processes, and shifting toward renewable energy sources. These strategies can lead to long-term cost savings and improved operational efficiency. Additionally, firms can participate in carbon trading to manage and reduce their carbon footprint. By buying and selling carbon credits, firms can meet their emissions targets and potentially profit from these transactions, further incentivizing emission reduction efforts.

Additionally, air pollution significantly affects labor productivity. Employees exposed to polluted environments may suffer from health issues, leading to increased sick leave and absenteeism, thereby reducing overall productivity. Moreover, air pollution can impact the mobility of high-skilled labor, as talented individuals may choose to avoid heavily polluted areas, posing challenges for firms in attracting and retaining top talent.

The marketing strategies of firms are also influenced by air pollution, as increasing consumer awareness of environmental issues can significantly alter purchasing behaviors. For instance, consumers are becoming more inclined to support brands that demonstrate a commitment to sustainability and environmental responsibility. This shift necessitates that firms adjust their marketing approaches to highlight their green initiatives and eco-friendly products. Additionally, firms can leverage certifications and eco-labels to build credibility and trust with environmentally conscious consumers. Marketing campaigns may also include transparency about environmental impacts and efforts to improve, using storytelling to connect with consumers on a deeper level. By aligning their marketing strategies with environmental values, firms can not only attract and retain customers but also differentiate themselves in a competitive market.

Considering policy risks, firms face significant challenges related to air pollution, as governments may implement stricter regulations and policies to combat environmental issues. In China, the practice of environmental courts provides a crucial legal framework for corporate environmental compliance. These courts enforce strict regulations and impose substantial fines, compelling firms to prioritize environmental protections. Firms must stay abreast of potential policy changes and proactively adjust their strategies to remain compliant. Some companies may attempt to influence policy decisions or evade penalties by increasing their connections with government officials.

However, this practice may pose significant ethical and legal risks. Possible regulatory evasion behaviors that firms may adopt include choosing to locate factories in areas with looser environmental regulations, increasing political connections, information opacity, transferring pollution, and short-term compliance strategies. These strategies face legal risks that may result in high penalties. Failure to adapt to policy changes may result in legal liability, financial losses, and damage to brand image.

In recent years, firms have increasingly invested in environmental, social, and governance (ESG) projects to demonstrate their commitment to sustainable development and attract socially responsible investors. ESG investments not only help improve corporate brand and reputation but also create a favorable marketing environment, ultimately establishing a positive corporate image. By actively engaging in ESG projects, firms can build consumer and investor trust, enhance market competitiveness, and achieve long-term sustainable development. Particularly in areas with severe air pollution, firms that invest in ESG to reduce emissions can meet regulatory requirements and significantly boost their public and market image. We will discuss this more in Chap. 19.

References

California Air Resources Board. (2012). *History|California air resources board*. Ca.gov. https://www.arb.ca.gov/about/history

China Research Data Services Platform (CNRDS). (2024). *Chinese research data services platform*. https://www.cnrds.com/Home

Evelyn, J. (1772). *Fumifugium*, or, *the inconvenience of the Aer and smoak of London dissipated: Together with some remedies humbly proposed*. W. Godbid.

Global Wildfire Information System. (2025). https://earthobservations.org/organization/work-programme/global-wildfire-information-system

Hultquist, T. (2015). *The great midwest wildfires of 1871*. https://www.weather.gov/grb/peshtigofire2

Health Effects Institute. (2024a). *State of global air 2024*. https://www.stateofglobalair.org

Institute for Health Metrics and Evaluation. (2024). *Global burden of disease 2021: Findings from the GBD 2021 study*. https://www.healthdata.org/research-analysis/library/global-burden-disease-2021-findings-gbd-2021-study

Lelieveld, J., Evans, J. S., Fnais, M., Giannadaki, D., & Pozzer, A. (2015). The contribution of outdoor air pollution sources to premature mortality on a global scale. *Nature, 525*(7569), 367–371.

Lelieveld, J., Klingmüller, K., Pozzer, A., Burnett, R. T., Haines, A., & Ramanathan, V. (2019). Effects of fossil fuel and total anthropogenic emission removal on public health and climate. *Proceedings of the National Academy of Sciences, 116*(15), 7192–7197.

National Aeronautics and Space Administration (NASA). (2009). *World of change: Antarctic ozone hole*. Nasa.gov; NASA Earth Observatory. https://earthobservatory.nasa.gov/world-of-change/Ozone

Natural Resources Canada. (2023). *Canada's record-breaking wildfires in 2023: A fiery wake-up call*. https://natural-resources.canada.ca/simply-science/canadas-record-breaking-wildfires-2023-fiery-wake-call/25303

Oatis, J. (2023, August 14). *Hawaii wildfires: How did the Maui fires start and are they still burning?* https://www.reuters.com/world/us/how-did-hawaii-wildfires-start-what-know-about-maui-big-island-blazes-2023-08-11/

Our World in Data. (2024). *Annual $PM_{2.5}$ emissions from wildfires*. https://ourworldindata.org/wildfires#all-charts

Roser, M. (2021). *Data review: How many people die from air pollution?* Our World in Data. https://ourworldindata.org/data-review-air-pollution-deaths

World Health Organization (WHO). (2021). *What are the WHO air quality guidelines?* World Health Organization. https://www.who.int/news-room/feature-stories/detail/what-are-the-who-air-quality-guidelines

World Health Organization (WHO). (2024). *Concept note: Second global conference on air pollution and health*. https://www.who.int/publications/m/item/concept-note--second-global-conference-on-air-pollution-and-health

United Nations Environment Programme (UNEP). (2018). *About Montreal protocol*. https://www.unep.org/ozonaction/who-we-are/about-montreal-protocol

United Nations Environment Programme (UNEP). (2023). *Air pollution note—Data you need to know*. https://www.unep.org/interactives/air-pollution-note/

Vohra, K., Vodonos, A., Schwartz, J., Marais, E. A., Sulprizio, M. P., & Mickley, L. J. (2021). Global mortality from outdoor fine particle pollution generated by fossil fuel combustion: Results from geos-chem. *Environmental Research, 195*(110754), 110754.

Chapter 11
Water Resource Management

Anyone who can solve the problems of water will be worthy of two Nobel prizes - one for peace and one for science.
John F. Kennedy

11.1 Water Resource: Current Status and Challenges

11.1.1 Global Water Resource Distribution and Utilization

The 2024 edition of the United Nations World Water Development Report highlights how developing and maintaining water security and equitable access to water services is essential to ensuring peace and prosperity for all. However, water resources are distributed unevenly across the globe, with significant disparities in availability. Approximately, 70% of the Earth's surface is covered by water, yet only about 2.5% is freshwater, and even less is readily accessible for human use. Of this freshwater, around 68.7% is locked in glaciers and ice caps, while only 0.3% exists in easily accessible surface water bodies like lakes and rivers.

From Fig. 11.1, we can see that there is a big difference in the quantity of renewable freshwater resources across different regions of the world. According to Koncagül and Connor (2023), the annual freshwater withdrawals of each social sector have grown rapidly over the past 120 years. This uneven distribution and water demand lead to critical challenges in water utilization, particularly in arid and semi-arid regions where demand often exceeds supply, necessitating effective management strategies to optimize usage and ensure sustainability.

Utilization of water resources also varies widely, influenced by many factors such as climate, geography, and economic development. Agriculture accounts for approximately 70% of global freshwater use, highlighting the critical need for efficient irrigation practices. Besides, industrial and domestic uses also place substantial pressure on available supplies, with urban areas increasingly competing for limited resources as the population grows quickly and the industrialization process accelerates.

L. Zhang, *Sustainability Economics*, https://doi.org/10.1007/978-3-031-87829-9_11

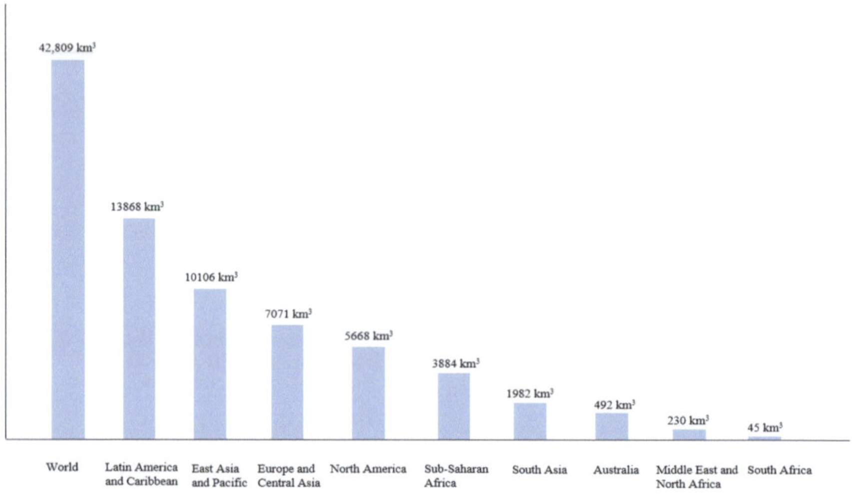

Fig. 11.1 Quantity of renewable freshwater resources by region. *Note* Renewable internal freshwater resources refer to the quantity of internal freshwater from inflowing river basins and recharging groundwater aquifers. *Data source* World Bank (2024)

Figure 11.2 illustrates the evolution of water use in agriculture, industry, and residential sectors over the past century. Water consumption in all three sectors has shown rapid growth, with industrial water use experiencing the most significant increase. In many developing countries, burdens on water resources are even more pronounced when rapid urbanization strains existing water infrastructure, leading to shortages and increased reliance on groundwater, which is often over-extracted. It can be easily concluded that those have countries experienced quick growth in freshwater withdrawals since the 1980s. This trend is typical of the world's freshwater consumption and use.

The uneven distribution of water resources is one of the most significant issues that human society is facing. The world has not yet faced an overall water crisis in recent years. However, economic and social development across different regions is still facing problems related to water stress. Water stress is defined in its simplest terms as occurring when total freshwater withdrawal substantiates a large share of available renewable freshwater resources (Koncagül & Connor, 2023). Although the world's water stress is around 20 percent, a figure that almost meets the upper bound of the no-stress category, many regions of the world are still suffering from problems caused by water stress.

People living in Africa, especially in Northern Africa, are the most affected by water stress, with the highest water stress that even exceeds 100%, the threshold of critical stress. The regions of Eastern, South-Eastern, and Northern America follow, falling into the low-stress category. This fact highlights that the uneven regional distribution of water resources and the concentration of population and economic activities in different regions lead to different water demands, which is the main reason for water resource stress in different regions.

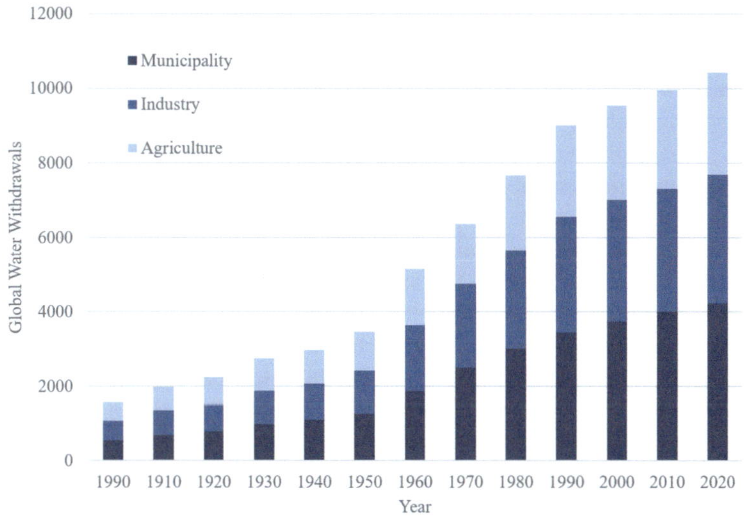

Fig. 11.2 Evolution of global water withdrawals, 1900–2018. *Data source* UN (2023a)

11.1.2 Externalities and Governance of Water Pollution

While global attention on water resources has focused primarily on water quantity, water quality remains equally important given the rise in water scarcity, drought, and related famine. It also poses a growing challenge as countries industrialize, urbanize, and their populations grow. While not fully quantified, the impacts of poor water quality on human development and the environment can be significant. Water pollution is now regarded as another significant challenge in the realm of water resource management, representing a critical issue that affects both environmental health and human well-being.

Economically, water pollution arises from market failures where the costs of pollution are not reflected in the prices of goods and services. For instance, agricultural runoff containing fertilizers and pesticides often leads to contamination of water bodies, creating negative externalities that impact communities downstream. These externalities occur when the polluter does not bear the full costs of their actions, resulting in the overuse of harmful practices. It is estimated that over 2 billion people globally lack access to safe drinking water, largely due to pollution, underscoring the urgent need for effective governance to mitigate these impacts (Koncagül & Connor, 2023).

In addition to negative externalities, information asymmetry plays a crucial role in exacerbating market failures related to water pollution. When consumers and regulators lack complete information about the sources and impacts of water pollution, it becomes challenging to hold polluters accountable. For example, industries may not disclose the full extent of their emissions or the potential risks associated with their operations, leading to uninformed decisions by policymakers and the public.

This lack of transparency can result in inadequate regulatory responses and insufficient investment in pollution control technologies, further perpetuating the cycle of contamination and harm to water resources.

Moreover, regulatory costs and unclear property rights contribute to the inefficiencies in managing water pollution (Olmstead, 2010). High costs associated with monitoring and enforcement can deter effective regulation, especially in developing regions where resources are limited. Sometimes, many companies adopt strategic water pollution, which to a certain extent strengthens information asymmetry and leads to higher regulatory costs for the supervision of water pollution. Reducing information asymmetry and the resulting regulatory costs related to water pollution remain important issues in the present practice of water resource management.

Additionally, when property rights over water resources are not well-defined, it creates confusion about accountability and responsibility for pollution. This ambiguity can lead to a "tragedy of the commons," where individuals exploit shared resources without regard for the long-term consequences. Establishing clear property rights and reducing regulatory burdens through efficient governance can help address these challenges, promoting sustainable water management practices that protect both human health and the environment.

11.1.3 Dynamic Impact of Climate Change on Water Supply and Demand

Climate change poses significant challenges to water supply and demand, creating a complex and dynamic relationship affecting ecosystems, agriculture, and urban areas simultaneously. On the supply side, adverse issues related to global climate change are rising rapidly. One notable phenomenon is that, as global temperatures rise, evaporation rates increase, leading to a reduction in the water levels of rivers, lakes, or reservoirs. This phenomenon is particularly pronounced in regions already faced with problems of water scarcity, where prolonged droughts can diminish available freshwater resources, greatly hindering the daily life of residents and economic and social development. One typical case is that the Colorado River Basin has seen a dramatic decline in water levels, affecting millions who rely on it for drinking water and irrigation (Shao, 2023).

Additionally, global climate change has largely induced changes in the water cycle in the atmosphere, which in turn leads to increasingly frequent abnormal precipitation patterns and may cause a water supply crisis. The changing precipitation patterns could result in more intense rainfall events, causing flooding and runoff that can further disrupt the natural replenishment of water supplies.

On the demand side, climate change is significantly impacting agricultural practices, which are heavily reliant on consistent water availability. When temperatures rise, crops such as wheat and corn may require up to 20% more water to thrive, while shifting growing seasons can lead to mismatches between water supply and agricultural needs. In some areas where water resources are not abundant, farmers are facing increased competition for water resources, leading to higher costs and potential crop failures. Agricultural failures caused by water shortages usually occur in the least developed countries or regions, where economic underdevelopment exacerbates the negative impact of water shortages.

A typical region is sub-Saharan Africa, where hunger and poverty are the main themes. Underdeveloped economies have difficulty in providing funds and technology to combat the many problems caused by climate change. As a result, the lack of water resources caused by extreme droughts has been greatly intensified in these regions. The lack of water resources has become one of the most important development issues in the region and even poses a huge threat to people's basic water supply.

On the other hand, the problem of water stress in urban areas should not be ignored either. Although climate change does not directly affect water demand in urban areas, it is exacerbating water stress in more subtle ways. Climate change can adversely affect water use in urban areas by altering the spatial and temporal distribution of water resources. Furthermore, urban areas are experiencing heightened water demand due to population explosion and increased heat, with cities like Phoenix and Las Vegas seeing a surge in water usage that strains local supplies and necessitates innovative conservation measures. The industry may also be affected by extreme weather. For example, extreme temperatures cause a reduction in water availability, leading power plant operators to switch from hydropower plants to nuclear power plants (An & Zhang, 2023).

The interplay between supply and demand under climate change necessitates adaptive and mitigative management strategies. An important aspect is to strengthen the supervision and control of water pollution in urban areas. Water stress in urban areas can be easily alleviated by using existing water resources more efficiently. In wider rural areas, policymakers and water managers must take a more comprehensive approach and implement integrated water resource management (IWRM) approaches that consider the variability of climate impacts in more general ways. This includes investing in infrastructure for water conservation, such as rainwater harvesting systems and efficient irrigation technologies, enhancing groundwater recharge through sustainable practices, and promoting sustainable agricultural practices like drought-resistant crops. By anticipating and responding to these dynamic changes, societies can better secure water resources for future generations while mitigating the adverse effects of climate change.

11.2 Water Resource Management: Practices

11.2.1 Water Resource Use Efficiency

Water resource use efficiency is a critical aspect of sustainable development, particularly in the context of increasing global water scarcity (Satija, 2024). Efficient water use involves maximizing the benefits derived from water resources while minimizing waste and environmental impact. This efficiency can be measured through various indicators, such as the ratio of water used for productive purposes to the total water withdrawn. By improving this ratio, economies can enhance agricultural productivity, support industrial growth, and ensure adequate water supply for domestic use, all of which are essential for sustainable economic development.

Water use efficiency in most regions of the world is experiencing an increment, as Fig. 11.3 shows. From 2015 to 2020, Oceania's water use efficiency increased to the greatest extent. However, areas including Northern Africa, Western Asia, Latin America, and the Caribbean experienced a slight drop in water use efficiency, uncovering the uneven fact of water use around the world.

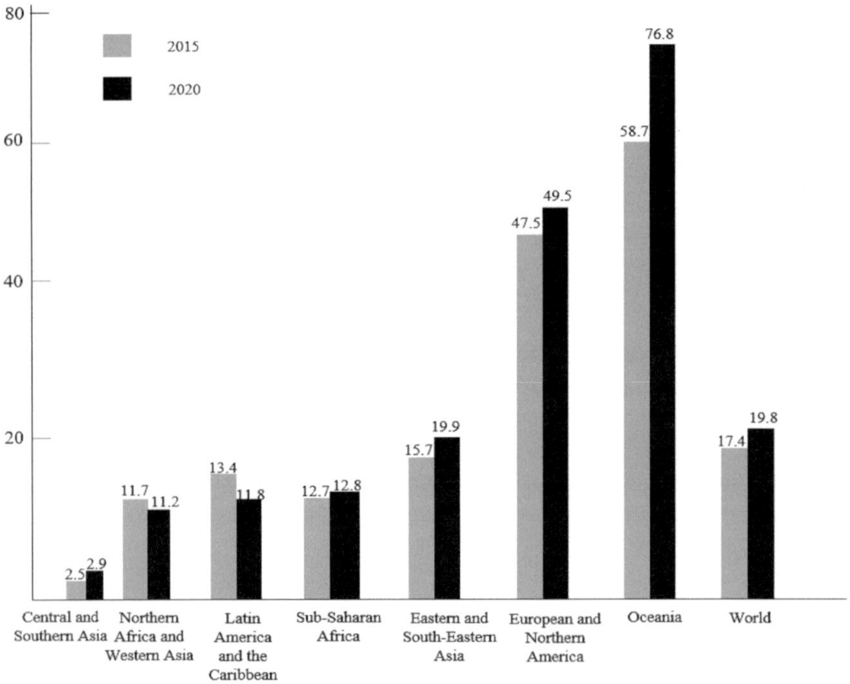

Fig. 11.3 Water use efficiency across regions in 2015 and 2020 (in US dollars per cubic meter). *Data source* UN (2023b)

Although China has abundant water resources, the contradiction between water supply and demand is becoming increasingly prominent due to its large population and rapid economic development. China's water resources are unevenly distributed, with relatively scarce water resources in the north and relatively abundant water resources in the south. In addition, industrial and agricultural water use accounts for a large amount of water resources, resulting in a certain degree of impact on domestic and ecological water use.

Since 1980, the proportion of industrial water use has increased rapidly, reaching a peak in 2010, as shown in Fig. 11.4. In terms of water resource use efficiency, in general, China has experienced rapid growth in the past four decades. As shown in Fig. 11.5, China's water productivity has risen rapidly. However, there are significant differences among provinces and regions in China. The spatial mismatch between water demand and distribution is an important issue. Many regions have long-term structural inefficiency. Although the government has taken a series of measures to improve water resource use efficiency, it still faces the challenge of addressing long-term structural inefficiency and regional differences to achieve sustainable water use for China.

To enhance water use efficiency, many comprehensive approaches must be adopted across agriculture, industry, and domestic use, each requiring specific strategies to address unique challenges.

In agriculture, irrigation technology can affect agricultural water use through different channels. Figure 11.5 provides an economic analysis of the economic channels that irrigation technology affects. There are four channels that could positively

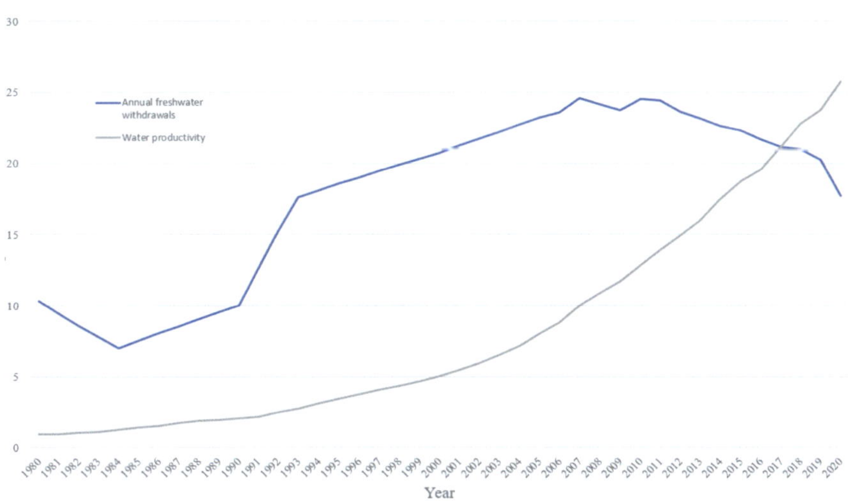

Fig. 11.4 Is water use sustainable and efficient in China? *Note* Water productivity is calculated as total gross domestic product (GDP), measured in constant 2015 US$ divided by annual total water. And industrial water withdrawals as a percentage of total water withdrawals (which is the sum of water used for agriculture, industry, and municipal purposes). *Data source* World Bank (2024)

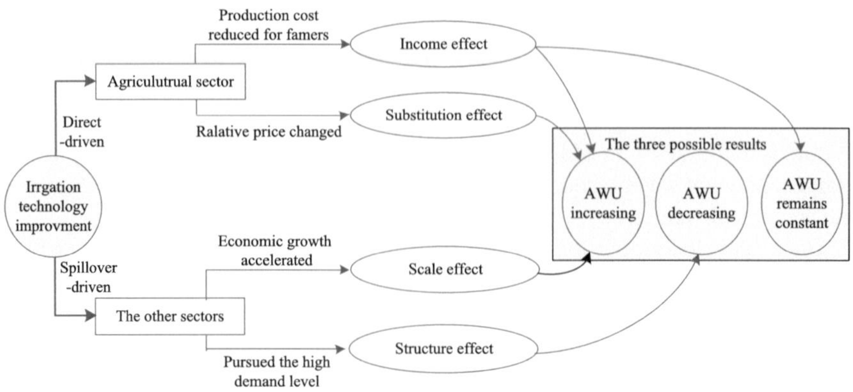

Fig. 11.5 Economic-wide channels affecting agricultural water use. *Source* Fang et al. (2020)

or negatively affect agricultural water use. Sometimes increasing water use efficiency could even lead to an increase in water use, which is called water rebound. Precision irrigation technologies utilize sensors and data analytics to monitor soil moisture levels, allowing farmers to apply water only when necessary. This targeted approach can reduce water usage by up to 30% while simultaneously improving crop yields. This reduction of water use is essential as water efficiency is negatively affected by the increasing frequency and severity of climate change (An et al., 2022). Additionally, practices such as drip irrigation minimize evaporation and runoff, further conserving water. Farmers can also benefit from adopting sustainable practices like crop rotation and planting drought-resistant varieties, which require less water and are more resilient to climate variability. Training programs focused on efficient water management techniques can empower farmers to optimize their water use, ensuring agricultural practices contribute to both productivity and sustainability.

It is worth noting that the development of irrigation technology does not always mean a decrease in total water use. Water rebound may occur in the process of technological development and agricultural expansion. Irrigation technology-induced water resource rebound refers to the phenomenon that in the process of agricultural irrigation, despite the use of water-saving irrigation technology, the actual water use increases due to increased production and expanded irrigation area. This phenomenon has been found in China's agricultural sector (Fang et al., 2020). Although water-saving technology has improved irrigation efficiency and reduced water use per unit area, farmers may expand the irrigation area or increase crop yields, resulting in an increase in total water use. The water rebound is particularly evident in areas where water resources are relatively scarce, posing challenges to water resource management and sustainable agricultural development.

In the industrial sector, investing in advanced wastewater treatment technologies is essential for improving water use efficiency (Rogers, 2002). Industries can implement systems that treat and recycle water, allowing them to reuse it for various processes, significantly reducing overall consumption. For example, sectors like textiles and

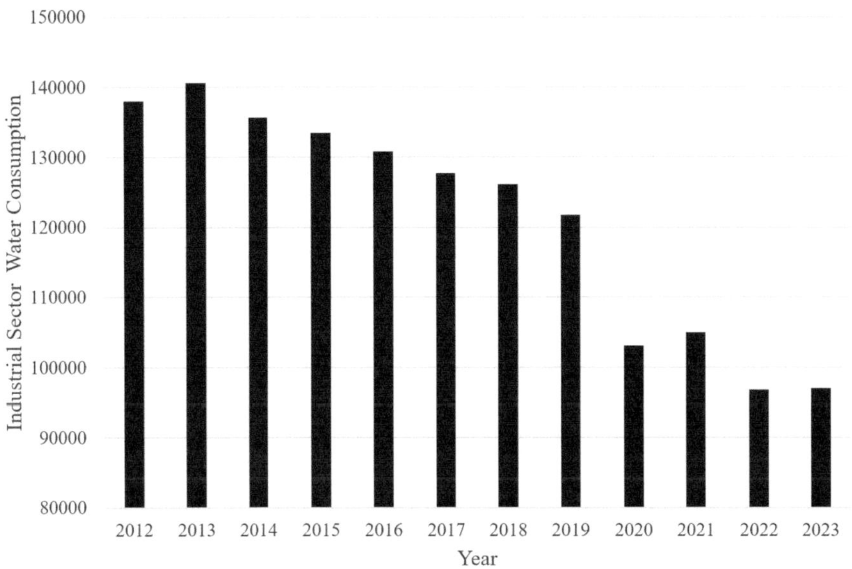

Fig. 11.6 Evolution of industrial sector's water consumption in China. *Note* The unit of water consumption is 100 million m^3. *Data source* China Water Resources Yearbook (2023)

food processing can recycle water used in production, minimizing waste and lowering operational costs. Furthermore, enforcing stricter regulations on effluent discharge ensures that industries do not pollute water sources, maintaining the quality of available water. Companies can also conduct water audits to assess their usage and identify areas for improvement, leading to more responsible water management practices that align with sustainability goals (Fig. 11.6).

For domestic use, public awareness campaigns play a vital role in promoting water conservation. Educating communities about the importance of water resources and practical methods for reducing consumption can lead to significant behavioral changes. Initiatives can include workshops on installing water-efficient fixtures, such as low-flow showerheads and dual-flush toilets, which can reduce household water use by up to 50%. Additionally, promoting rainwater harvesting systems allows households to collect and store rainwater for non-potable uses, further decreasing reliance on municipal water supplies. Implementing tiered pricing structures can also encourage households to conserve water by making excessive use more expensive, motivating families to adopt more sustainable practices. By integrating these strategies across agriculture, industry, and domestic use, we can significantly improve water resource management and promote sustainable practices that benefit both the economy and the environment.

While significant progress has been made in enhancing water resource use efficiency, several challenges remain that must be addressed. First, the lack of technology and infrastructure is a significant barrier, particularly in developing countries. For instance, many rural areas still rely on outdated irrigation methods, which can waste up to 50% of water due to evaporation and runoff. Second, deficiencies in policy and management frameworks hinder effective water use. In many regions, fragmented policies lead to inconsistent water allocation; for example, agricultural users may receive preferential treatment over urban needs, exacerbating water scarcity in cities. Finally, insufficient public awareness and participation restrict efforts to protect and utilize water resources efficiently. Many individuals do not recognize the value of water. Campaigns in some communities have shown that increasing awareness can lead to a 20% reduction in water usage, highlighting the potential for greater engagement. Addressing these challenges requires collaborative efforts from governments, businesses, and communities to achieve sustainable water resource management.

11.2.2 Water Rights and Trade System

Due to the water stress driven by the uneven distribution of water resources and population explosion, many regions are now faced with water scarcity issues. This necessitates the motivation for improving water use efficiency. Market mechanisms, although still in their infancy, are a promising approach. Establishing water rights and trading systems is one of the crucial market mechanisms for water resource management (Rosegrant & Binswanger, 1994).

Water rights refer to the legal entitlements of individuals, communities, or organizations to use water resources. These rights should be established through laws, policies, economic conditions, and traditional customs. The trading system is a market built by the government to exchange water quotas, in which the price of the quota determines the supply and demand of water quotas. By allocating water quotas and allowing these quotas to be traded in the market, the power of prices can make the social marginal benefits of both parties equal to the social marginal costs, thus achieving the socially optimal utilization rate. In this case, the efficiency of water resource use improves.

Water rights trading systems allow holders of water rights to buy, sell, or lease their rights on the market. This system can enhance the efficiency of water resource use by transferring water rights from less efficient users to more efficient ones. The economic principle behind water rights trading is that market mechanisms reflect the scarcity and value of water resources, incentivizing users to conserve and use water more efficiently.

The prerequisite for the successful establishment of the trading system to improve water usage efficiency begins with assigning water rights properly. The allocation of water rights is typically based on the availability and demand for water resources, ensuring fair and sustainable use. By appropriately granting different water rights to high-water-consuming and low-water-consuming industries and establishing a water rights trading market, the price mechanism can adjust the water consumption of both parties, improve water resource efficiency, and achieve the socially optimal state. This market design not only directly improves efficiency by regulating water consumption but also provides economic incentives for companies to invest in research and development to promote water conservation (Fang & Zhang, 2020).

However, water rights trading also involves institutional costs and information asymmetry issues. Institutional costs include the expenses of establishing and maintaining market mechanisms and regulating and enforcing trading rules. Information asymmetry refers to the problem where trading parties may not fully understand each other's water usage and the actual value of water rights, leading to market failures. More importantly, the initial allocation and clarification of water rights are critical issues in the water rights trading system, as clear ownership is necessary for legal transactions and management. Sometimes, negative externalities, such as ecological damage and environmental pollution caused by the overuse of water resources, are not reflected in market prices and require additional policies and measures to address. The implementation of water rights trading systems requires transparent market mechanisms and strict regulations to prevent abuse and ensure fair transactions.

Example

Demand for Smartphones

In China, the groundwater rights trading in Suqian City, Jiangsu Province, is a typical example. Suqian is a hub for the liquor industry, with a high demand for groundwater resources. However, it is also an area of groundwater over-extraction, with strict limits on new water withdrawals. To balance the development needs of the liquor industry with groundwater conservation, Suqian implemented water rights reforms, using market mechanisms to allocate groundwater resources. Specifically, the Suqian Water Resources Bureau developed a pilot reform plan, using market mechanisms within the legal water withdrawal limits to adjust groundwater resource allocation. The state-owned water rights company repurchased groundwater withdrawal quotas from Squian Yanghe Co., Ltd. and released them to the market, activating the stock. This not only protected groundwater resources but also ensured the legal water use of liquor enterprises, effectively solving the "water vein" problem of enterprises using groundwater. Although the water rights trading system is still in a rapid development stage and faces a series of practical challenges, it has gradually demonstrated its ability to improve the efficiency of water resource use and will surely be further developed in the future.

11.2.3 Integrated Watershed Management

To address the water resource crisis, it is essential to manage regional water resources with a systematic approach. Given the complexity of water resources within a river basin, the implications of one single strategy or policy must be considered from an overall perspective. Integrated River Basin Management (IRBM) takes a systematic view aimed at coordinating and optimizing the use of water resources within a river basin (Watson, 2004). It takes the river basin as a unit, comprehensively considering the natural, cultural, and ecological aspects of water. Such an integrated framework emphasizes managing the river basin as a whole, considering the interconnections between upstream and downstream areas. Through comprehensive planning and coordination, it ensures the sustainable use of water resources, reducing flood risks and water shortages.

IRBM involves a variety of practices aimed at ensuring sustainable water resource use and environmental protection. These practices include protecting upstream vegetation to prevent soil erosion and maintain water quality, and promoting water conservation among households and businesses by installing water-efficient fixtures and adopting water-saving habits. Encouraging cross-sector and regional cooperation and communication helps in sharing information and coordinating actions. It is worth noting that governments play a crucial role by fostering public participation and ensuring that planning and policies fully consider environmental water needs.

IRBM also emphasizes water-saving agricultural practices, such as drip irrigation and selecting drought-resistant crops, and ensuring access to clean drinking water for all. Protecting and sustainably using wetlands, along with comprehensive wastewater treatment and reuse, are essential components. Eco-friendly hydraulic engineering maintains environmental flows, while sustainable fisheries and the health of estuaries and coastal areas support biodiversity and ecosystem services. Through these integrated efforts, IRBM aims to balance human needs with ecological sustainability, ensuring the long-term health and productivity of river basins.

One important feature is that IRBM focuses on multistakeholder participation and cooperation, involving governments, communities, businesses, and NGOs in the decision-making process to ensure fairness and effectiveness. Active participation of all stakeholders, achieving information exchange, and ensuring transparency in planning and decision-making processes are crucial for the successful implementation of IRBM. Enhancing decision-making transparency and promoting equal dialogue among stakeholders, including all water users, are the best ways to resolve water conflicts.

For example, the EU Water Framework Directive emphasizes the need for active public participation in its implementation, requiring three rounds of written consultations during the planning process and providing channels for the public to access essential information (European Commission, 2023). The mechanisms for stakeholder participation vary depending on the content and requirements of river basin management. It includes many aspects, for example, participating in river basin decision-making bodies, management institutions, or advisory committees, contributing to the formulation of plans, engaging in consultations, attending public hearings, and promptly informing affected communities.

Despite the many advantages of IRBM, such as improving water resource use efficiency and promoting ecological protection, there are still numerous challenges. These include the uneven distribution of water resources, fragmented management systems, difficulties in stakeholder coordination, and information asymmetry. Existing management mechanisms often lack unified planning and coordination, leading to resource waste and environmental degradation. In the future, integrated river basin management needs to strengthen multistakeholder cooperation, establish transparent and efficient management systems, and utilize advanced technological means for monitoring and management. Public participation and education are also crucial to raising awareness of water resource protection across society. Through comprehensive governance and scientific planning, we can achieve sustainable water resource use and environmental improvement.

11.2.4 Industrial Wastewater Management

Industrial water use has always accounted for a considerable share of water consumption, especially in developing and developed regions. The water used in industry is often not directly consumed and converted into other forms of substances. Instead, it produces a large amount of industrial wastewater, which is discharged into rivers, lakes, and seas after a series of sewage treatment processes. In fact, cities in developed areas are now suffering from water pollution, which is largely caused by industrial sewage. Directly discharging untreated sewage into nature not only causes great damage to the ecological environment, seriously pollutes the survival and development of fish and birds, but also has an immeasurable impact on the economy and life of human society. This calls for industry and governments to invest more resources and attention in wastewater control to reduce the irreversible impact of wastewater on the economy, society, and ecosystems.

Industrial wastewater management involves the treatment and disposal of wastewater generated by industrial activities to prevent environmental pollution and protect public health. This treatment process includes several key steps. First, identifying and segregating different types of wastewater based on their chemical composition and potential hazards. Then, employing various treatment methods such as physical, chemical, and biological processes to remove contaminants and reduce toxicity. Finally, ensuring the treated water meets regulatory standards before it is discharged into the environment or reused in industrial processes.

In practice, the processes involved in industrial wastewater treatment mainly include the following:

- *Pre-treatment*, where large particles and debris are removed using screens and grit chambers.
- *Primary treatment*, which further removes suspended solids and organic matter through sedimentation tanks.
- *Secondary treatment*, where biological methods such as activated sludge or biofilters degrade organic matter with the help of microorganisms.
- *Tertiary treatment*, which targets specific pollutants like nitrogen, phosphorus, and heavy metals through chemical precipitation, filtration, and adsorption.
- *Disinfection*, where pathogens are killed using methods like chlorination, UV irradiation, or ozonation.

- *Sludge treatment*, where the sludge generated during the process is treated and disposed of through methods like thickening, digestion, dewatering, and incineration.

These processes work together to ensure that wastewater meets environmental standards before being discharged, reducing pollution and protecting the environment. Currently, significant achievements have been made in the technological advancements of industrial wastewater treatment processes. Advanced treatment technologies, such as membrane bioreactors (MBR), advanced oxidation processes (AOPs), and biochar adsorption technology, have been applied in some industrial sectors. These technologies have shown excellent performance in improving efficiency, reducing energy consumption, and decreasing pollutant emissions. Additionally, the introduction of intelligent and automated control systems has made the wastewater treatment process more efficient and reliable. Industrial wastewater treatment processes are continuously improving, making positive contributions to environmental protection.

Figure 11.7 shows the changes in China's industrial wastewater management over the past 20 years. The following trends can be clearly seen: First, the total discharged wastewater volume has fluctuated upward; second, the amount of COD removed from industrial wastewater has shown a continuous downward trend. These two significant characteristics are consistent with China's consistent implementation of strict industrial water pollution control policies during its rapid industrialization and the improvement of wastewater treatment technology.

Despite the tremendous development in the process of industrial wastewater treatment, effective management of industrial wastewater still faces many challenges.

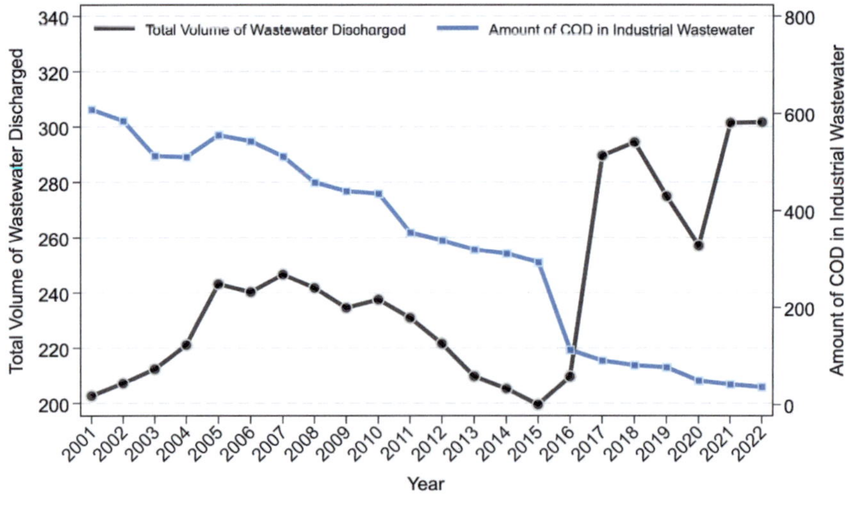

Fig. 11.7 Wastewater management in China's industry. *Data source* China Environmental Statistical Yearbook (2022)

These challenges go beyond the technical level and require concerted efforts from the government and industry. Intuitively, a large amount of wastewater containing highly toxic substances is discharged into rivers and lakes, which will further cause huge ecological disasters in the surrounding areas.

Environmental degradation does not incur direct costs for the factories that produce and discharge this wastewater, leading the effort of factories for wastewater treatment to be lower than the socially optimal level. Firms may adopt strategic behaviors to discharge wastewater in order to avoid the high cost of purifying the wastewater. These strategies include releasing highly polluted water at night, over-discharging when the monitoring system fails, and locating plants in areas with weaker environmental regulation (Schwarzenbach et al., 2010).

To deal with strategic wastewater emissions, governments and water-intensive industries should take effective actions to regulate wastewater. The most effective measure is to enact regulation laws (An et al., 2019). Powerful measures to promote wastewater reduction include providing financing incentives for greener enterprises, providing tax incentives for R&D expenditures to promote wastewater treatment, and adopting more effective monitoring systems, such as online wastewater monitoring systems. In addition, governments can use increasingly rich sensor data to monitor water environments with high precision around the clock.

Through multipronged initiatives from industry and government, firms can not only be encouraged to manage industrial wastewater more effectively to reduce the impact on the environment but also be promoted to a more sustainable development path.

11.3 Future Water Resource Management

11.3.1 Technologies in Water Management

Economic and social development has led to an increasing demand for sustainable water resources, which has further intensified concerns about the already pessimistic efficiency of water resource use and posed new challenges. Fortunately, the emergence of new technologies can help improve water use efficiency and alleviate the crisis of water shortage.

New technologies in water resource management are revolutionizing our view of sustainable development. Advanced process technologies, such as membrane filtration, reverse osmosis, and advanced oxidation processes, have significantly improved the efficiency and effectiveness of water purification. These technologies not only improve the quality of treated water but also reduce energy consumption and operating costs, making them more sustainable and environmentally friendly.

Membrane filtration technology removes tiny particles and pollutants from water by using a semi-permeable membrane. Reverse osmosis technology separates water molecules from pollutants by high pressure, and advanced oxidation processes use strong oxidants to decompose organic pollutants.

In addition to treatment technologies, smart water management systems are also gaining more and more attention. These systems use sensors, data analysis, and artificial intelligence to monitor and manage water resources in real time. By providing accurate and timely information, smart water management systems can optimize water consumption, reduce waste, and make better decisions.

Smart water meters can monitor household and industrial water use in real time, detect water leaks in time, and optimize water use strategies through data analysis. Additionally, smart irrigation systems can automatically adjust irrigation volume according to soil moisture and weather conditions to improve water use efficiency. The application of these smart systems not only improves the efficiency of water resource use but also reduces water waste.

Innovative approaches such as water recycling and reuse are becoming more common. By treating and reusing wastewater for non-drinking purposes, such as irrigation and industrial processes, we can significantly reduce the demand for fresh water. This not only saves precious water resources but also reduces the impact of wastewater discharge on the environment. For example, wastewater can be used for cooling systems and cleaning equipment after treatment, while urban sewage can be used for green irrigation and road washing after treatment.

As society continues to develop and implement these new technologies, it is closer to achieving sustainable water resource management and ensuring safe water resources for all.

When managing water resources, governments can also adopt a variety of new technologies to improve efficiency and effectiveness. Integrated urban water resource intelligence technology can optimize urban water management and reduce waste through real-time monitoring and data analysis. This allows for the prediction of pollution diffusion in advance, enabling timely preparation and response to potential pollution incidents. By integrating various data sources, such as weather forecasts and water usage patterns, it can provide a comprehensive overview of urban water systems, facilitating better decision-making and resource allocation.

Another technology, called the online pollution monitoring system, can promptly detect and deal with water pollution emissions to ensure water quality safety. It uses sensors and data analytics to continuously monitor water quality parameters, such as turbidity and contaminant levels. When abnormal readings are detected, the system can automatically alert authorities and trigger immediate response actions.

Additionally, basin integrated water resource management technology promotes the sustainable use of water resources by comprehensively managing water resources in the basin and coordinating the interests of all parties. This involves the use of hydrological models, geographic information systems (GIS), and remote sensing to assess and manage water resources at the basin level. By considering factors such as water availability, demand, and ecological needs, it ensures that water resources are allocated and used efficiently. Furthermore, it also facilitates collaboration among different stakeholders, such as governments, communities, and industries, to achieve balanced and sustainable water management.

Given the advantages of applying new technology in water resource management, there are still numerous challenges. First, data acquisition and processing remain significant issues. Although the Internet of Things and big data technologies can provide vast amounts of data, effectively collecting, storing, and analyzing these data still requires further improvement.

Second, the application and promotion of new technologies may encounter obstacles. Although many new technologies show good performance in laboratory settings, various issues, including high costs and complex maintenance, have emerged in practice. Additionally, significant differences in terms of technology and management capability between regions and countries increase the difficulty of cross-regional cooperation.

Finally, public acceptance and participation in new technologies are also hindering promotion and application. Only by continuously overcoming these challenges can we fully harness the potential of new technologies in water resource management and achieve sustainable water management goals.

In the future, new technologies will play an even more powerful role in the field of water resource management. As emerging technologies, such as big data, artificial intelligence, and the Internet of Things, continue to develop, monitoring and managing water resources in a more accurate and efficient way seems promising. With the application of these technologies, water resource management will become more intelligent, efficient, and sustainable, providing strong support for the protection and utilization of global water resources (Fig. 11.8).

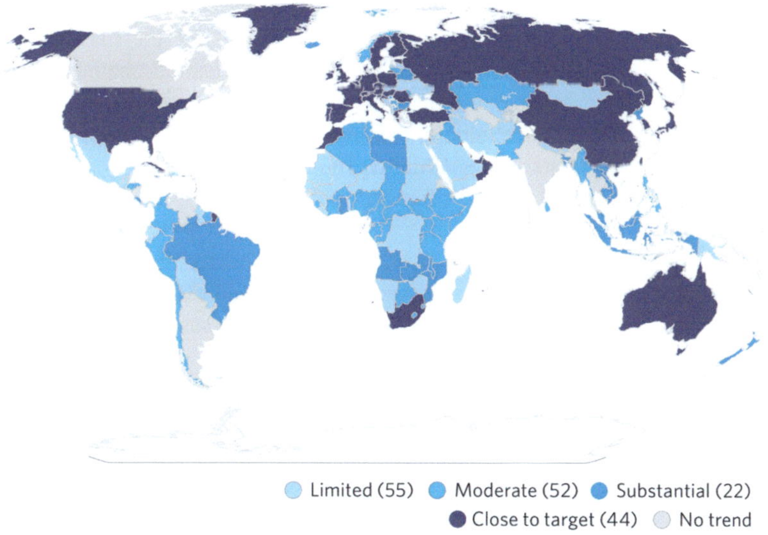

Fig. 11.8 Integrated water resources management implementation progress, 2017–2020. *Note* Numbers in the parentheses refer to the number of relevant policies. *Source* UN (2023b)

11.3.2 Social Capital and Public Participation

Firms are direct producers and emitters of industrial wastewater and are the main participants in industrial wastewater management activities. Sometimes, firms may fail to conduct effective water management actions due to the negative externality of wastewater and a lack of sufficient economic incentives. To achieve a socially optimal effort, the government, as the regulator of society, could play an irreplaceable role in improving water use efficiency by applying multiple measures that correct market failures.

However, in some situations, such as untrustworthy public policies, information asymmetry, and conflicting objectives, government regulation may not be able to fully consider the interests of various aspects of society. Firms and governments may make decisions on water management based on the goal of maximizing their own interests, which harms the interests of the public. In this context, third-party participation in water resource management is particularly important. Social capital and public participation can serve as effective supplementary forces to express the demands of various parties, convey information, and promote the implementation of corresponding water resources management policies.

Social capital refers to the networks, trust, and cooperative abilities within a region or community, which could facilitate collaboration and coordination among stakeholders. In water resource management, social capital helps to establish effective communication channels, promote information sharing, and integrate resources, thereby improving management efficiency and outcomes.

A typical example is China's River Chief System, which designates government officials at various levels as "river chiefs" responsible for the management and protection of rivers. By establishing effective communication channels, the River Chief System enhances cooperation and coordination among government agencies, communities, and stakeholders, thereby strengthening social capital. Besides, the implementation of the River Chief System not only improves water resource management efficiency but also promotes public participation in water resource protection.

Public participation is an indispensable part of water resource management. Public participation refers to the active involvement and contribution of the public in the decision-making process, which can enhance the transparency and fairness of decisions and build public trust and support for water resource management.

For instance, the Water Resource Protection Zone project in the United States involves public participation in the decision-making process, thereby increasing the transparency and fairness of decisions (Water Resource Protection Plan, 2014). The public participants provide valuable local knowledge and experience, helping to formulate management plans that are more in line with actual conditions. Additionally, public participation raises environmental awareness and encourages active involvement in water resource protection and management. In California, the public participated in the planning and management of water resource protection zones, actively engaging in activities such as tree planting and river cleaning, effectively improving local water resource conditions. More importantly, public participation can make up for environmental costs caused by government supervision, which is crucial for the supervision of sewage discharge.

The combination of social capital and public participation can significantly enhance the effectiveness of water resource management. By building strong social networks and trust relationships, communities and stakeholders can better collaborate to address the challenges in water resource management. Public participation can provide diverse perspectives and solutions, helping to formulate more comprehensive and effective management strategies.

To sum up, social capital and public participation can serve as supplementary forces for promoting information exchange, policy transparency, and law enforcement in water resource management, contributing to the achievement of sustainable water resource management goals. In the future, the power of social capital and public participation could be further released as more relevant institutions and designs are introduced.

11.3.3 International Cooperation in Cross-Border Water Management

International cooperation in cross-border water management is crucial for ensuring the sustainable use and protection of shared water resources. As global water scarcity issues intensify, managing transboundary water resources has become increasingly important. Many countries share rivers, lakes, and groundwater resources, necessitating cooperation and coordination among nations. International cooperation can help address issues such as water allocation, pollution control, and ecological protection, promoting regional stability and sustainable development.

The reasons for cross-regional cooperation are multiple. First, the natural flow of water resources is not confined by national borders, making it difficult for a single country to manage an entire basin independently. Second, transboundary water management involves the interests and needs of many countries, and only through cooperation can fair and reasonable resource allocation be achieved. Finally, environmental issues in transboundary water management, such as pollution and ecological degradation, often have cross-border impacts, requiring joint efforts from all countries to address and resolve.

One notable example is the Nile Basin Initiative (NBI), which covers ten countries that share the Nile River's water resources. The goal of NBI is to promote cooperative management and development of the Nile's water resources, ensuring equitable use and benefits for all people along the Nile River. The basic principles of cross-regional cooperation include information sharing, joint decision-making, and benefit-sharing.

Through information sharing, countries can understand the water resource status and management measures within the basin, promoting transparency and trust. Joint decision-making ensures equal participation and responsibility-sharing among countries in water resource management, avoiding conflicts and issues arising from unilateral decisions. Benefit-sharing, through cooperative projects and joint development, ensures that all countries benefit from water resource utilization, enhancing the sustainability of cooperation. By implementing joint projects, capacity building, and policy dialogue, the NBI has fostered collaboration and trust among the riparian states, helping to address water-related challenges and conflicts. For example, the NBI has implemented several cross-border water management projects, such as the *Nile Basin Integrated Water Resources Management Project*, which has improved water resource sustainability within the basin.

The success of the NBI demonstrates the importance of international cooperation in cross-border water management. By establishing cooperative mechanisms and common goals, countries can better manage and protect shared water resources, achieving regional sustainable development and stability. In the future, international cross-regional cooperation in water resources will become even more important and urgent. As water resource pressures from climate change and population growth continue to increase, there are an increasing number of countries and regions that need to cooperate more closely to address challenges related to water resource management.

The emerging new technologies will provide more tools and means for cross-regional cooperation, helping countries better monitor, manage, and protect water resources. Additionally, international organizations and multilateral cooperation mechanisms will play a greater role in promoting cross-regional water resource management cooperation by providing technical support, financial assistance, and policy coordination, helping countries achieve sustainable water resource management goals.

Future international cooperation in water resources will also focus more on equity and inclusion, ensuring that all stakeholders, especially vulnerable and marginalized communities, can participate in and benefit from water resource management fairly. By strengthening public participation and building social capital, countries can jointly develop and implement more comprehensive and effective water resource management strategies, ensuring the sustainable use and protection of water resources.

In summary, future international cross-regional cooperation in water resources will rely on technological innovation, support from international organizations, and equitable and inclusive cooperation mechanisms. Through joint efforts, countries can better address water resource management challenges, achieve sustainable development goals, and ensure that everyone has access to safe and sufficient water resources.

References

An, Y., Li, Q., & Zhang, L. (2022). Managing agricultural water use in a changing climate in China. *Sustainable Production and Consumption, 33*, 978–990.

An, Y., & Zhang, L. (2023). The thirst for power: The impacts of water availability on electricity generation in china. *The Energy Journal, 44*(2), 205–240.

An, Y., Zhang, L., & Adom, P. K. (2019). Economics of wastewater management in China's industry. *Environment and Development Economics, 24*(5), 457–478.

China Environmental Statistical Yearbook. (2022). *National Bureau of Statistics*. https://www.stats.gov.cn/

China Water Resources Yearbook. (2023). *Ministry of Water Resources of China*. http://www.mwr.gov.cn/

European Commission. (2023). *Water Framework Directive*. https://environment.ec.europa.eu/topics/water/water-framework-directive_en

Fang, L., Wu, F., Yu, Y., & Zhang, L. (2020). Irrigation technology and water rebound in China's agricultural sector. *Journal of Industrial Ecology, 24*(5), 1088–1100.

Fang, L., & Zhang, L. (2020). Does the trading of water rights encourage technology improvement and agricultural water conservation? *Agricultural Water Management, 233*, 106097.

Koncagül, E., & Connor, R. (2023). *The United Nations world water development report 2023: Partnerships and cooperation for water; facts, figures and action examples*. https://www.unwater.org/publications/un-world-water-development-report-2023

Olmstead, S. M. (2010). The economics of managing scarce water resources. *Review of Environmental Economics and Policy, 4*(2), 179–198.

Rogers, P. (2002). Water is an economic good: How to use prices to promote equity, efficiency, and sustainability. *Water Policy, 4*(1), 1–17.

Rosegrant, M. W., & Binswanger, H. P. (1994). Markets in tradable water rights: Potential for efficiency gains in developing country water resource allocation. *World Development, 22*(11), 1613–1625.

Satija, P. (2024). *Water use efficiency key to achieving multiple SDGs | policy circle*. https://www.policycircle.org/opinion/water-use-efficiency-key-to-sdgs/

Schwarzenbach, R. P., Egli, T., Hofstetter, T. B., Gunten, V., & Wehrli, B. (2010). Global water pollution and human health. *Annual Review of Environment and Resources, 35*, 109–136.

Shao, E. (2023). The Colorado River is shrinking. See what's using all the water. *The New York Times*. https://www.nytimes.com/interactive/2023/05/22/climate/colorado-river-water.html

United Nations (UN). (2023a). *The sustainable development goals report 2023: Special edition*. https://unstats.un.org/sdgs/report/2023/

United Nations (UN). (2023b). *UN World Water Development Report 2023*. https://www.unwater.org/publications/un-world-water-development-report-2023

Water Resource Plan. (2014). *Water resource protection plan*. https://dlnr.hawaii.gov/cwrm/planning/hiwaterplan/wrpp/

Watson, N. (2004). Integrated river basin management: A case for collaboration. *International Journal of River Basin Management, 2*(4), 243–257.

World Bank. (2024). *FAO's global Information system on water and agriculture*. https://www.fao.org/aquastat/en/databases/

Chapter 12
Land Use Management

We abuse land because we regard it as a commodity belonging to us. When we see land as a community to which we belong, we may begin to use it with love and respect.
Aldo Leopold (1887–1948)

12.1 Land Use: Current Status and Challenges

12.1.1 Land Use and Its Trend

Throughout human history, the expansion of human activities has been closely linked to changes in land use. As civilizations grew and populations increased, the demand for land led to significant modifications of natural landscapes, impacting ecosystems and biodiversity. The global expansion of human activities is accompanied by an urgent need for comprehensive land utilization to maintain the survival and development of civilization.

With societal development, people gradually began to view land use from a scientific perspective, developing a series of scientific theories. From a modern perspective, land use refers to the management and modification of natural environments into built environments such as fields, pastures, and settlements. It encompasses a wide range of activities, each with its own unique impact on the environment and society.

One of the most important types of land use is agriculture, which involves the cultivation of crops and the raising of livestock. If not managed sustainably, this can lead to soil degradation and deforestation. Another significant type is forestry, which includes the management and harvesting of forests for timber and other resources. This requires careful planning to prevent habitat loss and maintain biodiversity.

In modern society, rapid urbanization calls for effective urban development management. Urban development refers to the expansion of cities and infrastructure, often resulting in the conversion of natural landscapes into built environments, which can contribute to pollution and reduce green spaces.

L. Zhang, *Sustainability Economics*, https://doi.org/10.1007/978-3-031-87829-9_12

Conservation efforts have emerged more recently and have become a popular practice aimed at protecting and restoring natural habitats. These efforts ensure the sustainable use of land resources while preserving ecosystems and biodiversity.

The aforementioned land use practices focus on different aspects and have their own emphasis. They are increasingly important in modern society and have attracted the attention of policymakers and the public to land development and utilization.

Figure 12.1 shows global land use and food production. Only about 29% of the Earth's surface is covered by land, approximately 148 million km^2. Of the land surface, habitable land accounts for 71% of the total land area. About 46% of the land is used for agricultural production. Agricultural land is divided into 77% for meat and dairy products and 23% for crops. The areas of water bodies and urban construction for human activities are almost the same, each accounting for about 1% of the habitable land area. These figures show that the current global land is still mainly used for agricultural activities. Forest areas account for only 38% of the habitable area, about 40 million km^2.

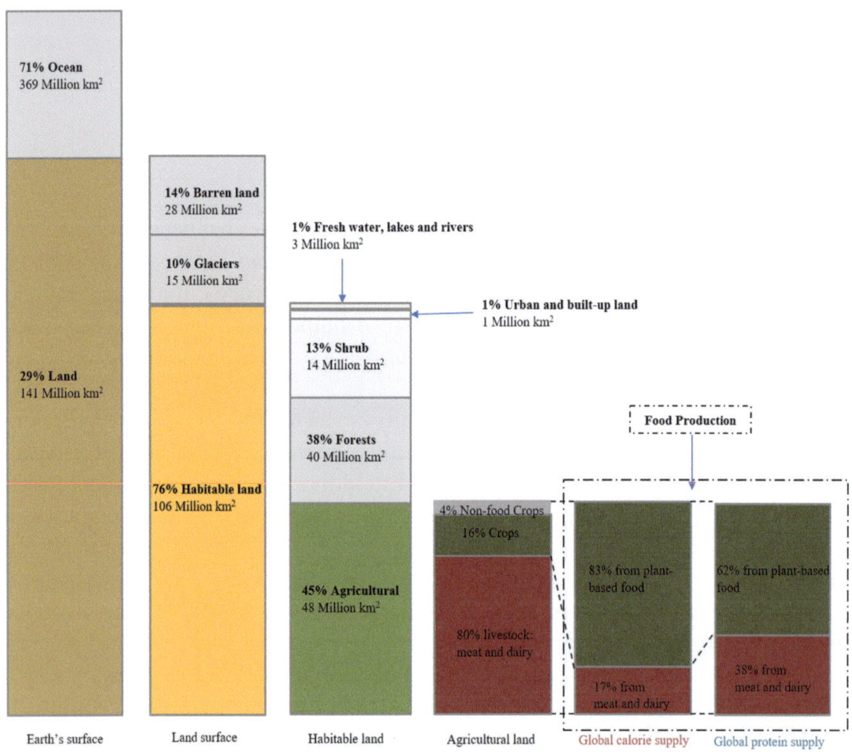

Fig. 12.1 Global land use and food production. *Data source* FAO (2023) and Poore and Nemecek (2018)

The way land is used has profound implications for the environment, economy, and society. Effective land use planning and management are crucial for sustainable development, as they help balance the competing demands for land resources while minimizing negative impacts on ecosystems and biodiversity. This involves integrating various land use activities, such as agriculture, urban development, and conservation, to ensure that each use is optimized without compromising the others. For instance, sustainable agricultural practices can maintain soil health and productivity, while urban planning can incorporate green spaces to enhance biodiversity and reduce pollution. Additionally, strategically placed conservation efforts can protect critical habitats and maintain ecosystem services, ensuring that land resources are used efficiently and sustainably for future generations. By adopting a holistic approach, land use planning can address the complex interactions between different land uses and promote balanced and sustainable development.

Improper land use can lead to severe consequences for the environment and society. Unsustainable agricultural practices often result in soil degradation and deforestation, while poor forestry management can cause habitat loss and reduce biodiversity. Additionally, urban development without proper planning contributes to pollution, reduces green spaces, and creates heat islands. Conservation efforts that are not well-implemented may fail to protect ecosystems and biodiversity. The impacts of improper land use are not only isolated but can interact with each other, leading to more serious negative effects. When both agriculture and forestry are mismanaged, the combined effects can exacerbate deforestation, soil erosion, and loss of wildlife habitats, significantly undermining ecosystem health and resilience. These interconnected impacts underscore the critical importance of sustainable land use practices to ensure the long-term health of our planet.

Over the past few decades, global land use patterns have been shaped by various factors, including population growth, economic development, and environmental policies, leading to significant changes and trends in how land is utilized and managed. Two prominent trends stand out: One of the most notable trends is the rapid urbanization occurring worldwide, particularly in developing countries. As populations grow and economies develop, more land is being converted from rural to urban use, leading to the expansion of cities and infrastructure. This urban sprawl often comes at the expense of agricultural land and natural habitats, resulting in significant changes to the landscape and ecosystem services.

Another important trend is the increasing emphasis on sustainable land management practices. Governments and organizations around the world are recognizing the need to balance development with environmental conservation. This has led to initiatives such as reforestation, the establishment of protected areas, and the promotion of sustainable agricultural practices. These efforts aim to mitigate the negative impacts of land use changes, such as deforestation, soil degradation, and loss of biodiversity, while ensuring that land resources are used efficiently and sustainably. The continuous expansion of human activities, the sustainable development goals of nature, and the dynamic changes in different land uses constitute the most distinctive opportunities and challenges for land use in today's world. Human society needs to pay enough attention to solve the prominent problems in the land use process and address related environmental issues.

12.1.2 Land Degradation and Desertification

Land degradation refers to the decline in land quality and productivity due to factors such as soil erosion, deforestation, and pollution. Desertification is a specific type of land degradation that occurs in arid, semi-arid, and dry sub-humid areas, leading to the transformation of fertile land into desert-like conditions. Land degradation and desertification are among the most direct and serious consequences of inappropriate land use. Unsustainable agricultural practices often result in soil degradation and deforestation, while poor forestry management can cause habitat loss and reduce biodiversity. These issues are further compounded by urban development without proper planning, which may contribute to pollution, reduce green spaces, and create heat islands. Additionally, conservation efforts that are not well-implemented may fail to protect ecosystems and biodiversity. When both agriculture and forestry are mismanaged, the combined effects can exacerbate deforestation, soil erosion, and loss of wildlife habitats, significantly undermining ecosystem health and resilience. These interconnected impacts underscore the critical importance of sustainable land use practices to ensure the long-term health of our planet.

Perhaps the most dramatic example of land degradation is the deforestation of the Brazilian rainforest. The Brazilian Amazon Rainforest, often referred to as the "lungs of the Earth", has been facing severe land degradation due to deforestation and other human activities. Over the past 50 years, the Amazon has lost approximately 17% of its forest cover, primarily driven by agricultural expansion, logging, and infrastructure development. Figure 12.2 depicts the evolution of Amazon deforestation in Brazil from 2001 to 2019. Although there has been a slight decline, deforestation remains severe. In 2022 alone, the Amazon lost over 1230 km^2 of forest, marking the highest level of deforestation in six years. This deforestation disrupts the habitat of countless species, leading to a decline in biodiversity, and releases significant amounts of carbon dioxide into the atmosphere, exacerbating climate change. The deforestation and burning of the Amazon contribute to nearly half of Brazil's annual carbon emissions. The situation in the Amazon underscores the urgent need for sustainable land management practices and stronger conservation efforts to protect this vital ecosystem.

Figure 12.3 depicts the desertified land due to high temperatures and a lack of rainfall, indicating the adverse consequences of desertification. Economically, land degradation reduces agricultural productivity, leading to lower crop yields and increased food insecurity. This can have a cascading effect on rural economies, where agriculture is often the primary source of income, exacerbating poverty and reducing the quality of life for millions of people. According to the United Nations, approximately 16.6 billion hectares of land are degraded globally due to human activities, with 60% of this degradation occurring on agricultural land. Additionally, the costs

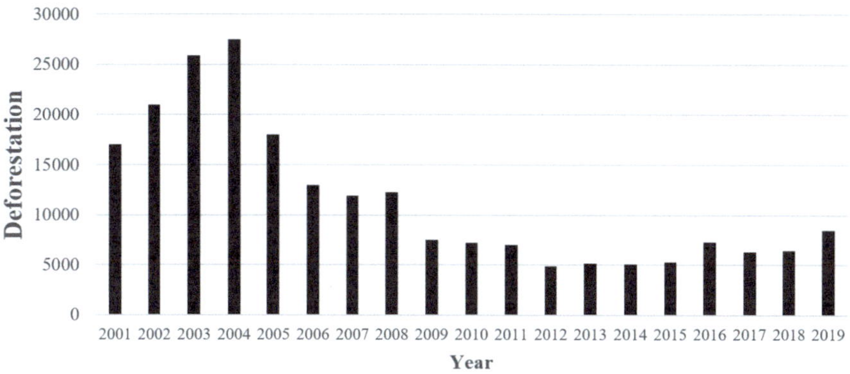

Fig. 12.2 Evolution of the Amazon deforestation. *Data source* NASA (2019)

Fig. 12.3 Decertified land

associated with restoring degraded land and implementing sustainable land management practices can be substantial, placing a financial burden on governments and communities.

Ecologically, land degradation and desertification result in the loss of biodiversity and the disruption of ecosystem services. Degraded lands often lose their ability to support diverse plant and animal species, leading to a decline in biodiversity. This loss can have far-reaching consequences, as ecosystems become less resilient to environmental changes and less capable of providing essential services such as water filtration, carbon sequestration, and soil fertility. The spread of desertification further exacerbates these issues, as once-productive lands turn into barren deserts, unable to support life. The United Nations reports that land degradation affects 40% of the world's land area, impacting the livelihoods of nearly 3 billion people.

The combined economic and ecological impacts of land degradation and desertification highlight the urgent need for sustainable land management practices. By adopting integrated approaches that consider both economic and ecological factors, people can mitigate the adverse effects of land degradation and desertification, ensuring the long-term health and productivity of our land resources. In general, tree planting and afforestation are the most effective weapons to fight land degradation and desertification. Specifically, measures to mitigate land degradation include adding organic fertilizers and soil conditioners to conserve water and soil. Engineering practices that effectively control land desertification include large-scale afforestation, planting drought-resistant plants, and using straw chessboard barriers. On the policy side, governments can implement regulations that promote sustainable land use, enforce anti-deforestation laws, and support conservation efforts. Collaboration among governments, communities, and the private sector is essential to develop and enforce these policies, ensuring that land management practices are both sustainable and effective.

Reading

Controlling the Mu Us Desert in China

The governance of the Mu Us Desert in China has employed a variety of engineering techniques and practices, reaching significant achievements. Large-scale afforestation and the planting of drought-resistant plants such as sand willow, caragana, and sea buckthorn have been used to effectively stabilize the sandy land and prevent wind erosion. The root systems of these plants could secure the soil, reduce water and soil loss, and improve soil fertility. Additionally, straw checkerboard barriers have been implemented, where straw grids are laid on the sandy land to physically stabilize the shifting sands and reduce wind erosion. Besides, techniques such as drip irrigation and sprinkler irrigation have been adopted to improve water use efficiency and reduce waste. By adding organic fertilizers and soil conditioners, the fertility and water retention capacity of the sandy soil are enhanced, improving soil structure and increasing soil stability and productivity. These comprehensive measures have significantly improved the ecological environment of the Mu Us Desert and proved effective in curbing desertification.

12.1.3 Land Use in the Process of Urbanization

Historically, urbanization has been a driving force behind economic development and societal progress. For example, during the Industrial Revolution, cities expanded rapidly as people moved from rural areas to urban centers in search of better job opportunities and living conditions. This trend has continued into the modern era, with urban areas now housing more than half of the world's population. Land use in the process of urbanization involves the transformation of natural and rural areas into urban spaces to accommodate growing populations and economic activities. This process significantly alters land use patterns, often leading to conflicts between different land use demands.

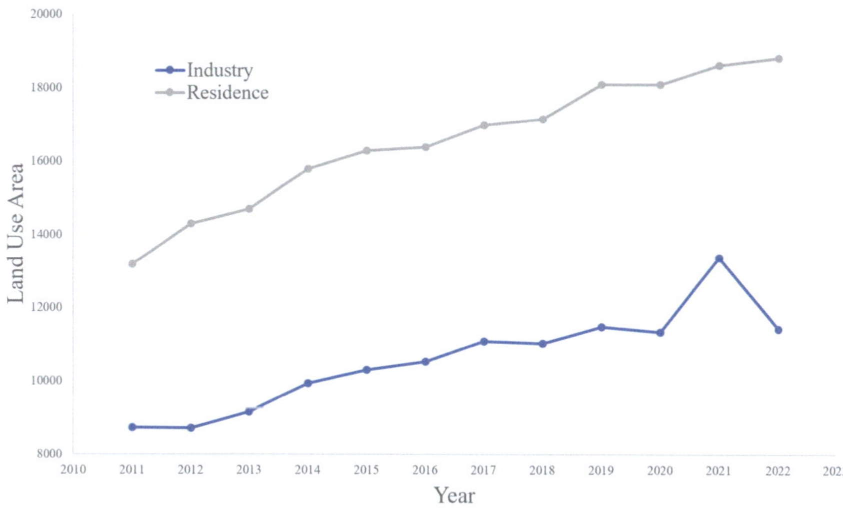

Fig. 12.4 Evolution of China's industrial and residential land use. *Note* The unit of land use area is 10,000 ha. *Data source* China Land and Resources Yearbook (2022)

As cities expand, agricultural lands and natural habitats are frequently converted into urban areas, resulting in the loss of arable land and biodiversity. For instance, between 2000 and 2018, the global urban population increased by 1.4 billion, leading to the conversion of approximately 1.2 million square kilometers of land to urban use. This rapid urban growth creates tensions between the need for housing and infrastructure development and the preservation of green spaces and agricultural productivity. Figure 12.4 illustrates a typical example of the expansion of industrial and residential land use in China. As depicted, China has experienced a quick expansion in industrial and residential land use. This brings concerns over food security related to agricultural land use. The competition for land resources can also lead to unsustainable practices, exacerbating environmental degradation and social inequalities.

Therefore, effective coordination in land use planning is essential to address these conflicts and promote sustainable urban development. Integrated land use planning approaches can help balance the competing demands for land by considering the needs of various stakeholders, including urban planners, environmentalists, and local communities. Strategies such as zoning regulations, green belts, and urban agriculture can be employed to ensure that urban expansion does not come at the expense of environmental sustainability and food security. Collaborative efforts among government agencies, the private sector, and civil society are crucial in developing and implementing these strategies.

Moreover, innovative solutions and technologies can play a vital role in enhancing land use coordination during urbanization. For example, smart city initiatives leverage data and technology to optimize land use, improve resource efficiency, and enhance the quality of urban life. By adopting a holistic approach that integrates economic, social, and environmental considerations, cities can achieve sustainable growth while minimizing the negative impacts of urbanization on land use. This requires continuous dialogue and cooperation among all stakeholders to ensure that urban development is both inclusive and sustainable.

12.2 Land Management System

To achieve sustainable land use, it is necessary to understand the institutional background of land use. The land management system is the sum of systems, institutions, and means for the state to manage, supervise, and regulate land ownership, land registration, land use, land market, and land taxes and fees. Countries around the world adopt different land management systems based on their basic political systems, historical processes, economic and social development levels, and cultural backgrounds. Land ownership refers to the legal and institutional arrangements for the ownership and use rights of land, which determine who can own, use, and manage land. The land market involves the market mechanism for the sale, lease, and transfer of land use rights, which determines land prices and resource allocation through supply and demand. There is a close connection between land ownership and the land transaction market. Land ownership determines the use rights of land, which directly affects the transaction and operation of the land market. Under private ownership, land can be traded to a great extent. The market mechanism plays a leading role in this situation. The price of land is determined by the relationship between supply and demand, and trading activities are active. Under public or collective ownership, the transaction of land is subject to more restrictions and supervision. The government or collective organizations play an important role in land allocation and use, and the role of the market mechanism is relatively weak compared with the system of private ownership of land. Regardless of the ownership of land, clear property rights and effective market mechanisms are key to ensuring the rational allocation and efficient use of land resources. The following discusses the land management systems of two major countries, each of which has distinct characteristics.

12.2.1 Land Management System in China

China's land management system is characterized by a combination of government control and market mechanisms. The government plays an important role in land allocation, supervision, and use, and controls land use and transactions through various policies and regulations. China's land ownership has distinct characteristics, stipulating that the state and collective are the owners of land. At the same time, institutional reforms are implemented to meet the needs of diversified land use scenarios. The diversity of land use rights meets different uses and needs and contributes to the stability and sustainable development of the land market. In terms of land ownership,

China has developed state-owned land rights in urban areas and collective land rights in rural areas.

State-owned land is mainly used for urban development, industrial projects, and infrastructure construction, while collective land is mainly used for agriculture and rural development. The state-owned land use process involves the government transferring land use rights to developers and other users through the "*Bidding, Auctioning, and Listing*" system. The government sets the reserve price and conditions for the land, and interested parties obtain land use rights through open competition. This process ensures transparency and fairness in land allocation and promotes efficient use and development of land. In addition, the government also implements strict land use planning and zoning regulations to control urban expansion, protect agricultural land, and ensure sustainable development.

In rural areas, China implements a new land transfer system that allows the separation of ownership, contracting rights, and management rights. The ownership of collective land belongs to the village collective, and farmers have the right to contract the land and can use the land in a long-term and stable manner. At the same time, farmers can transfer the management rights of the land to other people or organizations and realize large-scale management and efficient use of the land through leasing, subcontracting, and equity. This system helps to optimize the allocation of rural land resources and promote agricultural modernization and rural economic development.

Reading

"Bidding, Auctioning, and Listing" in China

China's "Bidding, Auctioning, and Listing" system originated in the 1990s, aiming to allocate land use rights in an open and transparent manner to prevent corruption and improper transactions. The implementation of this system has effectively improved the transparency and fairness of the land market and promoted the rational allocation and efficient use of land resources.

The process of land use rights transactions through "Bidding, Auctioning and Listing" includes the following steps: First, the local government issues a land transfer announcement, detailing the basic situation, use, planning conditions, transfer period, starting price and other information of the land to be transferred. Then, the units or individuals interested in participating in the bidding submit registration materials and pay a deposit, and the government reviews the qualifications of the applicants. Next, open competition is carried out through bidding, auctioning or listing, and the highest bidder obtains the land use right. After the successful bidding, the winner signs a contract with the government, pays the land transfer fee in accordance with the contract, and completes the land use rights delivery procedures. Finally, the winner takes the contract and payment receipts to the land management department to register the land use rights and obtain a certificate. Through the above steps, the openness, fairness and transparency of land use rights transactions are ensured, which helps prevent corruption and promote the efficient use and rational allocation of land.

12.2.2 Land Management System in the U.S.

The land management system in the United States is mainly driven by market mechanisms, supplemented by government supervision and policy support. Land prices are determined by supply and demand, and the market plays a leading role in the allocation of land resources. The government regulates the land market by formulating and implementing land use planning, environmental protection regulations, and tax policies to ensure the sustainable use of land resources. Land ownership in the United States is mainly private, with individuals and companies owning most of the land. The government also owns a large amount of land, especially land for public use, protection, and national parks. Clear property rights and legal protection allow land to be traded and transferred freely, promoting an active land market and economic development.

In urban areas, land is mainly used for residential, commercial, and industrial development. The government controls urban expansion and protects the environment and public interests through land use planning and zoning regulations. The land transaction market is active, land prices are determined by market supply and demand, and the government regulates the market through taxes and policies to prevent over-development and environmental damage.

In rural areas, land is mainly used for agriculture and natural resource protection. The government supports agricultural development and land protection through agricultural subsidies, land protection programs, and environmental regulations. Private landowners can buy, sell, and lease land freely, and the market mechanism plays an important role in the allocation of land resources. At the same time, the government also protects natural resources and the ecological environment through the establishment of protected areas and national parks. The characteristics of the U.S. land management system are its high degree of marketization and free trade. At the same time, the government ensures the sustainable use of land resources and environmental protection through policies and regulations. This system not only promotes economic development but also considers public interests and environmental protection.

12.2.3 Land Use Planning and Policy Analysis

Based on the existing land management system, how can the authorities implement sustainable land use? To achieve this goal, the government should conduct land use planning and policy analysis. Land use planning and policy analysis are critical components of sustainable land use management.

Land use planning involves systematic assessment and allocation of land resources to meet the needs of society while minimizing the negative effects of human activities on the environment. The process of planning includes evaluating land suitability, identifying potential conflicts, and developing strategies to mitigate these conflicts. Effective land use planning requires the integration of spatial data, environmental impact assessments, and community input to ensure that decisions are informed and inclusive.

Land use policy analysis, different from planning, involves evaluating existing policies and developing new ones to address emerging challenges and opportunities in land use management. It includes processes that assess the effectiveness of current land use policies, identify existing gaps, and formulate recommendations for improving the efficiency of current land use. Sometimes, analysis of current policies can provide a good reference for the next step of land use planning. By adopting comprehensive approaches for land use planning, policymakers can promote sustainable land use practices, reduce environmental degradation, and enhance the quality of life for communities.

Land use planning involves several methods to ensure sustainable and efficient use of land resources. One common method is Land Suitability Analysis (LSA), which assesses the suitability of different land areas for various uses based on factors such as soil quality, topography, and climate. LSA focuses on the optimal use of land within the constraints of limited environmental pollution.

However, to better plan land use, authorities must have a deep understanding of the environmental impact of different types of land use. This necessitates Environmental Impact Assessments in land use planning, which evaluate the potential environmental effects of proposed land use changes, ensuring that development projects do not cause undue harm to the environment. This process involves a thorough analysis of the potential impacts on air and water quality, wildlife habitats, and other environmental factors.

Community input and stakeholder engagement are also essential components of effective land use planning. By involving local communities and stakeholders in the planning process, planners can ensure that land use decisions reflect the needs and priorities of those who will be most affected by them. This approach helps to build consensus and support for land use plans, making them more likely to be successful in the long term. These methods, combined with a comprehensive approach to understanding land use policies, help create informed and balanced land use plans that promote sustainable development, reduce environmental degradation, and enhance the quality of life for communities.

Policy analysis of land use involves multidisciplinary approaches that incorporate insights from environmental science, economics, law, and social sciences. Cost-benefit analysis is a commonly used method that evaluates the economic efficiency of policies by comparing the costs and benefits related to different policies, helping to determine the most economically viable solution. This method emphasizes multidimensional economic evaluations of different aspects from a general perspective.

To actively participate in the decision-making process and ensure that all stakeholders receive appropriate returns from land use policy, it is necessary to include relevant stakeholders in the decision-making process. Therefore, policymakers are required to make decisions from the perspective of stakeholders. Stakeholder analysis provides a powerful solution as it identifies and assesses the interests and influences of various groups affected by policies, ensuring that the perspectives and needs of all relevant parties are considered in the decision-making process. This method identifies these people before the land use policy begins, groups them according to their levels of participation, interest, and influence in the process, and determines how best to involve and communicate with each of these stakeholder groups throughout the land use policy. Through this method, policymakers can maximize the enthusiasm of stakeholders in land use policies and promote policy implementation.

Under some conditions, policymakers and the public want to evaluate the consequences of various potential land use policies that have not yet been implemented, to make the best decision. In this case, scenario analysis of land use policy is a good choice. Scenario analysis explores the potential outcomes of different land use policy options under various future conditions, allowing policymakers to anticipate and plan for a range of possible scenarios, thereby enhancing the resilience and adaptability of land use policies. This method involves creating detailed narratives or models that describe how different variables, such as economic trends, environmental changes, and technological advancements, might evolve over time. By examining these scenarios, policymakers can identify potential risks and opportunities, develop strategies to mitigate adverse impacts, and capitalize on positive developments. Scenario analysis helps ensure that land use policies are robust and flexible, capable of adapting to a wide range of future conditions and uncertainties.

Case

The Land Intensification Allowance in Singapore

Singapore is a country with extremely limited land resources. Faced with rapid urbanization, how to efficiently use land resources has become an important issue for the government. In order to meet this challenge, the Singapore government has formulated a series of land use policies aimed at optimizing land use efficiency and promoting sustainable development. Among them, the Land Intensification Allowance (LIA) policy is a successful example. The policy encourages enterprises to optimize the layout of industrial facilities and increase building density by providing tax incentives, thereby improving the productivity and economic benefits of the land. Eligible enterprises can enjoy an initial allowance of 25% of the development project cost, and an additional allowance of up to 5% per year for 13 years, with a total allowance of up to 100% of the project's eligible expenditure.

The LIA policy is mainly aimed at enterprises that require a large amount of industrial land, such as manufacturing, logistics and warehousing, food and beverage processing, and data centers. When applying for LIA subsidies, companies must meet the following conditions: First, the project must meet the government's land use density standards. Then, the company needs to carry out value-added production or service activities, such as high-tech manufacturing, advanced logistics and data management. Finally, company's capital expenditures are required to include land development, building renovation and other costs directly related to intensive land use. Through this policy design, the Singapore government not only helps companies reduce land development costs, but also improves the intensiveness of industrial facilities and economic added value, which helps Singapore better achieve sustainable use of land resources and high-quality economic growth.

12.3 Land Use by Sector

12.3.1 Agricultural Land Use

Agriculture is the foundation of social and economic development. It is the basis for food production and guarantees basic human needs, supporting the development of rural economies. In many regions, agriculture is the main source of employment and income for farmers. Additionally, agricultural land plays an important role in maintaining ecosystems, promoting biodiversity, and ensuring environmental sustainability.

According to the Land Statistics (FAO, 2024), the world's total agricultural land in 2022 was 4,781 million hectares, more than one-third of the global land area. This figure highlights the important role of agricultural land in economic and social development and underscores the need to promote the sustainable development of agricultural land.

In Fig. 12.5, the growth of agricultural land is uneven around the world. There is a small proportion of areas that experience agricultural land growth over 10%, mainly located in less developed countries, especially in Africa and Southeast Asia. Agricultural land use growth in Central Africa, North America, and South America was less than 10%. Most other regions experienced negative growth in agricultural land.

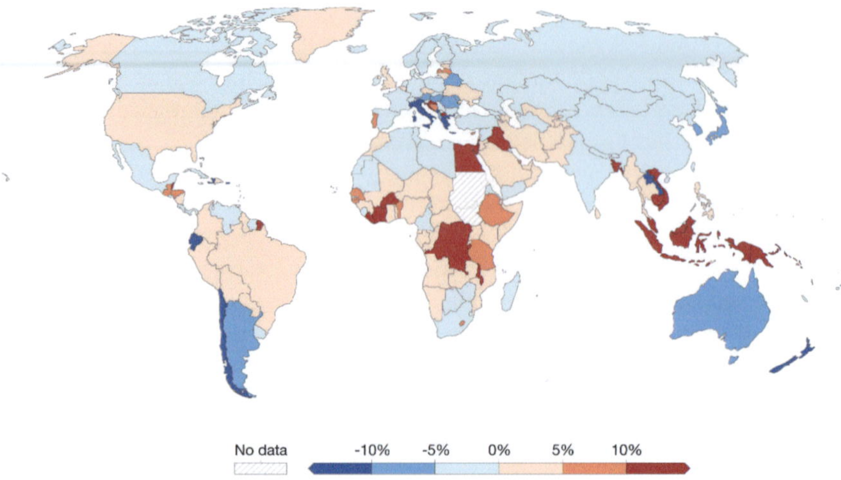

Fig. 12.5 Agricultural land change over the prior decade, 2021. *Note* Agricultural land area is compared to its area 10 years before. Positive values (shown in red) indicate an expansion of agriculture. Negative values (in blue) show a decline. Agricultural land is the sum of croplands and pasture used for livestock grazing. This figure is plotted by Our World in Data (2024) and the original data source is FAO (2024)

Rapid population growth, expansion of industrial activities, and climate change are likely the main reasons for the uneven changes in agricultural land use around the world. The explosive growth of the population has led to a huge demand for agricultural products and is the primary reason for the growth of agricultural land. This phenomenon is particularly likely to occur in developing countries with low agricultural productivity and rapid population growth.

As a large number of people gather in cities, the size of cities rapidly expands, and industrial and residential land expands accordingly, forming competition for agricultural land. This situation is more common in countries with a high level of industrialization, such as China. Additionally, climate change has an important impact. Some lands that were previously unsuitable for agricultural development have become suitable for cultivation under the influence of climate change and have therefore been rapidly developed.

The impact of the rapid advancement of agricultural technology cannot be ignored either. Modern agricultural production, such as drip irrigation technology and greenhouse agriculture, has transformed land that was originally considered unsuitable for agricultural production into modern agricultural land, greatly improving land utilization.

Although the total area of agricultural land has been growing in the past few decades, the development of agriculture is still facing tremendous pressure. Population growth and the demand for improving living standards have put pressure on the production and variety of agricultural products. This has triggered concerns about food security and an urgent need for sustainable agricultural development.

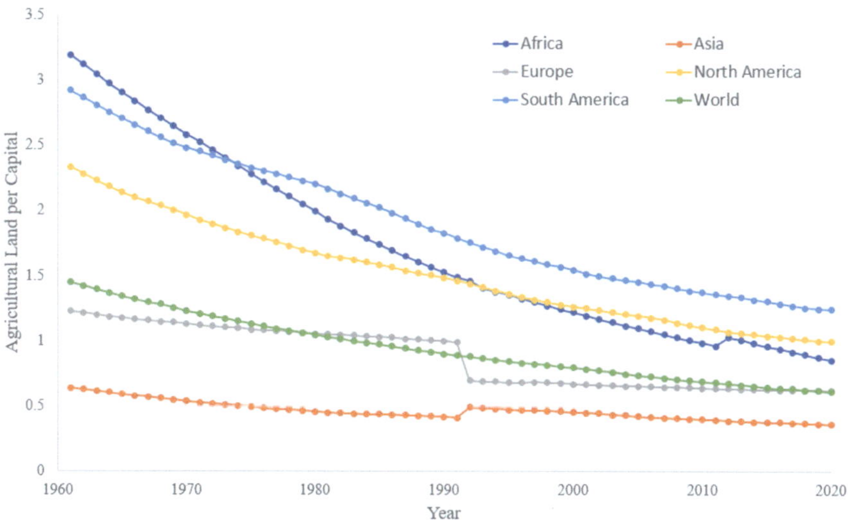

Fig. 12.6 Agricultural land per capita of the world. *Note* Agricultural land is the sum of cropland and land used as pasture for grazing livestock. The data is drawn from FAO (2024)

There is a clear pattern in agricultural land per capita, as shown in Fig. 12.6. Global agricultural land per capita has experienced a rapid decline in the past few decades. The decline in Africa is particularly obvious, primarily due to rapid population growth. In North America and Europe, the decline has been relatively gradual, likely because of slow population growth. In these places, the erosion of agricultural land by industrial expansion is the main reason for the decline in per capita agriculture.

At the same time, agricultural productivity has increased over the past few decades. The reasons for the huge improvement in agricultural productivity include modern agricultural production technology, the use of fertilizers, and improvements in farming methods. The most striking example is the productivity of wheat, which is now more than three times what it was in 1960. This improvement in agricultural productivity can not only meet the demand for agricultural products of the world's huge population but also ease the tension between agricultural land and industrial and residential land.

While agricultural productivity has increased, some adverse effects have inevitably emerged. First, excessive use of fertilizers and pesticides has led to soil and water pollution, damaging the ecological environment. Second, agricultural mechanization and large-scale planting have led to a decrease in biodiversity and the destruction of many animal and plant habitats. Additionally, intensive agricultural management may lead to land degradation and a decrease in soil fertility, affecting long-term agricultural production.

Government policies play a vital role in mitigating agricultural environmental pollution. By formulating and implementing reasonable policies, governments can

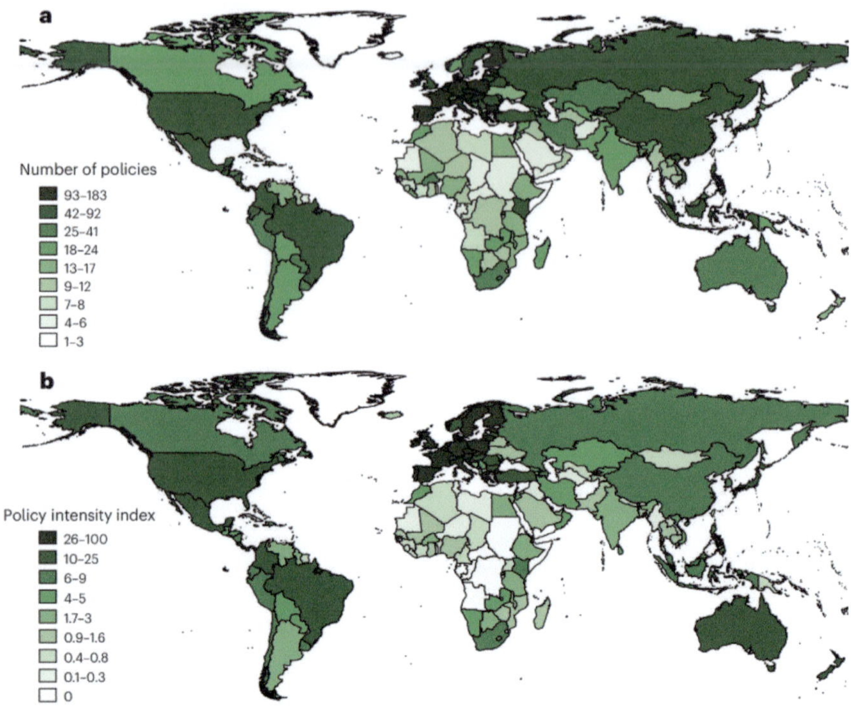

Fig. 12.7 Agri-environmental policies, 2022. *Note* This figure depicts the number of policies related to reducing the environmental impacts of agricultural environmental pollution. *Source* Wuepper et al. (2024)

effectively guide agricultural production in a sustainable direction, protect the ecological environment, and ensure the long-term healthy development of agriculture. In the past few decades, governments around the world have implemented a large number of policies aimed at reducing agricultural environmental pollution, as Fig. 12.7 depicts (Wuepper et al., 2024).

The relevant policies contain different regulations, frameworks, and payment plans. These measures aim to reduce the excessive use of fertilizers and pesticides, protect forests, and promote biodiversity. Through the efforts of governments in terms of enforcement, strictness, and accuracy, these detections can not only reduce environmental pollution but also improve the efficiency and quality of agricultural production and promote the sustainable development of the rural economy.

12.3.2 Industrial Land Use

The growth of land use in large cities around the world today is attributed to the rapid development of industrial areas. The spatial structure of urban industrial distribution shapes the urban spatial form and is closely related to other types of land use, such as transportation, economic activities, and housing. At the same time, the growth and expansion of urban population and land use reconfigures the spatial structure of industrial distribution.

Industrial land is not only an engine for economic development but also provides a large number of job opportunities, which promotes social stability. The development of industrial areas also promotes the construction and improvement of infrastructure and improves the overall service of the city. Besides, the intensive use of industrial land helps to improve the utilization efficiency of land resources and reduce the negative impact of urban expansion on the surrounding environment.

Industrial land use in cities is difficult to separate from residential, commercial, and transportation areas. Measuring the industrial land area within a city is a difficult task. Fortunately, the development of remote sensing technology has made this work possible. Using a collaborative approach involving multiple satellite datasets and machine learning techniques, Fig. 12.8 shows the changes in industrial and non-industrial land use within cities in ten major industrial countries over the past 20 years (Yoo et al., 2024).

Between 2000 and 2019, China's industrial land area grew most significantly, especially in coastal areas such as around Shanghai, while the United States, Japan, and developed European countries saw relatively small increases. By 2019, China's industrial land area had surpassed that of the United States, showing a significant trend in industrial land expansion between China and other Asian countries. The results reveal that the growth of industrial land in developed countries has slowed down, while that in developing countries has expanded rapidly.

Although the rapid expansion of industrial land can bring great benefits for economic development, its disadvantages, such as polluting the environment, discharging wastewater and exhaust gas, and creating harsh noise, should not be ignored. The negative externalities of urban industrial land expansion, such as air pollution and water pollution, are difficult to deal with through market mechanisms. The choice of residential areas and industrial land in cities requires government intervention.

Since the type of land use within a city is endogenously determined to some extent, the government should mitigate the negative externalities of urban industrial land when the situation worsens. The local government should plan the overall functions and zoning of the city to minimize the negative effects of industrial land. Policy tool sets, including environmental taxes and specific subsidies, are also desirable. Location-specific environmental taxes and labor subsidy policies can reduce the negative externality of industrial land and improve overall social welfare (Kyriakopoulou & Xepapadeas, 2016).

In addition, some specific administrative measures can help alleviate the negative impact of industrial land concentration. Specifically, the government can set up a special environmental protection fund to support enterprises in environmental protection technology transformation and pollution control projects. The government

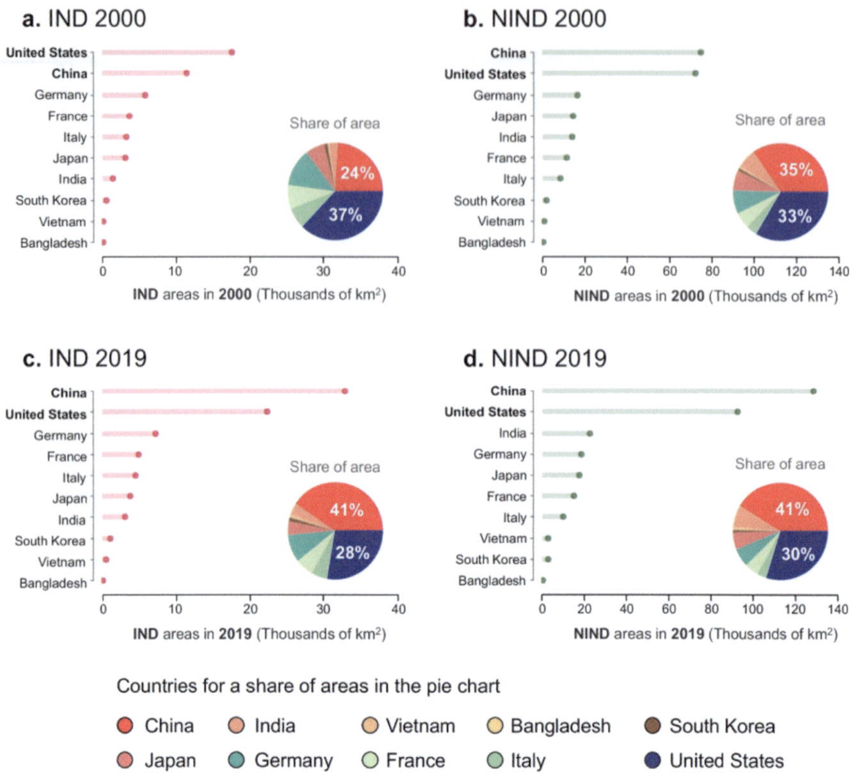

Fig. 12.8 Ten countries' urban industrial land and urban non-industrial land from years 2000 to 2019. *Note* It depicts the Total areas of urban industrial land (IND) and urban non-industrial land (NIND) for between year 2000 and 2019 across ten major countries. *Source* Yoo et al. (2024)

can ensure that the emissions of industrial enterprises meet environmental protection requirements by formulating strict emission standards and monitoring mechanisms. The government can also promote the establishment of centralized treatment facilities in industrial parks, such as sewage treatment plants and waste gas treatment equipment, to centrally treat industrial pollutants and reduce the impact on the environment. Through these specific measures, the government can effectively reduce the environmental pollution problems caused by the expansion of industrial land and promote the coordinated development of the economy and the environment.

12.3.3 Nature Reserve

The establishment of nature reserves around the world can be traced back to the late nineteenth century when some countries began to realize the importance of protecting natural resources. In 1872, the United States established the world's first national park, Yellowstone National Park, marking the birth of modern nature reserves. Since then, many nature reserves have been established around the world to protect rare animals and plants and unique natural landscapes.

In the mid-twentieth century, with the increase in environmental awareness, the international community began to promote the establishment and management of nature reserves, and organizations such as the United Nations Environment Program (UNEP) and the World Conservation Union (IUCN) played an important role in this process. Entering the twenty-first century, the number and area of nature reserves around the world have increased significantly, and the management and protection quality of nature reserves has gradually improved.

Recently, the coverage area of nature reserves around the world has continued to increase. At present, the protected area of terrestrial and inland water ecosystems has reached 22.5 million km^2, accounting for 16.64% of the global land area. The protected area of coastal waters and oceans has reached 28.1 million square kilometers, accounting for 7.74% of the global ocean area (Protected Planet, 2020). Since 2010, the area of newly added protected areas has exceeded 21 million km^2, accounting for 42% of the current total area of protected areas.

Despite the increase in coverage, the quality of protected area management and the number of marine protected areas still need to be improved. In addition, one-third of the world's key biodiversity areas have no protection at all, and less than 8% of the land is both protected and connected.

The main difficulties in protecting natural reserves include inadequate management quality, lack of funding, human interference, and climate change. These difficulties require global cooperation and coordination to ensure that nature reserves can effectively protect biodiversity and ecosystem services.

With the progress of industrialization and urbanization, the impact of nature reserves is growing, and people are gradually becoming aware of their importance. Nature reserves play an important role in protecting biodiversity by safeguarding habitats and species and maintaining ecological balance. They provide safe habitats for various plants and animals, preventing species from becoming extinct. By protecting these areas, the balance and health of the ecosystem are maintained, ensuring the continuation of biodiversity.

In addition, nature reserves help conserve water and purify the air, contributing to a healthy ecosystem. Ecosystems such as forests and wetlands in nature reserves help conserve water, regulate the water cycle, and prevent soil erosion. At the same time, vegetation in these areas can absorb invisible pollutants to purify the air.

Nature reserves also provide space for the public to relax and have fun, promoting the development of ecotourism. They offer the public an opportunity to get close to nature and promote the development of ecotourism. Visitors can engage in activities such as hiking, bird watching, and camping in these areas, which not only allows them to enjoy the beauty of nature but also raises awareness of environmental protection.

Based on the economic benefits and ecological functions of nature reserves described above, our society should invest sufficient resources to expand the area of nature reserves to protect biodiversity. When constructing nature reserves, governments or enterprises make decisions based on economic analysis. The marginal benefit of adding one hectare of nature reserve must equal the marginal cost to reach the socially optimal size. Intuitively, this is a simple calculation of economic theory. However, in practice, the social optimal status is hard to achieve because the economic benefits of nature reserves exist in the form of a positive externality.

It is easy to measure the economic cost of building nature reserves. The costs include the land cost of nature reserves, the cost of maintaining the operation of the reserves, and the opportunity cost of building reserves. The benefits, including the direct economic benefits generated by nature reserves through tourism and other means and the ecological and economic benefits of protecting ecological diversity, unfortunately, are difficult to measure directly.

Researchers have developed many methods to measure the economic value of natural reserves indirectly. For example, the contingent valuation method (CVM) can estimate the economic value of natural reserves by surveying individuals' willingness to pay for preservation or their willingness to accept compensation for loss, capturing non-use values like existence and bequest values. The travel cost method (TCM) assesses recreational value by analyzing visitors' travel expenses and visitation rates, while the hedonic pricing method (HPM) evaluates how proximity to natural reserves influences property prices, reflecting the value of environmental amenities. Additionally, ecosystem service valuation quantifies benefits such as carbon sequestration, water purification, and flood regulation, assigning monetary values to these services.

Through these methods, governments can conduct precise value assessments when planning natural reserves, ensuring that both ecological and economic dimensions of sustainability are integrated into decision-making processes.

12.4 Land Use Management: What to Do Next?

12.4.1 New Technologies in Land Use Management

With the acceleration of global urbanization and industrialization, land use management faces unprecedented challenges and opportunities. To achieve sustainable use of land resources, advanced technologies should be applied to improve land use efficiency and reduce negative environmental impacts. Several new technologies that play an increasingly important role in land use management are introduced below.

Remote sensing obtains surface information through equipment such as satellites and drones and can monitor land use changes in real-time and provide accurate data support. This technology can help governments and enterprises better plan and manage land resources and promptly discover and solve problems in land use. Governments around the world are using remote sensing technology to monitor deforestation, urban expansion, and changes in agricultural activities to formulate corresponding land use policies.

Smart city systems combine Internet of Things (IoT) and Geographic Information System (GIS) technologies to achieve intelligent management of land resources. Through sensors and data analysis, the system can monitor land use in real-time, optimize land use efficiency, and reduce resource waste.

Artificial intelligence and big data technologies are also playing an increasingly important role in land use management. By analyzing a large amount of land use data, AI can predict future land demand and change trends and help decision-makers develop scientific land use plans. Big data technology can integrate multi-source data, provide comprehensive land use information, and support precise management and decision-making. Through big data analysis, potential problems and opportunities in land use can be identified, thereby optimizing land use strategies.

Although new technologies have broad prospects in land use management, it is important to note that these new technologies also have some disadvantages. Table 12.1 summarizes the advantages and disadvantages of new technologies to help us understand the prospects of technology in land use management.

Table 12.1 Pros and cons of new technology in land use management

Technology	Advantages	Disadvantages
Remote Sensing	• Provides wide-range, high-resolution land cover and use data • Real-time monitoring of environmental changes, supports disaster warning and emergency response • Supports long-term monitoring and historical data analysis, aids in land use planning	• Data processing and analysis require high technical skills and equipment • High cost, especially for high-resolution satellite imagery • Affected by weather and atmospheric conditions, data may be incomplete or inaccurate
Smart City	• Achieves efficient management and optimization of urban land resources • Improves land use efficiency, reduces resource waste • Optimizes land use and environmental protection through smart sensors and data analysis	• Requires significant initial investment and technological infrastructure • Data privacy and security issues need special attention • Requires continuous maintenance and updates, outdated technology may affect system performance
AI and Big Data	• Provides deep data analysis and prediction, supports scientific decision-making • Identifies potential problems and optimization opportunities, enhances land management efficiency • Supports automation and intelligent management, reduces human errors	• Data quality and source reliability directly affect analysis results • Requires large data storage and computing resources • High technical complexity, implementation and maintenance require expertise

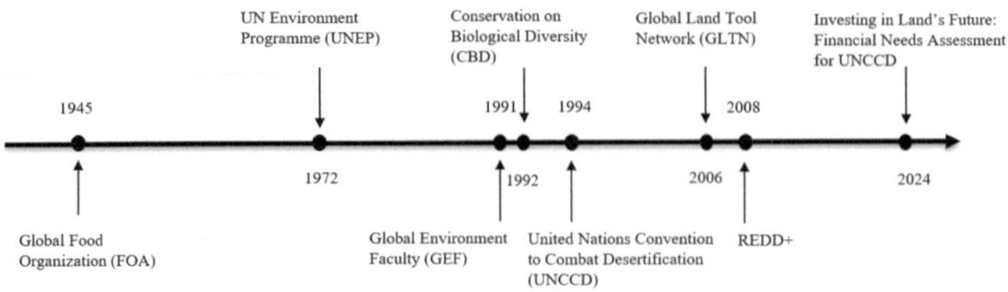

Fig. 12.9 International organizations and initiatives in land use management

12.4.2 Global Horizon in Land Use Management

Land use management requires cooperation. Even with abundant resources, it is still difficult for a single country, region, or organization to achieve sustainable land use. The transboundary issues related to land use, pollution externalities, and climate change are global and far beyond the capabilities of a single country or region.

Land use management from a global perspective is designed to meet the complex challenges brought about by globalization. Transboundary issues such as water management, air quality, and biodiversity conservation require international cooperation to solve. Through global coordination, countries can share best practices and advanced technologies to improve the efficiency and effectiveness of land management.

In addition, global changes such as climate change, population growth, and urbanization have a significant impact on land use. Land management from a global perspective can help countries jointly respond to these changes, formulate adaptable policies and measures, and ensure the sustainable use of land resources.

Global cooperation on land use management can help reduce pollution and improve land use efficiency. Fortunately, the international community has made a series of efforts. Since 1945, organizations and initiatives to promote sustainable land use development at the global level have gradually been established, as shown in Fig. 12.9. Land use management from a global perspective involves multiple international agreements, initiatives, and organizations, such as the Convention on Biological Diversity and the REDD + Initiative. These cooperation frameworks aim to promote sustainable land use and environmental protection through cross-border cooperation.

References

China Land and Resources Yearbook. (2022). *National Bureau of Statistics*. https://www.stats.gov.cn/

Food and Agriculture Organization of the United Nations (FAO). (2023). *FAO statistics*. https://www.fao.org/statistics/data-collection/general/en

Food and Agriculture Organization of the United Nations (FAO). (2024). *Land statistics 2001–2022. Global, regional and country trends.* https://www.fao.org/statistics/highlights-archive/highlights-detail/land-statistics-2001-2022.-global--regional-and-country-trends/en

Kyriakopoulou, E., & Xepapadeas, A. (2016). Atmospheric pollution in rapidly growing industrial cities: Spatial policies and land use patterns. *Journal of Economic Geography, 17*(3), 607–634.

National Aeronautics and Space Administration (NASA). (2019). *Tracking Amazon deforestation from above.* https://earthobservatory.nasa.gov/images/145988/tracking-amazon-deforestation-from-above

Our World in Data. (2024). *Agricultural production.* https://ourworldindata.org/grapher/oil-production-by-country

Poore, J., & Nemecek, T. (2018). Reducing food's environmental impacts through producers and consumers. *Science, 360*(6392), 987–992.

Protected Planet. (2020). *Protected Planet report 2020.* https://protectedplanetreport2020.protectedplanet.net/

Wuepper, D., Wiebecke, I., Meier, L., Vogelsanger, S., Bramato, S., Fürholz, A., & Finger, R. (2024). Agrienvironmental policies from 1960 to 2022. *Nature Food, 5*(4), 323–331.

Yoo, C., Xiao, H., Zhong, Q., & Weng, Q. (2024). Unequal impacts of urban industrial land expansion on economic growth and carbon dioxide emissions. *Communications Earth & Environment, 5*, 203.

Part IV
Society-Related Sustainability Issues

Chapter 13
Poverty and Inequality

To progress, we have to abandon the habit of reducing the poor to cartoon characters and take the time to really understand their lives, in all their complexity and richness.
Banerjee and Duflo (2011)

13.1 Understanding Poverty and Inequality

13.1.1 Definitions and Concepts

Poverty and inequality are critical issues that significantly impact the trajectory of sustainable development. Poverty is a situation where an individual does not have enough financial resources and necessities to maintain a basic standard of living. The first Sustainable Development Goal (SDG1) is dedicated to "End poverty in all its forms everywhere." This goal encompasses not only monetary poverty but also includes aspects such as access to basic services, ownership of assets, and vulnerability to disasters and environmental shocks.

There are two primary measures to evaluate poverty: absolute poverty and relative poverty. Absolute poverty is defined by the lack of basic human needs such as food, clean drinking water, sanitation facilities, and adequate shelter. It is typically measured against a fixed standard or poverty line, often based on income and basic needs.

For example, the international poverty line—which is presently set at $2.15 per day in 2017 purchasing power parity—is an essential indicator to track and evaluate extreme poverty worldwide (Jolliffe et al., 2022). This standard is based on the poverty lines of the world's poorest countries, which are typically established to cover the cost of a diet sufficient to satisfy basic daily caloric requirements. An individual living on less than $2.15 per day is classified as existing in extreme poverty, a designation that carries significant implications for global poverty alleviation efforts.

L. Zhang, *Sustainability Economics*, https://doi.org/10.1007/978-3-031-87829-9_13

The international poverty line is not just a statistical benchmark. It is a tool for policymakers, NGOs, and international organizations to track progress, allocate resources, and design strategies aimed at lifting the most vulnerable out of poverty.

On the other hand, relative poverty measures income inequality in relation to the average income of a society, often set at a threshold such as 50% or 60% of the median income. In other words, relative poverty is often measured by income-group-specific poverty lines that account for the varying needs and living standards across countries at different levels of economic development.

In simple terms, a person is considered to be living in relative poverty if nearly everyone else in the country resides in modern brick houses, while that person can only afford an open tent. However, this is not the case if everyone else also lives in open tents, such as in a nomadic tribe. Due to the lower levels of absolute poverty in wealthy nations, relative poverty is regarded as the most useful measure for determining poverty rates in these developed countries (Ravallion & Chen, 2011).

For instance, as countries like China have evolved from low-income to upper-middle-income status, their national poverty lines have increased, reflecting a relative concept of poverty. In 1998, when China was a low-income country, it used the international poverty line of $2.15 to measure poverty. However, as its economy grew, by 2010, when China became an upper-middle-income country, the poverty line was raised to $6.85. This adjustment reflects the changing standards of living and the relative poverty experienced by the population as the country's economic status improved. This approach allows for a more nuanced understanding of poverty that is sensitive to the economic context of each country, capturing those who, despite not living in absolute poverty, may still face significant economic challenges relative to their societal peers.

Figure 13.1 depicts the income-group-specific poverty line based on statistics from 2017. Countries classified as Low-Income Countries (LIC) have a per capita gross national income below $996, with a corresponding poverty line set at $2.15. Lower-Middle-Income Countries (LMIC) have a per capita gross national income ranging from $996 to $3896, with a poverty line of $3.65. The poverty line in Upper-Middle-Income Countries (UMIC) is $6.85, with per capita gross national income ranging from $3896 to $12,055. High-Income Countries (HIC), exceeding a per capita gross national income of $12,055, are designated a poverty line of $24.35.

The distribution of global poverty across various regions is illustrated in Fig. 13.2. Overall, global poverty is concentrated in Sub-Saharan Africa and Eastern and Southern Africa, where the number of poor people is much higher than in other regions. Additionally, South Asia and Western and Central Africa also have a high number of poor people.

Notably, poverty and sustainability are deeply interconnected. Poverty often forces individuals and communities to overexploit natural resources for survival, leading to environmental degradation. Conversely, environmental degradation can exacerbate poverty by reducing the productivity of natural resources that are essential for the livelihoods of the poor. This creates a vicious cycle that hinders sustainable development.

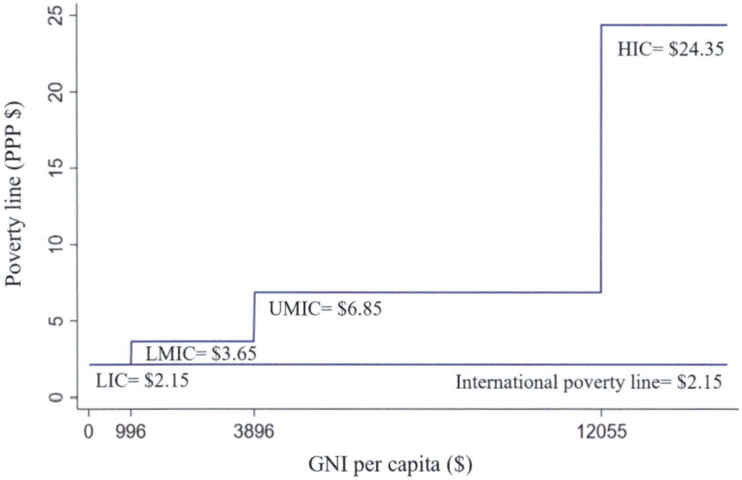

Fig. 13.1 The international poverty line and income-group-specific poverty line

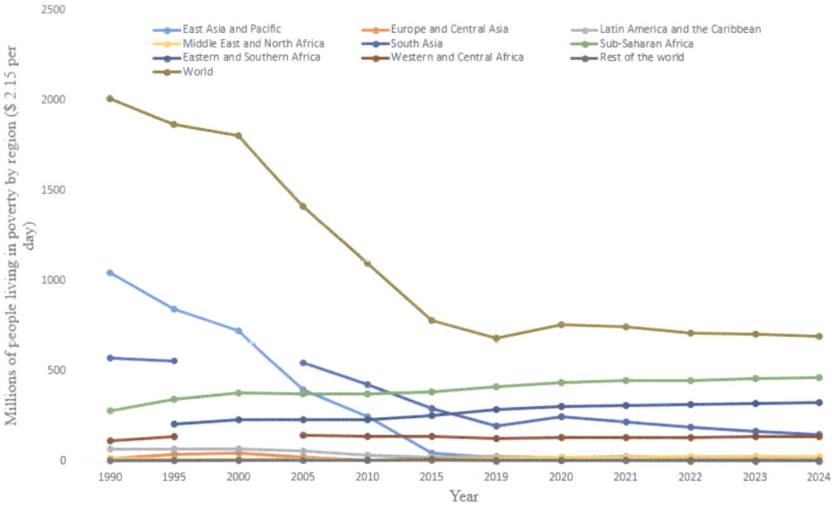

Fig. 13.2 Millions of people living in poverty by region ($2.15 per day). *Data source* Poverty, prosperity, and planet report (2024)

Inequality refers to the unequal distribution of resources, opportunities, and outcomes among individuals or groups within a society or between different societies. It encompasses both economic and social inequality. The unequal distribution of income and wealth within a population is referred to as economic inequality. In contrast, social inequality extends beyond economics to include disparities in power, opportunities, and social status, often affected by factors such as race, gender, and education.

Inequality can manifest in various aspects of life, including access to education, healthcare, political representation, and social services. It often perpetuates cycles of disadvantages that limit individual potential and societal progress. In addition, inequality is a pervasive issue that undermines long-term social and economic development. It can also hinder poverty reduction and erode individuals' sense of accomplishment and self-worth. Historically, the incomes of the bottom forty percent of the population have grown more slowly than the national average in the majority of nations.

The current state of global inequality is a pressing issue that demands our attention. The COVID-19 epidemic has had a huge influence on global inequality, reversing the good trend of lowering within-country inequality and resulting in the highest increase in between-country disparity in three decades. The incomes of the most impoverished 40% of the population have been disproportionately affected. Furthermore, one out of every six people worldwide has experienced discrimination, with women and people who are disabled being especially vulnerable. In 2022, the number of refugees reached a record high of 34.6 million, with an estimated 7000 migrant deaths reported worldwide. Addressing intra- and inter-country inequality is a difficult task that requires distributing resources equally, investment in education and skill acquisition, the implementation of social protection regulations, the fight against discrimination, the assistance of marginalized groups, and the promotion of international cooperation for equitable trade and financial systems.

13.1.2 Measurement and Indicators

Poverty and inequality require precise measurement for effective policymaking and intervention assessment. Scholars have created kinds of ways for measuring poverty, with the most used being the poverty line and the multidimensional poverty index (MPI) (Alkire & Foster, 2011; Gillie, 1996).

First, poverty can be identified by assessing which individuals or households fall below a certain poverty line (see Fig. 13.1), a threshold that represents the minimum income or resources needed to secure basic needs such as food, clean water, sanitation, and shelter. Then, poverty rates are calculated by determining the proportion of the population living below the defined poverty line. Poverty rates provide a snapshot of the economic well-being of a population and are a key indicator for tracking progress against poverty. They are used to inform policy decisions, allocate resources, and measure the success of poverty alleviation efforts over time.

As mentioned previously, poverty lines can be absolute, reflecting a fixed standard of living necessary to meet basic needs. Poverty lines can also be relative, reflecting a proportion of the median income or consumption expenditure of a population. Absolute poverty lines are set in real terms throughout time and geography, capturing deprivations in fundamental requirements like food, clothes, and housing. Relative

Table 13.1 Global Multidimensional Poverty Index Indicators and weights

Dimensions	Indicator	Weight
Health	Nutrition	1/6
	Child mortality	1/6
Education	Years of schooling	1/6
	School attendance	1/6
Standard of living	Cooking fuel	1/18
	Sanitation	1/18
	Drinking water	1/18
	Electricity	1/18
	Housing	1/18
	Assets	1/18

Source United Nations Development Programme (2024)

poverty lines, on the other hand, account for a lack of means to participate fully in one's society as it evolves, reflecting the unequal component of poverty.

Given the dynamic nature of economies and the cost of living, poverty lines must be regularly updated to maintain their relevance. The adoption of the 2017 Purchasing Power Parity (PPP) measures led to a revision of global poverty lines in 2022. The international poverty line, used to define extreme global poverty, was revised to US$2.15 from US$1.90. Similarly, poverty lines for LMIC moved to US$3.65 from US$3.20, and for UMIC to US$6.85 from US$5.50. These updates reflect changes in living standards and economic conditions, ensuring that poverty measures remain accurate and actionable.

In addition, the MPI offers a more comprehensive approach to poverty measurement. Unlike income-based measures, the MPI considers a range of dimensions that contribute to an individual's quality of life as illustrated in Table 13.1. These include:

- Health: Nutrition and child mortality rates.
- Education: Years of schooling and whether individuals have completed a certain level of education.
- Living Standards: Access to cooking fuel, sanitation, drinking water, electricity, housing, and assets such as telephones and bicycles.

By considering these three dimensions, the MPI provides a more nuanced picture of poverty that goes beyond income alone. It offers a more detailed understanding of the various deprivations faced by the poor.

Accurate identification of poverty is essential for several reasons. It allows for the targeting of resources and interventions to those most in need, ensuring that assistance reaches the populations where it is needed most. It also enables the monitoring and evaluation of poverty reduction strategies, providing a clear picture of which approaches are working and which need to be adjusted. Furthermore, accurate poverty identification is crucial for holding governments and organizations accountable for their commitments to reducing poverty and improving living standards.

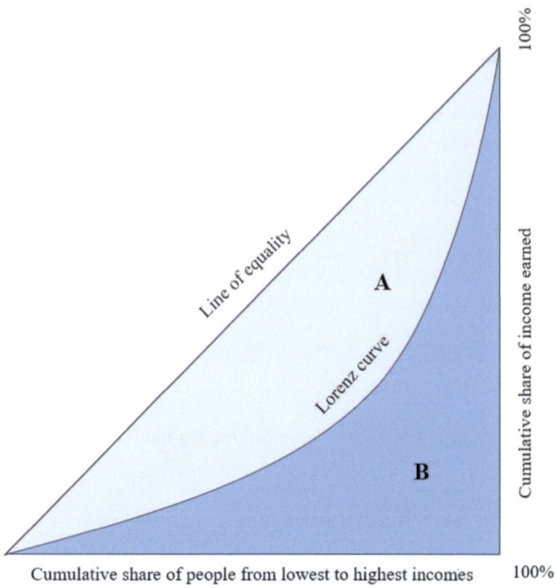

Fig. 13.3 The Gini coefficient

In summary, poverty identification is a multifaceted process that involves measuring income against established poverty lines and considering a range of deprivations that affect quality of life. By using a combination of the poverty line and the MPI, policymakers can gain a comprehensive understanding of poverty in their societies and develop targeted, effective strategies to address it.

Inequality can be quantified through various indices, with the Gini coefficient being the most widely recognized. The Gini coefficient is typically defined mathematically using the Lorenz curve, which illustrates the cumulative distribution of income within a population. As shown in Fig. 13.3, the y-axis represents the proportion of total income, while the x-axis represents the cumulative percentage of the population. A 45-degree line on this graph indicates perfect income equality.

The Gini coefficient is calculated as the ratio of the area between the line of equality and the Lorenz curve (denoted as area A) to the total area under the line of equality (areas A and B combined). In other words, the Gini coefficient is given by the formula:

$$G = \frac{A}{A + B} \tag{13.1}$$

The Gini coefficient measures the distribution of income or wealth within a population, ranging from 0 to 1. A Gini coefficient of 0 indicates perfect equality, where everyone has the same income or wealth, while a coefficient of 1 signifies perfect inequality, with one person possessing all the income or wealth and everyone else having none. As the Gini coefficient increases from 0 to 1, it reflects a rise in

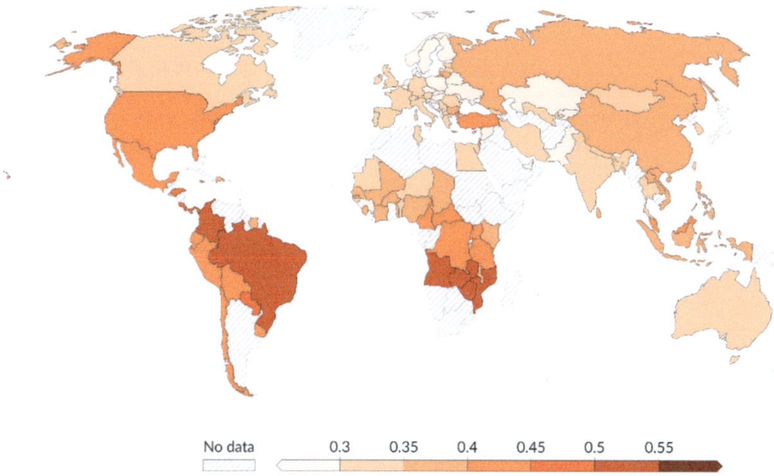

Fig. 13.4 Global Gini coefficient in 2023. *Note* The map is plotted by Our World in Data (2024). *Data source* World Bank (2024)

inequality. Higher Gini coefficients indicate a more unequal distribution, with larger disparities between the wealthy and the less wealthy.

Figure 13.4 shows the global Gini coefficient in 2023. The Gini coefficient can be used to understand the degree of economic inequality in a society and how it evolves over time, making it a useful tool for identifying trends in income concentration and assessing an economy's overall health in terms of equitable resource distribution.

Other indicators include the Palma ratio, which compares the income share of the top 10% to the bottom 40%, offering insights into the concentration of income at different ends of the spectrum. This ratio is particularly useful for understanding the extent of income polarization and for tracking changes in income distribution over time.

In all, these measurement tools are crucial for policymakers as they provide quantitative evidence of the extent and nature of poverty and inequality. They enable the assessment of the effectiveness of policies and programs aimed at reducing disparities and improving the well-being of all members of society. Monitoring these indicators allows governments and international organizations to alter their strategy to ensure that economic growth is inclusive and that development gains are distributed more equally across the population.

13.1.3 Evolution of Poverty and Inequality Over Time

The evolution of poverty and inequality is a complex narrative that unfolds across historical, economic, and social landscapes. Understanding how these issues have changed over time is crucial for developing effective strategies to address them.

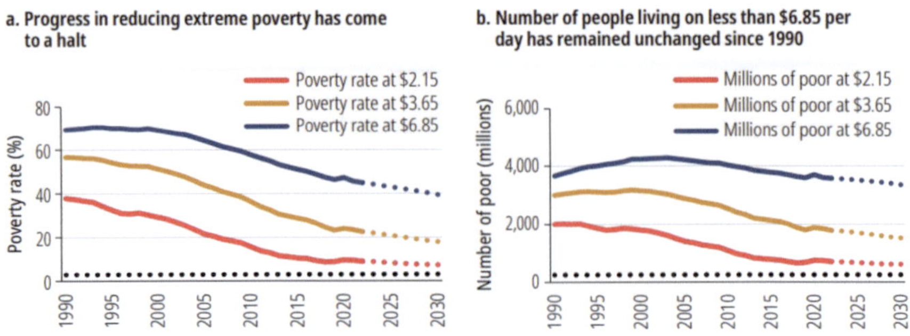

Fig. 13.5 Global extreme poverty reduction. *Source* World Bank (2024)

The history of poverty is as old as civilization itself. Throughout most of human history, poverty was a universal condition for the majority of the population. It was only with the onset of the Industrial Revolution that significant changes began to alter the economic landscape, leading to increased productivity and improvements in living standards. However, this progress was uneven. Poverty remained a persistent challenge, particularly in regions that were slow to industrialize or lacked access to the benefits of economic growth.

As shown in Fig. 13.5, efforts to reduce poverty and income inequality have achieved significant progress in recent decades. Extreme poverty has been reduced at a global level, with the most significant improvements seen in South Asia and parts of Africa. As for multidimensional poverty, the 2024 Global MPI report reveals that 1.1 billion out of 6.3 billion people live in acute multidimensional poverty (United Nations Development Programme, 2024), with over half of them being children. Common deprivations include inadequate housing, sanitation, electricity, cooking fuel, nutrition, and school attendance. However, there is still much work to be done to address poverty and ensure everyone has the chance to lead healthy and fulfilling lives.

Inequality, too, has seen a dynamic evolution. In many societies, the shift from agrarian to industrial economies led to a temporary increase in inequality as wealth and power became concentrated in the hands of a few. Over time, as labor movements gained traction and social policies were implemented, inequalities were somewhat mitigated in many developed countries. However, in recent decades, there has been a trend toward increasing income and wealth inequality, both within and between countries.

Several factors contribute to the evolution of poverty and inequality. First, globalization has played a significant role in this process. On one hand, it has facilitated economic growth and lifted millions out of poverty, especially in emerging economies. On the other hand, it has also led to increased competition and job displacement, contributing to income instability and wage stagnation for certain segments of the workforce, particularly in developed countries.

Second, technological advancements have been a double-edged sword in the context of poverty and inequality. On the one hand, technology has increased productivity and created new industries, which can drive economic growth and reduce poverty. On the other hand, it has also led to job displacement in sectors that are unable to adapt, and it has the potential to exacerbate inequality if the benefits of technological progress are not widely shared.

Third, government policies have been instrumental in shaping the trajectory of poverty and inequality. Social security nets, fair taxation, and improvements in education and healthcare have been effective in reducing poverty and mitigating inequality in many countries. Conversely, inadequate policy responses, corruption, and weak governance can entrench poverty and exacerbate inequality.

In recent years, the COVID-19 pandemic has posed new challenges to the fight against poverty and inequality. The pandemic has disproportionately affected the most vulnerable populations, reversing gains in poverty reduction and widening income disparities. It has also highlighted the deep inequalities in access to healthcare, digital technology, and social protection.

As we look to the future, the evolution of poverty and inequality will continue to be shaped by a range of factors, including economic growth patterns, technological change, demographic shifts, and policy decisions. To effectively address these issues, it is essential to learn from the past and to adapt strategies in response to emerging trends and challenges. This includes investing in inclusive growth, ensuring that the benefits of technological progress are broadly shared, and implementing policies that promote social protection and reduce inequalities.

The evolution of poverty and inequality is a story of human progress and struggle. It is a tale of how societies have grappled with the fundamental questions of fairness and well-being. As we move toward a more equitable and sustainable society, understanding this history and learning from it will be key to shaping a future where everyone can thrive.

13.2 Causes of Poverty and Inequality

13.2.1 Environmental Factor

Environmental factors play a crucial role in perpetuating poverty and inequality. Less-favored agricultural lands, natural resource dependency, and the impacts of natural disasters and climate change create significant barriers for poor populations. These challenges are compounded by issues such as water scarcity and air pollution, which disproportionately affect the most vulnerable, trapping them in a cycle of poverty and environmental degradation.

Less-favored Agricultural Lands: Poor rural populations in developing countries often reside in areas with severe biophysical constraints on production and limited market access. These regions, termed less-favored agricultural lands, are prone to low productivity and degradation, which can entrap populations into poverty (Barbier & Hochard, 2019).

Natural Resource and Ecosystem: The poor are often concentrated in fragile environments where livelihoods are intimately dependent on natural resource use and ecosystem services, leading to overexploitation. This overexploitation can further degrade the environment, creating a cycle of poverty and environmental decline (Duraiappah, 1998). For example, the loss of critical ecosystems such as mangroves, coral reefs, and tropical forests, which have been significantly degraded or lost in many developing regions, directly impacts the livelihoods of the rural poor who depend on these ecosystems for subsistence and income.

Natural Disasters and Climate Change: Natural disasters like earthquakes, droughts, and floods can lead to property damage and hardship. Climate change also poses a fundamental risk to poverty reduction efforts, with more frequent and intense weather events. These events can severely hinder livelihoods and significantly impede poverty reduction efforts. Notably, the nexus between carbon emissions and job creation, particularly in manufacturing, presents a challenge for developing countries aiming for rapid economic growth. While these jobs can lift people out of poverty, they often come with high carbon emissions that contribute to climate change, which in turn can exacerbate poverty.

Water Scarcity and Air Pollution: As shown in Fig. 13.6, water scarcity and air pollution also disproportionately affect the poor, limiting their access to clean water and increasing health risks, which can trap them in poverty (Rentschler & Leonova, 2023). Both indoor and outdoor air pollution have significant health impacts, particularly on the poor who may not have the means to mitigate exposure, leading to increased health costs and reduced productivity, which can perpetuate poverty.

These environmental factors are interconnected and often exacerbate one another, creating a complex web of challenges that contribute to poverty. Addressing these factors requires a multifaceted approach that considers the unique environmental and geographical conditions of disadvantaged regions, as well as the need for sustainable resource management and climate change mitigation strategies.

Fig. 13.6 Drought and agriculture

13.2.2 Social Factor

Social factors are deeply intertwined with poverty and inequality, influencing individuals' access to opportunities and resources. Limited education, poor health, and gender inequality are significant contributors, as they restrict economic mobility and perpetuate disadvantage. Additionally, social exclusion, cultural norms, and the dynamics of migration and urbanization further exacerbate the challenges faced by marginalized communities.

Education and Human Capital: Limited access to education is a significant social factor contributing to poverty. Education empowers individuals, enhances earning potential, and fosters social mobility. Conversely, a lack of education traps people in a cycle of poverty, as they are unable to access higher-paying jobs or engage in income-generating activities that require specific skills or knowledge.

Health and Nutrition: Poor health and malnutrition can directly lead to poverty by reducing an individual's ability to work and earn income. Health issues can also result in significant medical expenses, which can push families into poverty.

Gender Inequality: Gender disparities can contribute to poverty, particularly in regions where women have limited access to resources, education, and decision-making power. Gender inequality can restrict women's ability to access economic opportunities, which in turn perpetuates poverty.

Social Exclusion and Discrimination: Social exclusion and discrimination based on race, ethnicity, caste, or social status, which limit access to resources and opportunities, are significant social factors contributing to poverty. Discrimination can prevent individuals from accessing education, jobs, and social services, thereby increasing the risk of poverty.

Migration and Urbanization: Social factors influencing migration and urbanization can contribute to poverty. Rural–urban migration, often driven by the search for better economic opportunities, can lead to overcrowding in urban areas, leading to slum conditions and limited access to basic services, thus trapping people in poverty.

Addressing these socioeconomic issues is critical to breaking the cycle of poverty, and it necessitates a multifaceted approach that includes education, health interventions, gender equality, and social inclusion policies. By tackling these social determinants of poverty, societies can create more equitable opportunities for all, reducing the risk of poverty and enhancing the well-being of vulnerable populations.

13.2.3 Political Factor

Political factors significantly impact the distribution of resources and opportunities, often exacerbating poverty and inequality. Poor governance, corruption, and political instability can divert resources away from poverty alleviation efforts. The lack of political representation for the poor and the capture of resources by elites further entrench disparities, while international relations and global politics can either mitigate or worsen these issues depending on their structure and implementation.

Governance and Corruption: Poor governance and rampant corruption can lead to the misallocation of resources and the diversion of public funds away from poverty alleviation efforts. Corruption can undermine the effectiveness of social programs and economic policies, as these programs may be hijacked by elites for personal gain.

Political Instability and Conflict: Political instability and conflict can disrupt economic activities, leading to a lack of investment in poverty-reducing initiatives. Conflicts often divert resources away from social welfare and can lead to long-term economic stagnation, which perpetuates poverty.

Lack of Representation: The lack of political representation and voice for the poor can result in policies that overlook the needs of the most vulnerable populations. When the poor are excluded from the political process, their concerns and interests are often ignored, leading to policies that do not address the root causes of poverty.

Capture by Elites: Political factors can influence how resources are allocated, often leading to the capture of resources by elites. This can result in the concentration of wealth and power, leaving the poor with limited access to resources and opportunities.

International Relations and Global Politics: International relations and global politics can also impact poverty. Trade policies, foreign aid, and international agreements can either alleviate or exacerbate poverty, depending on how they are structured and implemented.

13.2.4 Poverty Trap and Middle-Income Trap

Understanding the causes of poverty and inequality is crucial, but it is equally important to recognize the specific economic mechanisms that can perpetuate these conditions. Two such mechanisms are the poverty trap and the middle-income trap. These traps highlight the structural and systemic challenges that can hinder economic development and social progress. By examining these traps, we can gain deeper insights into the complex dynamics that keep individuals and countries from achieving sustainable growth and prosperity.

A poverty trap is a situation where individuals or communities are unable to escape poverty due to a combination of factors that reinforce their impoverished state. This cycle is self-reinforcing, making it extremely hard for people to break out of poverty. Key factors contributing to poverty traps include lack of access to education, poor health, limited assets, and financial constraints.

First, individuals are unable to acquire the skills needed for higher-paying jobs without access to quality education, leading to limited economic opportunities and perpetuating poverty. Second, health issues can reduce an individual's ability to work and earn income, leading to higher medical expenses and further entrenching poverty. Third, individuals in poverty often lack the financial resources to invest in assets that could increase their productivity, such as tools or machinery. Fourth, the poor often

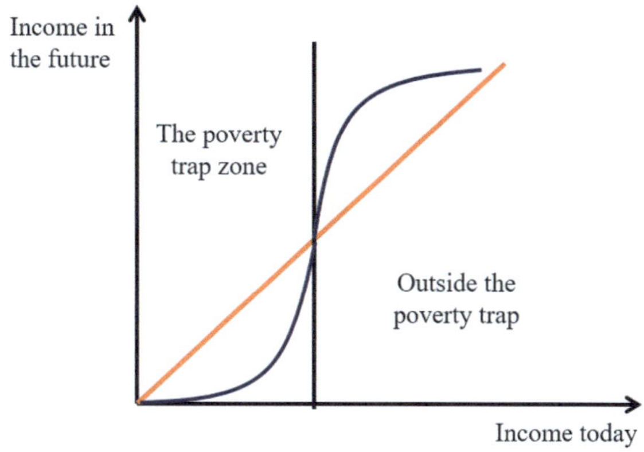

Fig. 13.7 The S-curved poverty trap illustrated in Banerjee and Duflo (2011)

lack collateral, making it difficult for them to obtain loans from banks, which further limits their economic opportunities.

For example, in a rural village where most residents live below the poverty line, children may forgo education to help their families by working in fields or doing household chores. Without education, these children grow up to have limited skills and job opportunities, often continuing to work in low-paying, labor-intensive jobs. This perpetuates the cycle of poverty. Banerjee and Duflo (2011) developed a S-curved shape to demonstrate the nutrition-based trap: In sub-Saharan African countries, Malaria-infested countries are stuck in the poverty trap zone (left of the curve in Fig. 13.7), as their malaria-depleted workers are less productive, and thus has no money to treat malaria.

The middle-income trap describes a situation where countries struggle to attain high-income status despite economic growth and falling poverty rates. These countries face significant challenges related to economic growth, wage competition, and innovation, often relying on policies that are based on superficial measures of economic efficiency.

Key characteristics of the middle-income trap include diminishing returns to capital investment, exhaustion of cheap labor and imitation gains, and lack of innovation. First, as countries grow wealthier, the returns on capital investment tend to diminish, making it harder to sustain high growth rates. Second, middle-income countries often rely on cheap labor and the imitation of technologies from high-income countries. However, as these advantages diminish, countries need to transition to more innovative and productive economic structures. Third, many middle-income countries struggle to move from imitation to innovation, which is necessary to achieve high-income status.

By understanding these traps, policymakers and stakeholders can develop more effective and targeted interventions to promote sustainable development and social equity.

13.3 Consequences of Poverty and Inequality

13.3.1 Environmental Impact

Poverty and inequality are closely linked to environmental degradation. Impoverished communities often lack the resources and opportunities to engage in sustainable practices. As a result, they may resort to environmentally harmful activities to meet their immediate needs. For example, deforestation for firewood, slash-and-burn agriculture, and overfishing are common practices among impoverished populations. These activities can lead to soil erosion, loss of biodiversity, and depletion of natural resources, which further entrench poverty and inequality.

Notably, when communities are trapped in poverty, they may overexploit natural resources to survive. This overexploitation can have several detrimental effects on the environment. First, impoverished communities often rely on forests for fuel, building materials, and agricultural land. Nevertheless, deforestation can lead to habitat loss, reduced biodiversity, and increased carbon emissions, contributing to climate change. Second, unsustainable agricultural practices, such as overgrazing and monocropping, can degrade soil quality. This reduces agricultural productivity and can lead to desertification, making it even harder for communities to sustain themselves. Third, overuse of water resources for irrigation and domestic needs can deplete local water supplies. This can lead to conflicts over water access and further exacerbate poverty and inequality. Fourth, impoverished communities may lack access to proper waste management systems, leading to pollution of air, water, and soil. This pollution can have severe health impacts and reduce the quality of life for these communities.

Economic and social inequalities can also contribute to environmental degradation. Wealthier individuals and corporations often have greater access to resources and political influence, allowing them to exploit natural resources without facing significant consequences. This can lead to environmental degradation that disproportionately affects poorer communities, who may lack the means to protect their environment or advocate for their rights.

Poverty and inequality exacerbate vulnerability to climate change. Poor populations frequently reside in locations prone to climate-related disasters such as floods, droughts, and hurricanes. They may lack the resources to adapt to these changes, leading to loss of livelihoods, displacement, and further entrenchment of poverty. Additionally, climate change can exacerbate existing inequalities by disproportionately affecting those who are already marginalized. As shown in Fig. 13.8, Cuni-Sanchez et al. (2025) investigate the influence of site and household wealth on the proportion of adaptation responses to climate change used by each household in ten African mountain regions. They find a clear disparity in the ability of households to adapt based on their wealth. Wealthier households have more resources to implement a variety of adaptation strategies, while poorer households are more limited in their adaptive capacity.

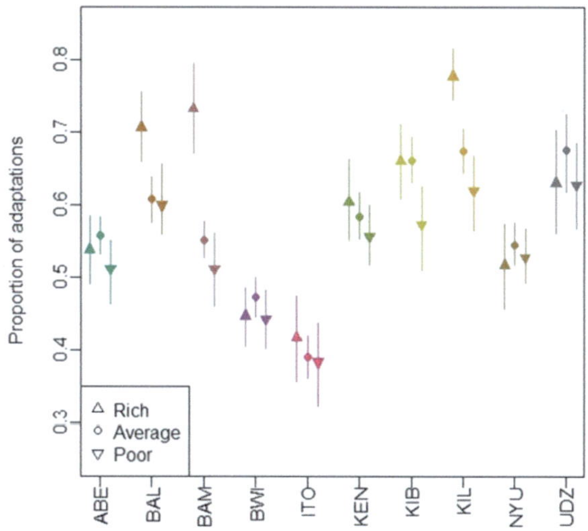

Fig. 13.8 Influence of site and household wealth on the proportion of adaptation responses to climate change. *Source* Cuni-Sanchez et al. (2025)

13.3.2 Social Impact

Poverty and inequality have profound social impacts that can hinder sustainable development. One of the most significant social consequences is the perpetuation of social exclusion and marginalization. Individuals and communities living in poverty often face barriers to accessing essential services such as education, healthcare, and social protection. This lack of access can lead to a cycle of poverty, where the disadvantaged are unable to improve their circumstances, further entrenching inequality.

Education is a critical area where the social impacts of poverty and inequality are evident. Children from impoverished families often have limited access to quality education, which affects their future employment opportunities and earning potential. This educational disparity contributes to the intergenerational transmission of poverty, as children who grow up in poverty are more likely to remain poor in adulthood. Additionally, the lack of education can limit individuals' ability to participate fully in society, reducing social cohesion and increasing the risk of social unrest.

Healthcare is another domain where the social impacts of poverty and inequality are stark. Impoverished communities often lack access to adequate healthcare services, leading to higher rates of illness and mortality. Poor health can reduce individuals' ability to work and support their families, exacerbating poverty. Furthermore, health disparities can lead to increased healthcare costs for society as a whole, straining public health systems and reducing the overall quality of life.

Social inequality also affects housing and living conditions. Impoverished individuals are more likely to live in substandard housing with inadequate access to clean water, sanitation, and other basic services. These poor living conditions can result in a range of health issues and reduce individuals' ability to participate in economic and

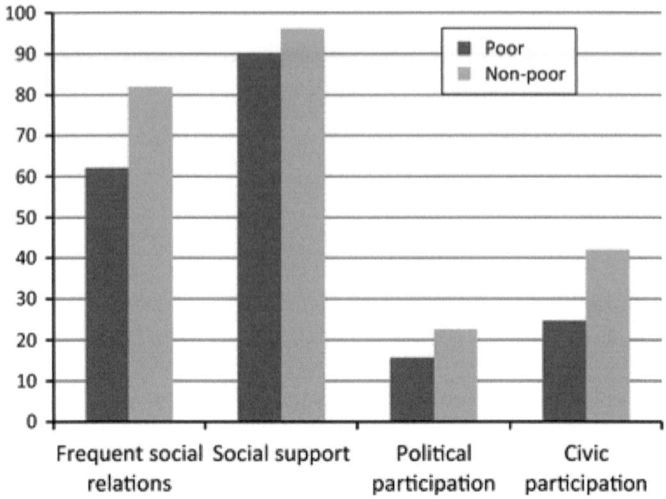

Fig. 13.9 The relation between poverty and social relations/participation in Sweden. *Source* Mood and Jonsson (2016)

social activities. For example, Mood and Jonsson (2016) investigate the social consequences of poverty using longitudinal data from the Swedish Level-of-Living Survey. As shown in Fig. 13.9, they find that impoverished individuals have poorer social relationships and lower participation in political life and organizational activities. In addition, inadequate housing can contribute to social instability, as marginalized communities may feel neglected and disenfranchised.

The social impacts of poverty and inequality extend to the labor market as well. Impoverished individuals often face discrimination and limited opportunities in the job market, leading to higher unemployment rates and lower wages. This economic marginalization can reduce social mobility and increase the gap between the rich and the poor. Additionally, job insecurity and poor working conditions can lead to stress and mental health issues, further exacerbating social inequalities.

In summary, poverty and inequality have far-reaching social impacts that can undermine sustainable development. By addressing these issues through inclusive policies and programs, societies can promote social cohesion, improve health and education outcomes, and create more equitable opportunities for all individuals. This, in turn, can contribute to a more sustainable and just world.

13.3.3 Economic Impact

Poverty and inequality are closely linked to economic growth. Marrero and Servén (2022) revisit the links between inequality, growth, and poverty. Analyzing a panel of 158 countries from 1960 to 2010, they find that the relationship between economic

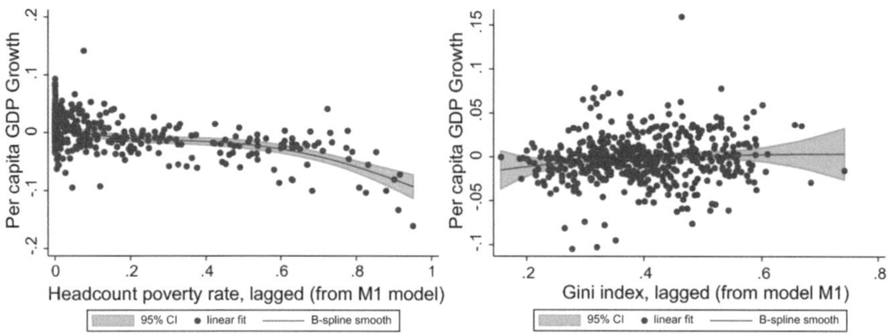

Fig. 13.10 Poverty, inequality and growth: nonlinear nonparametric estimates. *Source* Marrero and Servén (2022)

growth and poverty is consistently negative, as shown in Fig. 13.10. Specifically, a 10-percentage-point decrease in the headcount poverty rate is associated with an increase in per capita GDP of between 0.5 and 1.2% per year. In contrast, the relationship between economic growth and inequality is empirically fragile, indicating that it can be either positive or negative.

One of the primary ways in which poverty affects the economy is through reduced productivity. Individuals living in poverty often lack access to education, healthcare, and other critical services, which limits their ability to develop skills and maintain good health. This, in turn, reduces their productivity and ability to contribute effectively to the economy. When a significant percentage of the population is unable to participate fully in economic activities, the overall economic output is diminished, slowing down growth and development.

Moreover, economic inequality can lead to inefficient allocation of resources. Wealth and resources are frequently concentrated in the hands of a small group of people in highly unequal countries, while the majority of the population struggles to meet basic demands. This concentration of wealth can result in underinvestment in critical areas such as education, healthcare, and infrastructure, which are essential for long-term economic growth. When resources are not distributed equitably, it can stifle innovation and limit the potential for economic diversification, making the economy more vulnerable to shocks and less resilient.

Another significant economic impact of poverty and inequality is the creation of barriers to social mobility. In societies with high levels of inequality, individuals from lower-income backgrounds often face significant obstacles in accessing opportunities for advancement. This lack of social mobility can perpetuate cycles of poverty, as individuals are unable to improve their economic status despite their efforts. The resulting economic stagnation can lead to a lack of dynamism in the labor market, reducing the overall competitiveness of the economy.

Furthermore, poverty and inequality can lead to increased public expenditure on social welfare programs. Governments may need to devote significant resources to support disadvantaged populations through social safety nets such as unemployment compensation, dietary assistance, and healthcare subsidies. While these programs are essential for alleviating immediate suffering, they can place a significant strain on public finances, limiting the government's ability to invest in other areas critical for economic development. High levels of public debt and fiscal deficits can also result from increased social spending, which can have long-term implications for economic stability.

The economic impacts of poverty and inequality also extend to the private sector. Businesses operating in highly unequal societies may face a limited consumer base, as a large portion of the population lacks the purchasing power to buy goods and services. This can constrain market growth and reduce profitability for companies. Investors may be reluctant to invest in regions with significant social and economic disparities, further limiting economic opportunities and growth.

Addressing these issues is essential for promoting inclusive economic growth and ensuring that all individuals have the opportunity to contribute to and benefit from economic development. By enacting policies that reduce poverty and inequality, societies can create a more fair and prosperous future for all.

13.4 Strategies for Reducing Poverty and Inequality

Reducing poverty and inequality is vital for sustainable development, and it requires a multifaceted approach that integrates policy, environmental, social, and political strategies.

First of all, inclusive economic policies, such as progressive taxation and social safety nets, play a vital role in redistributing wealth and providing financial support to vulnerable populations. Furthermore, investing in education and skill development provides individuals with the tools they need to obtain higher-paying jobs, thus breaking the cycle of poverty. Universal healthcare access further alleviates the financial burden of medical expenses, ensuring that individuals can maintain their health and productivity.

Environmental strategies are equally important. Sustainable resource management practices in agriculture, forestry, and fisheries help preserve natural resources while providing livelihoods for impoverished communities. Programs promoting sustainable farming techniques, reforestation, and responsible fishing can mitigate environmental degradation and support long-term economic stability. Transitioning to renewable energy sources not only creates job opportunities but also reduces energy costs for low-income households, with government support through subsidies and training programs being essential for this transition.

Social strategies, such as social protection programs and community development projects, provide a safety net for individuals experiencing economic difficulty and uplift entire communities. Social protection programs, including unemployment benefits and pensions, prevent people from falling into poverty during crises and support their recovery. Community development projects improve infrastructure, housing, and public services, creating jobs and fostering social cohesion.

Political strategies are fundamental to ensuring the effectiveness of these efforts. Good governance and anti-corruption measures are essential for transparent and accountable resource allocation, ensuring aid reaches those in need. Participatory decision-making processes that involve marginalized communities ensure that their needs and perspectives are considered in policy development, resulting in more effective and equitable solutions.

In summary, tackling poverty and inequality demands a comprehensive strategy that integrates economic, environmental, social, and political dimensions. These combined efforts pave the way for a sustainable and just future.

Case

Solar Photovoltaic Interventions for Poverty Alleviation in China

Since 2013, China has implemented a large-scale initiative to deploy solar photovoltaic (PV) projects in rural areas to alleviate poverty. This strategy, known as the Solar Energy for Poverty Alleviation Program (SEPAP), seeks to add more than 10 GW of solar power by 2020, benefiting over 2 million people. The program promotes solar installations in high-poverty rural villages through three types of projects: village-level arrays, village-level joint construction arrays, and rooftop installations targeted at poor villagers.

Zhang et al. (2020) observed that the PV poverty alleviation pilot strategy boosts per-capita disposable income in a county by roughly 7%-8%, based on a panel dataset of 211 pilot counties that received targeted PV investments between 2013 and 2016. The effect is favorable and considerable in the first year of policy implementation, and it more than doubles in the next two to three years. Notably, the PV poverty reduction effect is more pronounced in poorer places, particularly in Eastern China.

The success of SEPAP can be attributed to several factors. First, the program provides a direct income effect by distributing earnings from solar installations to eligible poor families. Second, it provides indirect benefits by improving access to knowledge and information, resulting in increased revenue potential and business growth. However, challenges such as rent-seeking behavior and inadequate data sharing need to be addressed to ensure the program's effectiveness.

This case highlights the potential of renewable energy interventions in reducing poverty and inequality, demonstrating that targeted policies can have a significant and lasting impact on rural communities.

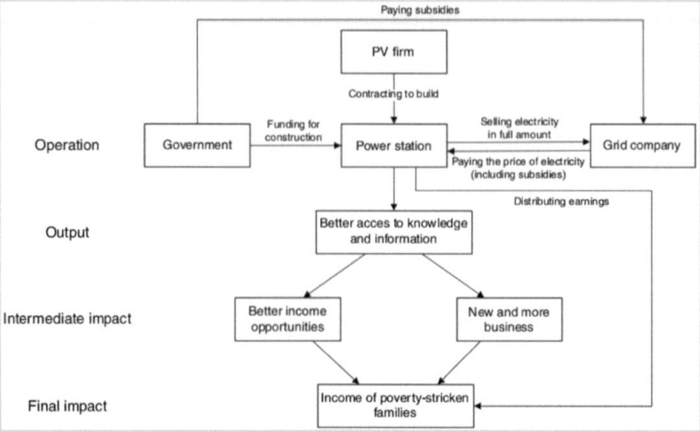

Fig. 13.11 The ways in which the SEPAP influences household income. *Source* Zhang et al. (2020)

References

Alkire, S., & Foster, J. (2011). Counting and multidimensional poverty measurement. *Journal of Public Economics, 95*(7–8), 476–487.

Banerjee, A. V., & Duflo, E. (2011). *Poor economics: A radical rethinking of the way to fight global poverty*. Public Affairs.

Barbier, E. B., & Hochard, J. P. (2019). Poverty-environment traps. *Environmental and Resource Economics, 74*(3), 1239–1271.

Cuni-Sanchez, A., Aneseyee, A. B., Baderha, G. K., Batumike, R., Bitariho, R., Imani, G., Jha, N., Kaganzi, K. R., Kaplin, B. A., Klein, J. A., & Zafra-Calvo, N. (2025). Perceived climate change impacts and adaptation responses in ten African mountain regions. *Nature Climate Change*, 1–9.

Duraiappah, A. K. (1998). Poverty and environmental degradation: A review and analysis of the nexus. *World Development, 26*(12), 2169–2179.

Gillie, A. (1996). The origin of the poverty line. *The Economic History Review, 49*(4), 715–730.

Jolliffe, D. M., Atamanov, A., Lakner, C., Mahler, D. G., & Baah, S. K. T. (2022). Assessing the impact of the 2017 PPPs on the international poverty line and global poverty. *World Bank Washington, DC.*

Marrero, G. A., & Servén, L. (2022). Growth, inequality and poverty: A robust relationship? *Empirical Economics, 63*(2), 725–791.

Mood, C., & Jonsson, J. O. (2016). The social consequences of poverty: An empirical test on longitudinal data. *Social Indicators Research, 127*, 633–652.

Our World in Data. (2024). *Income inequality: Gini coefficient, 2023*. https://ourworldindata.org/grapher/economic-inequality-gini-index

Ravallion, M., & Chen, S. (2011). Weakly relative poverty. *Review of Economics and Statistics, 93*(4), 1251–1261.

Rentschler, J., & Leonova, N. (2023). Global air pollution exposure and poverty. *Nature Communications, 14*(1), 4432.

United Nations Development Programme. (2024). *2024 Global Multidimensional Poverty Index (MPI).* https://hdr.undp.org/content/2024-global-multidimensional-poverty-index-mpi#/indicies/MPI

World Bank. (2024). *Poverty, prosperity, and planet report 2024.* https://www.worldbank.org/en/publication/poverty-prosperity-and-planet

Zhang, H., Wu, K., Qiu, Y., Chan, G., Wang, S., Zhou, D., & Ren, X. (2020). Solar photovoltaic interventions have reduced rural poverty in China. *Nature Communications, 11*(1), 1969.

Chapter 14
Environmental Justice

No man is an island, entire of itself; every man is a piece of the continent, a part of the main.
"Devotions upon Emergent Occasions", John Donne (1623)

14.1 History of Environmental Justice

Despite the global endorsement of the Sustainable Development Goals (SDGs), there has been a notable increase in environmental justice struggles across various regions. These struggles are not merely isolated instances of injustice; rather, they reflect interwoven systems of oppression that benefit a select few while imposing suffering on the majority and jeopardizing ecological integrity. This chapter aims to illuminate the often-overlooked social dimension of sustainability and its intrinsic connections to human rights and environmental justice. Through an examination of diverse case studies from around the globe, this chapter underscores the critical need to transcend the fragmentation present within these social movements. By doing so, we can formulate comprehensive solutions that advance justice and safeguard the planet's ecosystems amid escalating economic and ecological crises.

Environmental justice is intrinsically linked to Sustainable Development Goal 16 (SDG 16), which seeks to foster peaceful and inclusive societies that promote sustainable development, ensure access to justice for all, and build effective, accountable and inclusive institutions at various levels. The principles of environmental justice align with SDG 16 by advocating for the fair distribution of environmental advantages and disadvantages, particularly emphasizing that marginalized communities should not bear a disproportionate burden of environmental hazards. This alignment underscores the importance of integrating justice and equity into sustainability efforts to achieve broader social and environmental goals.

Environmental sustainability is increasingly becoming a central determinant of the success or failure of comprehensive sustainability efforts. However, there is a growing recognition of the limitations that ecosystems place on economic activities and human existence, rooted in the principles of sustainability. This recognition

L. Zhang, *Sustainability Economics*, https://doi.org/10.1007/978-3-031-87829-9_14

brings to complex and inevitable issues related to the equitable treatment and meaningful involvement of all individuals, irrespective of income, race, color, national origin, tribal affiliation, or disability, in decision-making processes and federal activities and initiatives that significantly influence human health and the environment, a concept known as "environmental justice" which has been adopted by the U.S. Environmental Protection Agency (EPA) (Edwards & Darnall, 2010).

Environmental justice includes a broad range of issues, including air and water pollution, hazardous waste disposal, access to green spaces, and the ramifications of climate change. It ensures that all communities are represented in environmental decision-making processes and that their concerns are adequately addressed. By promoting the fair distribution of environmental advantages and disadvantages, environmental justice aims to create healthier, more sustainable communities for all. The notions of "equity" and "justice" are consistently listed as indispensable components for sustainability alongside environmental and economic considerations (Holdren, 2008). Equity involves the fair distribution of resources and opportunities, while justice encompasses the broader ethical and moral dimensions of fairness and rights. Although the justice element is frequently supported by the sustainability stool in public policy, in practice, it remains insufficiently explored or addressed.

To sum up, achieving environmental justice is crucial for the broader success of sustainability efforts. By recognizing and addressing the limitations that ecosystems place on economic activities and human existence, and by ensuring the fair treatment and meaningful involvement of all individuals, we can create more equitable and sustainable environments.

14.1.1 Early Beginnings and Civil Rights Roots

The notion of environmental justice is deeply intertwined with the civil rights movement of the 1960s and 1970s. Activists from low-income communities of color emerged as vocal proponents against perceived environmental injustices, citing the threats posed by hazardous waste and toxic chemicals in their neighborhoods. This period marked a critical awakening to the disproportionate environmental burdens experienced by marginalized groups, particularly those defined by race and ethnicity. The civil rights movement, which aimed to dismantle racial segregation and discrimination, laid the foundational principles for the environmental justice movement by exposing systemic inequalities that permeated various facets of life, including environmental quality and public health.

One of the earliest and most significant milestones in the history of environmental justice was the 1982 protests in Warren County, North Carolina (Slate, 2024). A seminal event in the history of environmental justice occurred in 1982 with the protests in Warren County, North Carolina. This predominantly African American community was chosen as the site for a hazardous waste landfill intended to house PCB-contaminated soil resulting from unlawful dumping along local roadways. Despite evaluating several potential locations, state officials ultimately selected this vulnerable community, igniting widespread protests and acts of civil disobedience among residents. These demonstrations captured national attention and brought the issue of environmental racism to the forefront, illustrating the extent to which marginalized communities bear a disproportionate share of environmental risks. The Warren County protests are frequently recognized as a pivotal moment for the environmental justice movement, inspiring activists and communities to mobilize against environmental inequities on a broader scale.

14.1.2 Emergence of Environmental Justice as a Movement

Following the protests in Warren County, the environmental justice movement began to gather significant momentum. A pivotal moment in this evolution came in 1987 with the publication of the landmark report Toxic Wastes and Race in the United States by the United Church of Christ's Commission for Racial Justice (Henderson & Wells, 2021; Slate, 2024). This report provided compelling empirical evidence demonstrating the disproportionate placement of hazardous waste facilities in communities of color, thereby lending credence to the concerns voiced by activists. The findings of this report highlighted the pervasive environmental injustices faced by marginalized groups and underscored the urgent need for policies aimed at combating environmental racism and advancing environmental justice.

In 1991, the environmental justice movement reached another critical milestone with the convening of the First National People of Color Environmental Leadership Summit in Washington, D.C. (Mendez, 2020). The summit united a diverse coalition, including activists, scholars, and community leaders from across the United States, to exchange ideas and formulate strategies for advancing environmental justice. One of the principal outcomes of the summit was the adoption of the Principles of Environmental Justice, a set of guiding values and objectives that articulated the movement's core goals. These principles emphasized the fundamental right of all individuals to inhabit a healthy environment, the necessity of fair distribution of environmental advantages and disadvantages, and the importance of ensuring community participation in environmental decision-making processes. The summit and its principles served as a crucial framework for the ongoing struggle for environmental justice, shaping the trajectory of future advocacy and policy efforts.

14.1.3 Institutionalization and Policy Development

The increasing recognition of environmental justice issues throughout the 1990s spurred notable policy advancements aimed at addressing the inequities faced by marginalized communities. In 1992, the U.S. EPA took a significant step by establishing the Office of Environmental Justice. This office was tasked with ensuring that environmental justice principles were integrated into the agency's programs and policies, with a particular focus on addressing the unique concerns of vulnerable populations in the context of federal environmental decision-making.

A landmark development in the institutionalization of environmental justice, titled Federal Actions to Address Environmental Justice in Minority Populations and Low-Income Populations, occurred in 1994 when President Bill Clinton signed Executive Order 12898. This executive order mandated that federal agencies evaluate and mitigate the disproportionately high and adverse effects of their actions on the health and environment of minority and low-income communities. Additionally, it underscored the critical role of public participation and community involvement in the decision-making processes that impact environmental outcomes. Executive Order 12898 represented a critical milestone in the formal integration of environmental justice principles into federal governance, establishing a framework for future policy development and ensuring that environmental justice became a key consideration in federal environmental practices.

14.1.4 Expansion and Globalization of the Environmental Justice Movement

The environmental justice movement, which originated in the United States, has gained substantial international traction, broadening its focus to encompass a diverse array of global environmental and social challenges (Perkins, 2022). Today, environmental justice issues are acknowledged across various regions, including Europe, Asia, Africa, and Latin America. The movement has significantly evolved from its initial parameters, integrating intersecting themes such as climate justice, food justice, and energy justice, all of which highlight the interconnectedness of environmental, social, and economic concerns.

Climate justice, a central theme in the expanded movement, highlights the disproportionate effects of climate change on vulnerable populations. These communities, often lacking the resources necessary for effective adaptation, suffer the most severe consequences of climate-related phenomena, such as extreme weather events, rising sea levels, and food insecurity. Advocates for climate justice strive to develop solutions that not only address the root causes of climate change but also prioritize the rights and needs of marginalized groups, ensuring they have access to essential resources and support to mitigate adverse impacts.

Food justice focuses on the systemic inequities in access to healthy, affordable food. It underscores the challenges faced by low-income and minority communities, which frequently encounter barriers to nutritious food, contributing to higher rates of diet-related diseases. The food justice movement advocates for policies and practices that enhance access to healthy food options, promote sustainable local food systems, and address the root causes of food insecurity.

Energy justice, similarly, emphasizes the need for equitable access to energy resources. It addresses issues such as energy poverty, where marginalized groups face difficulties accessing reliable and affordable energy, and the negative environmental consequences of energy production, including pollution from fossil fuel extraction and consumption. Energy justice advocates for a transition toward clean, renewable energy sources, ensuring that the benefits of such transitions are equitably distributed and that no community is left behind.

In 2000, the United Nations adopted the Environmental Justice Strategy Plan, which sought to embed principles of environmental justice into global governance frameworks. This strategy highlighted the importance of international collaboration to address environmental injustices and promote sustainable development. It called for the active inclusion of marginalized communities in environmental decision-making processes and emphasized the fair distribution of environmental benefits and burdens. The adoption of this strategy was a pivotal milestone, recognizing environmental justice as a global concern and incorporating it into international policy initiatives.

The Paris Agreement, finalized in 2015 and endorsed by 196 countries, marked another important advancement in the recognition of environmental justice principles. Aiming to limit global warming to well below 2 degrees Celsius above pre-industrial levels, the agreement emphasizes the necessity of addressing climate change in ways that foster sustainable development and eradicate poverty. Notably, it explicitly acknowledges the importance of considering the rights of marginalized communities, ensuring that climate action does not exacerbate existing inequalities. This inclusion reinforces the global acknowledgment of the intricate connections between climate change and social justice issues.

In 2020, the introduction of the Environmental Justice for All Act in the U.S. Congress represented a significant legislative development. This comprehensive measure aims to rectify environmental and health disparities affecting marginalized communities by bolstering legal protections, enhancing community participation in environmental decision-making, and increasing funding for environmental justice initiatives. The act aspires to guarantee that all communities have equitable access to clean air, water, and land, while also protecting them from the disproportionate impacts of environmental hazards. Its introduction reflects an increasing political commitment to addressing environmental justice issues at the national level and underscores the urgency for systemic transformation.

The environmental justice movement has achieved several key milestones, reflecting the growing recognition of environmental justice issues and the progress made in addressing these challenges. The following timeline of Table 14.1 highlights some of the most notable accomplishments in the movement's evolution:

Table 14.1 Key milestones of the environmental justice

Year	Key milestones
1982	North Carolina PCB Protest
1987	Toxic Wastes and Race Report
1991	First National People of Color Environmental Leadership Summit
1992	Establishment of the EPA's Office of Environmental Justice
1994	Executive Order 12898
2000	United Nations Environmental Justice Strategy Plan
2015	Paris Agreement
2020	Environmental Justice for All Act

14.2 Key Theories and Principles

14.2.1 Distributive Justice

Distributive justice is a fundamental principle within the broader framework of environmental justice, concerned with the fair distribution of both environmental benefits and burdens across society (Schlosberg, 2007). It asserts that no group should disproportionately shoulder the negative consequences of environmental degradation, such as exposure to pollution or hazardous waste, while others benefit from environmental goods like clean air and water.

This concept is embedded in the wider discourse of social justice, which seeks to ensure equitable distribution of resources and opportunities. In environmental justice, distributive justice specifically addresses the spatial and social dimensions of environmental disparities, highlighting how environmental harms and advantages are unequally shared across different communities. Often, these inequalities expose patterns of environmental racism and classism.

For instance, marginalized communities, particularly those with low income or predominantly made up of racial minorities, are frequently situated near polluting industrial facilities, waste disposal sites, and other environmental hazards. These communities often lack the political influence or economic resources to oppose the placement of such facilities, resulting in higher exposure to detrimental environmental conditions. Distributive justice advocates for policies that rectify these inequities by promoting equal access to a safe and healthy environment for all communities.

14.2.2 Procedural Justice

Procedural justice underscores the necessity of fair, transparent, and inclusive decision-making processes within environmental governance (Coolsaet, 2021). It emphasizes that all stakeholders, particularly those from marginalized communities, must have a meaningful voice in the environmental decisions that affect their well-being. This principle is grounded in the notion that equitable and just outcomes can only be achieved through participatory, accountable, and transparent procedures.

The core components of procedural justice include:

Access to Information: Ensuring that all relevant parties have timely access to necessary data concerning environmental issues. This encompasses information on pollution levels, potential health risks, and the implications of proposed projects.

Public Participation: Facilitating active engagement of affected communities in the decision-making process, such as through public hearings, community consultations, and opportunities for public commentary.

Accountability: Holding policymakers and decision-makers accountable for their actions, ensuring they incorporate and address the concerns of all affected stakeholders. This involves establishing mechanisms for monitoring compliance with environmental regulations.

Procedural justice plays a critical role in building trust between communities and policymakers, fostering a collaborative environment where marginalized groups can advocate for their rights. It guarantees that their voices are not only heard but also considered in shaping environmental policies and actions.

14.2.3 Recognition Justice

Recognition justice emphasizes the importance of acknowledging and valuing the diverse identities, cultures, and lived experiences of different communities (Whyte, 2011). This principle calls for recognizing the historical and ongoing injustices faced by marginalized groups and ensuring that their perspectives are integrated into environmental policymaking.

Recognition justice is especially crucial for Indigenous peoples, whose deep cultural, spiritual, and historical connections to the land are often overlooked in environmental decision-making. It advocates for the recognition of these unique relationships and the respect of Indigenous rights to self-determination and sovereignty over their territories.

This principle also addresses the intersectionality of environmental injustices, considering how various forms of social stratification, such as race, gender, and class, interact to create distinctive experiences of oppression and privilege. By recognizing these intersecting identities, policymakers can develop more comprehensive and targeted strategies for addressing environmental inequities in a manner that is both just and inclusive.

14.2.4 Capabilities Approach

The capabilities approach, articulated by Amartya Sen and Martha Nussbaum, centers on expanding individuals' capabilities to lead the lives they value and have reason to value (Fei, 2022). In the realm of environmental justice, this approach stresses the importance of ensuring that all individuals have the necessary resources and opportunities to achieve well-being, particularly through access to a healthy and sustainable environment.

Rather than focusing solely on the distribution of material resources, the capabilities approach emphasizes the actual freedoms and opportunities available to individuals to realize their full potential. It considers a broad spectrum of factors influencing well-being, including health, education, and social inclusion.

Incorporating the capabilities approach into environmental justice entails:

Assessing Capabilities: Evaluating the extent to which individuals and communities possess the capabilities required to achieve well-being, including access to vital environmental resources such as clean air, water, and safe living conditions.

Addressing Barriers: Identifying and mitigating the structural barriers that hinder individuals from realizing their capabilities. These may include environmental hazards, inadequate healthcare, and social discrimination.

Promoting Empowerment: Empowering individuals and communities to take control over their environments and improve their quality of life. This can be achieved through capacity-building efforts, community organizing, and advocacy for policy changes.

The capabilities approach offers a holistic framework for addressing environmental justice issues by focusing on creating the conditions that enable individuals and communities to flourish. It stresses the importance of removing barriers that restrict individuals' freedom to achieve their potential and ensuring that all people have the opportunity to lead fulfilling lives.

14.2.5 Intersectionality

Intersectionality, a framework developed by Kimberlé Crenshaw, examines the ways in which multiple forms of social stratification, such as race, gender, and class, intersect to shape individuals' experiences of both privilege and oppression (Jampel, 2018). In the context of environmental justice, this lens helps to better understand how overlapping identities influence an individual's susceptibility to environmental harm.

The concept of intersectionality acknowledges that environmental injustices are not uniformly experienced across society. For example, women of color may face distinct environmental challenges compared to their white counterparts or men of color, due to the compounded effects of racial discrimination, gender inequality, and socio-economic disadvantages. By considering these intersections, policymakers are better equipped to design more nuanced and effective strategies that address the complex ways in which environmental inequalities manifest.

Additionally, intersectionality underscores the importance of inclusive and diverse leadership within environmental movements. Representation of marginalized communities in decision-making processes is essential for ensuring that policies are equitable and that the environmental concerns of all social groups are adequately addressed.

14.2.6 Environmental Racism

Environmental racism refers to the systemic and disproportionate exposure of communities of color to environmental hazards, alongside their exclusion from environmental decision-making processes (Henderson & Wells, 2021). This concept highlights the intersection between racial discrimination and environmental justice, revealing how racial inequalities contribute to the unequal distribution of environmental burdens. Environmental racism manifests in several key ways, including:

Siting of Hazardous Facilities: Communities of color are more likely to be selected as locations for environmentally hazardous facilities, such as toxic waste dumps, industrial plants, and landfills. This practice results in higher exposure to pollution, posing significant public health risks for these populations.

Disproportionate Enforcement of Regulations: In many cases, environmental regulations are applied less rigorously in communities of color, leading to higher levels of pollution and environmental degradation compared to predominantly white communities.

Exclusion from Decision-Making: Often, communities of color are marginalized in environmental governance processes, which limits their ability to influence decisions that directly affect their health and environment. This exclusion perpetuates the environmental disparities they face.

To address environmental racism, it is crucial to advocate for racial equity within environmental policies and practices. This includes ensuring that communities of color are meaningfully engaged in environmental decision-making and that their rights to a safe and healthy environment are fully recognized and protected.

14.2.7 Social Justice and Economic Equity

Environmental justice is intrinsically linked to broader concepts of social justice and economic equity. These intersections are crucial for understanding the full scope of environmental justice and for developing comprehensive solutions.

Social Justice: Social justice seeks to create a society where all individuals have equal rights, opportunities, and access to resources. Environmental justice is a subset of social justice, focusing specifically on the equitable distribution of environmental benefits and burdens. It addresses how environmental harms disproportionately affect marginalized communities and seeks to rectify these inequities through inclusive and fair policies.

Economic Equity: Economic equity involves the fair distribution of economic resources and opportunities. Environmental justice intersects with economic equity by addressing how economic policies and practices contribute to environmental inequalities. For example, low-income communities often lack the financial resources to mitigate environmental hazards or to relocate from polluted areas. Addressing economic inequities is therefore essential for achieving environmental justice.

Health Equity: Environmental justice also intersects with health equity, as environmental hazards often lead to adverse health outcomes. Marginalized communities frequently experience higher rates of health issues, such as asthma, cancer, and cardiovascular diseases, due to their proximity to pollution sources. Ensuring equitable access to healthcare and addressing environmental determinants of health are critical components of environmental justice.

Example

Flint Water Crisis

The Flint water crisis serves as a poignant illustration of environmental injustice, highlighting significant failures in both distributive and procedural justice. In 2014, the city of Flint, Michigan, made a cost-saving decision to switch its water source from Lake Huron to the Flint River. Unfortunately, the river water was highly corrosive, leading to widespread lead contamination in the city's drinking water. This public health crisis disproportionately affected the predominantly African American and low-income residents of Flint, exposing them to toxic levels of lead and resulting in significant health problems.

The crisis underscores a profound lack of procedural justice, as the concerns of residents were ignored for months by government officials, and crucial information was withheld from the public. Moreover, it illustrates a failure of distributive justice, as the environmental burden of lead contamination fell disproportionately on marginalized communities. The Flint water crisis highlights the urgent need for more inclusive decision-making processes and for ensuring that environmental benefits and risks are distributed equitably, particularly among vulnerable populations.

In conclusion, the Flint water crisis underscores the urgent need for more inclusive and transparent decision-making processes. It also highlights the importance of equitable distribution of environmental advantages and risks, ensuring that all populations, especially those most vulnerable, are safeguarded against environmental injustices.

14.3 Environmental Inequality

14.3.1 Methods for Measuring Environmental Inequality

Assessing the distribution of environmental advantages and disadvantages across various communities is essential for understanding and addressing environmental inequality. Several methodologies are employed to quantify these disparities, each providing distinct insights into the spatial and social dimensions of environmental justice:

Geospatial Analysis: Geospatial analysis uses Geographic Information Systems (GIS) to map and assess the spatial distribution of environmental hazards in relation to different demographic groups (Tariqi & Naughton, 2021). This method enables researchers to identify and visualize patterns of environmental inequality. Tools such as the Environmental Justice Screening and Mapping Tool (EJScreen), developed by the U.S. EPA, integrate environmental and demographic data to highlight areas facing significant environmental burdens and vulnerable populations. By combining data on pollution sources, health outcomes, and socio-economic indicators, geospatial analysis offers a comprehensive understanding of environmental disparities.

Cumulative Impact Assessment: Cumulative impact assessment evaluates the combined effects of multiple environmental hazards on a community. This approach considers various environmental stressors, such as air and water pollution, toxic waste sites, and industrial facilities, alongside social determinants of health like income, education, and access to healthcare. Tools like the Environmental Justice Index (EJI) employ cumulative impact assessments to provide a holistic view of environmental inequality. By integrating diverse data sources, cumulative impact assessments help pinpoint communities most affected by environmental risks, thereby informing targeted policy interventions.

Health Impact Assessment: Health Impact Assessment (HIA) is a systematic process used to evaluate the potential health effects of policies, programs, or projects, particularly focusing on vulnerable populations. HIA integrates data on environmental exposures, health outcomes, and social determinants to assess the overall impact on community health. This method helps policymakers understand the potential health consequences of their decisions and develop strategies to mitigate adverse effects. HIA is particularly valuable in ensuring that health equity is incorporated into environmental and urban planning processes.

Environmental Justice Indexes: Environmental justice indexes combine environmental and socio-economic data to generate composite scores that reflect the overall environmental burden on communities. Prominent examples include the EJI and EJScreen. These indexes provide essential insights into the distribution of environmental risks and identify areas in need of intervention. By quantifying environmental burdens and highlighting disparities, environmental justice indexes support evidence-based policymaking and guide the allocation of resources to

	Modules	Domains	Indicators
Environmental Justice Rank	Social Vulnerability	Racial/Ethnic Minority Status	Minority Status
		Socioeconomic Status	Poverty; No High School Diploma; Unemployment, Renters; Housing Cost Burden; Lack of Health Insurance; Lack of Internet Access
		Household Characteristics	Age 65 and Older; Age 17 and Younger; Civilian with a Disability; English Language Proficiency
		Housing Type	Group Quarters; Mobile Homes
	Environmental Burden	Air Pollution	Ozone; Particulate Matter 2.5 (PM2.5); Diesel Particulate Matter; Air Toxics Cancer Risk
		Potentially Hazardous & Toxic Sites	National Priority List Sites; Toxic Release Inventory Sites; Treatment, Storage, and Disposal Sites; Risk Management Plan Sites; Coal Mines; Lead Mines
		Built Environment	Lack of Recreational Parks; Houses Built Pre-1980; Lack of Walkability
		Transportation Infrastructure	High Volume Roads; Railways; Airports
		Water Pollution	Impaired Surface Water
	Health Vulnerability	Pre-existing Chronic Disease Burden	Asthma; Cancer; Coronary Heart Disease; Diabetes; Poor Mental Health

Fig. 14.1 Environmental Justice Index indicators. *Source* ATSDR (2024a)

communities most in need of protection. The indicators included in the EJI (see Fig. 14.1) help users identify communities that may be disproportionately affected by socio-economic, environmental, and health burdens.

In conclusion, these methods and tools play a critical role in revealing and addressing environmental inequality. By employing a range of approaches: geospatial analysis, cumulative impact assessments, health impact assessments, and environmental justice indexes, researchers and policymakers can gain a more nuanced understanding of environmental burdens and design interventions that promote greater equity and justice in the distribution of environmental benefits and risks.

14.3.2 Examples of Environmental Injustices

Environmental injustices manifest in various forms, often disproportionately impacting marginalized communities. These injustices are reflected in the uneven distribution of pollution, resource depletion, and the effects of climate change. Below are detailed examples that illustrate these critical issues.

Resource Depletion: Resource depletion, manifested through deforestation, water scarcity, and soil degradation, often adversely affects marginalized communities that rely on natural resources for their livelihoods. These groups face heightened vulnerability to environmental changes and diminished access to essential resources.

In various regions worldwide, Indigenous communities bear the brunt of deforestation and land degradation. For instance, in the Amazon rainforest, Indigenous peoples have been engaged in ongoing struggles against illegal logging and territorial encroachment, threats that jeopardize their traditional ways of life and the region's biodiversity. The loss of forest resources not only undermines their cultural heritage but also threatens their food security and health. This case highlights the imperative for policies that safeguard Indigenous land rights while promoting sustainable resource management practices.

Climate Change: Climate change amplifies pre-existing environmental injustices, disproportionately affecting vulnerable communities that often have limited resources and adaptive capacity. These populations are typically the most severely impacted by climate-related consequences, such as extreme weather events, rising sea levels, and food insecurity.

Small Island Developing States (SIDS) are particularly susceptible to the adverse impacts of climate change. Rising sea levels, an increase in hurricane frequency, and coral reef degradation pose substantial threats to their economies, infrastructure, and livelihoods. Despite contributing minimally to global greenhouse gas emissions, SIDS face disproportionate risks and challenges related to climate change adaptation. This situation underscores the need for international cooperation and support to effectively address global environmental injustices.

Industrial Contamination: Industrial contamination often afflicts communities located near manufacturing plants and waste disposal sites. These areas are typically home to low-income and minority populations who suffer from increased exposure to toxic chemicals and pollutants.

Energy Injustice: Energy injustice refers to unequal access to clean and affordable energy. Marginalized communities often face higher energy costs and limited availability of renewable energy sources, thereby exacerbating economic and social inequalities. In many rural regions, particularly in developing countries, communities often lack access to reliable and affordable energy. This energy poverty hampers economic opportunities, educational attainment, and overall quality of life. Addressing energy injustice necessitates efforts aimed at expanding access to renewable energy sources and improving energy infrastructure in underserved areas.

14.4 Environmental Justice Across the World

Environmental justice is a pressing global issue, manifesting in diverse forms across different regions and cultures. The fundamental principles of environmental justice, namely, fair treatment and meaningful involvement of all individuals, irrespective of race, ethnicity, national origin, or income, are universally applicable. However, these principles must be adapted to address specific regional challenges and contexts. Figure 14.2 shows the overall environmental justice index in the U.S. We examine the state of environmental justice across various continents, highlighting key issues and illustrative case studies.

North America

In North America, movements advocating for environmental justice have historically concentrated on mitigating the disproportionate exposure of minority and low-income communities to environmental hazards. In the United States, significant activism and policy advancements have emerged since the 1980s. A prominent example is the Flint water crisis in Michigan, which exposed stark failures in both distributive and procedural justice. Predominantly affecting African American and

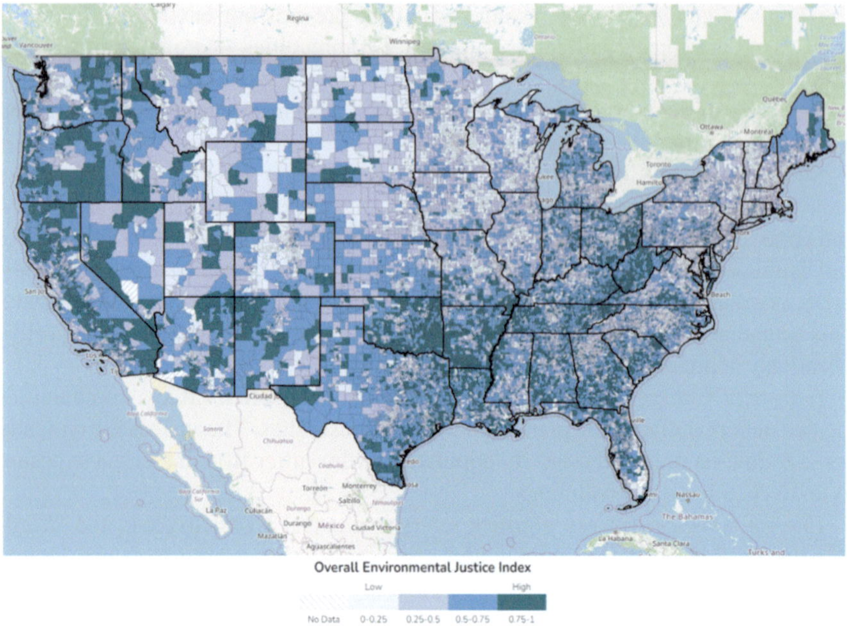

Fig. 14.2 Overall environmental justice index in the U.S. *Note* The map is generated by ATSDR (2024b)

low-income residents, this crisis arose from a cost-saving decision to switch the city's water source, resulting in toxic levels of lead contaminating drinking water. This situation has underscored the urgent need for inclusive decision-making processes and equitable distribution of environmental benefits and risks.

In Canada, environmental justice issues intersect significantly with Indigenous rights. Indigenous communities have taken a leading role in advocating for environmental justice, championing their land rights and opposing the detrimental impacts of climate change on their traditional lifestyles. For instance, the resistance of the Wet'suwet'en Nation against the Coastal GasLink pipeline project exemplifies the ongoing struggle for Indigenous sovereignty and environmental conservation. This pipeline, designed to transport natural gas, poses threats to the ecological integrity of Wet'suwet'en territory and jeopardizes the cultural heritage of its people.

Latin America

In Latin America, environmental justice issues predominantly revolve around the exploitation of natural resources and the rights of Indigenous peoples. The Amazon rainforest, home to numerous Indigenous communities, faces severe threats from illegal logging, mining, and agricultural expansion. These destructive activities not only degrade the environment but also undermine the livelihoods and cultural identities of Indigenous populations. Movements advocating for environmental justice in the region emphasize the protection of Indigenous land rights and promote sustainable resource management practices.

Example

The Amazon Rainforest

The Amazon rainforest is often referred to as the "lungs of the Earth" due to its vast capacity to absorb carbon dioxide. However, rampant deforestation driven by logging, mining, and agriculture poses a significant threat to this critical ecosystem. Indigenous communities, who have coexisted harmoniously with the forest for centuries, are disproportionately affected by these incursions. The degradation of forest resources significantly impacts their cultural heritage, food security, and health. Environmental justice initiatives in the Amazon focus on safeguarding Indigenous land rights while promoting sustainable practices to ensure the preservation of the rainforest for future generations.

Africa

Africa confronts unique environmental justice challenges, including land degradation, water scarcity, and the repercussions of climate change. Many African nations rely heavily on agriculture, rendering them particularly vulnerable to environmental fluctuations. Furthermore, the extraction of natural resources, such as oil and minerals, frequently leads to environmental degradation and social conflict. Efforts aimed at achieving environmental justice in Africa prioritize ensuring equitable access to resources, protecting the rights of local communities, and fostering sustainable development.

Example

The Niger Delta

The Niger Delta in Nigeria is one of the most oil-rich regions globally, yet it is simultaneously one of the most environmentally degraded. Decades of oil extraction have resulted in widespread pollution, including oil spills and gas flaring, which have devastated local ecosystems and communities. Predominantly inhabited by low-income and marginalized populations, the Niger Delta suffers from severe health issues, loss of livelihoods, and social unrest. Environmental justice movements in the area advocate for corporate accountability, environmental remediation, and the safeguarding of community rights.

Asia

Asia's rapid industrialization and urbanization have led to significant environmental justice dilemmas. Issues such as air and water pollution, deforestation, and displacement due to infrastructure projects disproportionately impact low-income and marginalized communities. In India, for example, large dam constructions have displaced millions, often without adequate compensation or resettlement plans. Environmental justice movements in Asia strive to uphold the rights of affected communities while seeking to balance economic development with environmental sustainability.

Example

The Narmada Valley Project

The Narmada Valley Project in India involves constructing a series of large dams on the Narmada River, intended to provide irrigation, drinking water, and hydroelectric power. However, this initiative has lead to the displacement of hundreds of thousands of individuals, primarily from Indigenous and economically disadvantaged backgrounds. Many of those displaced have not received sufficient compensation or resettlement assistance. Activisits, including the Narmada Bachao Andolan (Save the Narmada Movement), have been instrumental in fighting for the rights of these communities, advocating for fair compensation, sustainable development, and the protection of human rights.

Europe

Europe has made notable advancements in promoting environmental justice through the establishment of comprehensive regulatory frameworks and fostering public participation in environmental decision-making. Nevertheless, challenges persist, particularly in mitigating the needs of marginalized communities, such as the Roma, who frequently reside in areas suffering from significant environmental degradation. Environmental justice initiatives across Europe emphasize the importance of ensuring that all communities benefit equitably from environmental protections and sustainable development efforts.

Example

Roma Communities in Eastern Europe

Roma communities in Eastern Europe often inhabit informal settlements that lack access to essential services, including clean water, sanitation, and waste management. These populations face considerable environmental health risks due to their proximity to industrial sites and waste disposal areas. In response, environmental justice initiatives in the region seek to enhance living conditions for Roma communities, promote their meaningful inclusion in environmental decision-making processes, and confront the systemic discrimination they endure.

Oceania

In Oceania, the issues surrounding environmental justice are closely tied to the effects of climate change on SIDS. Rising sea levels, increased frequency of extreme weather events, and coral reef degradation pose existential threats to the survival of these islands. Movements advocating for environmental justice in Oceania emphasize the necessity of international collaboration and support to tackle these urgent challenges while promoting climate resilience among affected communities.

Reading

Climate Change in the Pacific Islands

The Pacific Islands represent one of the regions most vulnerable to climate change's impacts. As rising sea levels threaten to submerge entire islands, the increasing frequency and intensity of cyclones significantly disrupt infrastructure and livelihoods. Despite their minimal contributions to global greenhouse gas emissions, Pacific Island nations disproportionately bear the burden of climate-related risks and challenges in adaptation efforts. Environmental justice initiatives in this region prioritize advocacy for international climate action, securing financial resources for adaptation strategies, and safeguarding the rights and cultural heritages of island communities.

14.5 New Challenges in Environmental Justice

14.5.1 Climate Change and Vulnerability

Climate change represents one of the most critical environmental challenges of our era, manifesting its impacts unevenly across different regions and communities (see Fig. 14.3). Vulnerable populations, often characterized by limited resources for adaptation, are disproportionately affected by extreme weather events, rising sea levels, and shifting agricultural conditions. These communities encounter increased threats to their health, livelihoods, and housing, thereby exacerbating pre-existing inequalities.

Coastal populations in low-lying areas face significant risks of displacement due to rising sea levels, with small island nations and coastal regions in countries such as Bangladesh and the Philippines being particularly susceptible. Frequently, these areas lack the essential infrastructure and resources necessary to effectively respond to and recover from climate-related disasters. Addressing climate justice necessitates the implementation of targeted policies that prioritize the needs of these vulnerable groups, ensuring they possess the resources and support required to adapt and thrive.

Fig. 14.3 Climate change

Moreover, climate change tends to magnify existing social and economic disparities. For instance, individuals in low-income communities may not have access to air conditioning or adequate healthcare services, rendering them more vulnerable to heatwaves and other climate-induced health challenges. Therefore, ensuring equitable access to resources and support systems is vital for alleviating the adverse effects of climate change on these populations.

14.5.2 Technological Advancements and the Digital Divide

While technological advancements present considerable opportunities for tackling environmental challenges, they also introduce new layers of difficulty. The digital divide—the disparity between those with access to modern information and communication technologies and those without—can further entrench environmental injustices. Access to real-time environmental data and early warning systems can significantly bolster community resilience against ecological hazards. However, marginalized communities often lack access to these critical technologies, increasing their vulnerability.

For example, early warning systems designed for natural disasters such as hurricanes, floods, and wildfires heavily depend on digital communication channels. Communities that lack reliable internet connectivity or mobile access may fail to receive timely alerts, placing them at heightened risk. Bridging the digital divide is therefore imperative to ensure all communities can benefit from technological innovations aimed at addressing environmental challenges.

Additionally, the introduction of new technologies, such as renewable energy infrastructure, may yield unintended consequences. For instance, the extraction of rare earth minerals for solar panels and wind turbines can lead to environmental degradation and human rights violations within mining communities. To facilitate a just and equitable transition to a green economy, it is crucial to thoroughly consider the social and environmental ramifications associated with emerging technologies.

14.5.3 Urbanization and Environmental Gentrification

Rapid urbanization has given rise to a phenomenon known as environmental gentrification, where efforts to enhance urban environments, such as the development of green spaces and pollution reduction initiatives, unintentionally displace low-income communities. While these improvements aim to benefit urban areas, they often lead to increased property values and higher living costs, effectively pushing out the very populations that these changes were intended to support.

For instance, the creation of urban parks and greenways often leads to a rise in property values in surrounding neighborhoods, making housing less affordable for long-standing residents. This displacement disrupts social networks and weakens community cohesion, contributing to adverse social and economic outcomes for those who are forced to relocate. Preventing such displacement and ensuring that urban environmental improvements are inclusive remain key challenges for advocates of environmental justice.

One potential solution to environmental gentrification is the implementation of policies that protect affordable housing and encourage community-led development. Community land trusts and inclusionary zoning policies are examples of strategies that can help ensure low-income residents benefit from urban improvements without being displaced. Additionally, involving local communities in planning and decision-making processes is vital to ensure that their needs and priorities are adequately addressed.

14.5.4 Biodiversity Loss and Ecosystem Services

The loss of biodiversity and the degradation of ecosystem services represent profound challenges to environmental justice. Ecosystem services, such as clean air and water, food production, and climate regulation, are integral to human well-being. However, the destruction of natural habitats, pollution, and climate change are driving unprecedented rates of biodiversity decline, thereby threatening these essential services.

For example, activities such as deforestation and habitat destruction contribute to the decline of species that provide crucial ecosystem services, including pollination and natural pest control. This loss can jeopardize food security and livelihoods, particularly for communities dependent on subsistence agriculture and natural resources. Additionally, diminished biodiversity reduces ecosystem resilience to environmental changes, making them more susceptible to climate change and other stressors.

Addressing biodiversity loss and protecting ecosystem services necessitates integrated approaches that encompass social, economic, and environmental dimensions of sustainability. Strategies should include promoting conservation and sustainable land-use practices, safeguarding natural habitats, and restoring degraded ecosystems. Furthermore, engaging local communities in conservation efforts and ensuring they benefit from the sustainable utilization of natural resources is vital for achieving meaningful environmental justice.

14.5.5 Environmental Migration and Displacement

Environmental migration and displacement represent emerging challenges closely linked to environmental justice. Climate change, natural disasters, and environmental degradation can compel individuals and communities to abandon their homes,

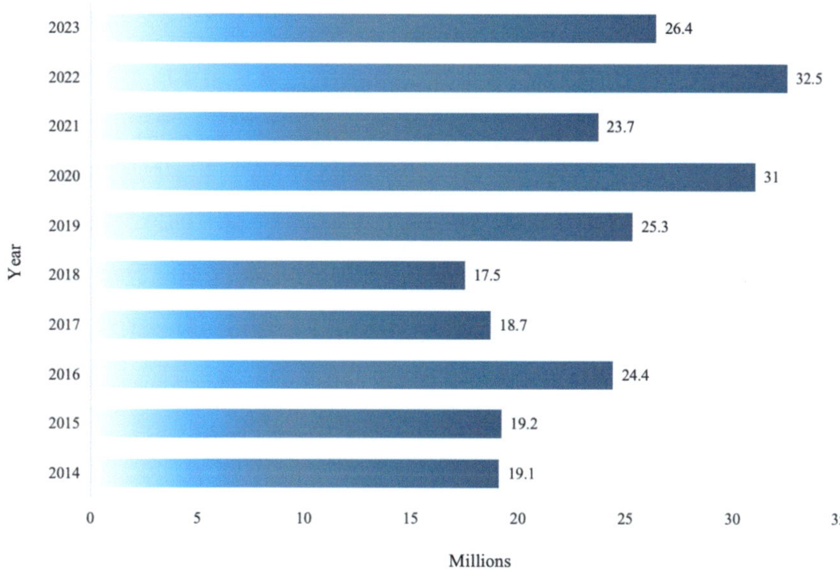

Fig. 14.4 New internal displacements due to disaster from 2014 to 2023. *Note* The figure is adapted from IDMC (2024)

resulting in displacement and migration. Vulnerable populations are often disproportionately impacted, as they typically lack the resources and support necessary to adapt to environmental changes and rebuild their lives.

Rising sea levels and the increasing frequency of extreme weather events, for example, can displace coastal communities, leading to the loss of homes, livelihoods, and cultural heritage. Similarly, environmental degradation, such as desertification and deforestation, can diminish the availability of natural resources, prompting individuals to migrate in search of improved living conditions.

Addressing the complexities associated with environmental migration and displacement requires comprehensive policies that focus on the rights and well-being of impacted communities. Such policies should provide support for adaptation and resilience-building initiatives, enabling affected individuals and families to navigate the challenges posed by environmental changes effectively. Figure 14.4 shows the new internal displacements due to disasters from 2014 to 2023.

References

Coolsaet, B. (2021). *Environmental justice: Key issues*. Routledge.

Edwards, D. D., & Darnall, N. (2010). Averting environmental justice claims? The role of environmental management systems. *Public Administration Review, 70*(3), 422–433.

Fei, W. (2022). The influence of Karl Marx's notion of justice on Martha Nussbaum's capabilities approach. *Educational Philosophy and Theory, 55*(8), 1–10.

Henderson, S., & Wells, R. (2021). Environmental racism and the contamination of black lives: A literature review. *Journal of African American Studies, 25*(1), 134–151.

Holdren, J. P. (2008). Science and technology for sustainable well-being. *Science, 319*(5862), 424–434.

Internal Displacement Monitoring Centre (IDMC). (2024). Global Report on Internal Displacement. https://www.internal-displacement.org/global-report/

Jampel, C. (2018). Intersections of disability justice, racial justice and environmental justice. *Environmental Sociology, 4*(1), 122–135.

Mendez, M. (2020). *Climate change from the streets*. Yale University Press.

Perkins, T. E. (2022). *Evolution of a movement: Four decades of California environmental justice activism*. University of California Press.

Schlosberg, D. (2007). Reconceiving environmental justice: Global movements and political theories. *Environmental Politics, 13*(3), 517–540.

Slate, N. (2024). "If it happens to them, it happens to you": The highlander folk school and the racial borders of environmental justice. *Environmental History, 29*(3), 474–499.

Tariqi, A. Q., & Naughton, C. C. (2021) Water, health, and environmental Justice in California: Geospatial Analysis of Nitrate Contamination and Thyroid Cancer. *Environmental Engineering Science, 38*(5) 377–388.

The Agency for Toxic Substances and Disease Registry (ATSDR). (2024a). *EJI indicators*. https://atsdr.cdc.gov/place-health/php/eji/eji-indicators.html

The Agency for Toxic Substances and Disease Registry (ATSDR). (2024b). *EJI map*. https://www.atsdr.cdc.gov/place-health/php/eji/eji-explorer.html

Whyte, K. P. (2011). The recognition dimensions of environmental justice in Indian country. *Environmental Justice, 4*(4), 199–205.

Chapter 15
Energy as Social Science

It is only thanks to cheap, plentiful, reliable energy that we live in an environment where the water we drink and the food we eat will not make us sick and where we can cope with the often hostile climate of Mother Nature. Energy is what we need to build sturdy homes, to purify water, to produce huge amounts of fresh food, to generate heat and air-conditioning, to irrigate deserts, to dry malaria-infested swamps, to build hospitals, and to manufacture pharmaceuticals, among many other things.
"The Moral Case for Fossil Fuels", Alex Epstein (2014)

15.1 Energy Poverty

15.1.1 Identification of Energy Poverty

The concept of energy poverty was first put forward in the United Kingdom and was originally known as "fuel poverty." It mainly focused on the situation where households could not afford the necessary energy costs to keep indoor temperatures at an adequate level. In the later part of the twentieth century, the notion of social equity gained prominence in the energy sector, and the concept of social justice was introduced to the energy sector. In the late twentieth century, the concept of social equity gained prominence in the energy sector. People began to recognize that access to energy should not be a privilege for a few, but a basic right for all. The rise of the concept of energy equity prompted scholars and policymakers to focus on those who did not have access to adequate energy. For example, international organizations such as the World Bank have begun to study energy poverty, defining it as the inability of households to access adequate, affordable, reliable, good quality and safe energy services to meet their basic needs. The spread of this idea has led to the gradual transformation of energy poverty from a phenomenon into a concept that is being widely studied and focused on, and is closely intertwined with concepts such as social equity and sustainable development.

L. Zhang, *Sustainability Economics*, https://doi.org/10.1007/978-3-031-87829-9_15

The definition of energy poverty has been refined progressively over time. The European Commission, in the Energy Efficiency Directive, has given a comprehensive definition. Energy poverty is defined as the inability of households to access energy services for fundamental living and health requirements, like heating, hot water, cooling, and lighting, and also involves cases where the electricity supply is insufficient or costly, making it unaffordable for households. Hence, energy poverty is typically considered in terms of two dimensions: energy accessibility and affordability. The energy accessibility dimension refers to a stable and reliable supply of energy, reflecting the integrity and infrastructure of the energy system. Accessibility encompasses not only the physical availability of energy supply facilities, but also the ability of residents to physically access these sources of energy. For example, in some urban–rural areas, although grid lines pass through the neighborhood, for various reasons (e.g., land title issues, high access costs, etc.), some residents cannot access the grid, resulting in their lack of access to electricity services, which is a case of energy poverty resulting from insufficient energy accessibility. The energy affordability dimension, on the other hand, is related to the ability to pay energy bills and involves energy prices, consumption habits, and household disposable income, which is reflected in the affordability of residents to pay for energy. Even if energy supply facilities are adequate, energy poverty can result if energy prices are too high and beyond the affordability of residents. For example, in some high-cost off-grid energy supply areas, such as some remote islands, residents need to use diesel generators to obtain electricity, and the high cost of transportation of diesel makes the price of electricity beyond the affordability of the average resident, who can only reduce the time of using electricity, which affects the normal quality of life, and this is energy poverty arising from the lack of affordability.

In recent years, the definition of energy poverty has shifted from being unidimensional to multidimensional. The Multidimensional Energy Poverty Index (MEPI) is a combined index used to assess energy poverty. It includes five dimensions: cooking, lighting, appliance services, recreation/education, and communication, and each of these dimensions has its own deprivation threshold. This multidimensional assessment approach helps in achieving a more comprehensive comprehension of energy poverty and is beneficial for the formulation of policies.

Table 15.1 shows the indicators commonly used to measure energy poverty. Indicators 1 to 5 connect household income levels with energy consumption habits from various angles. Indicator 6 is designed to uncover hidden energy poverty that stems from households' energy-saving practices. In the literature, the kind of cooking fuel used by households (for instance, depending on solid fuels like firewood or straw) is a common measure for assessing energy poverty (Pachauri et al., 2021). But this indicator is debatable, because the use of biomass fuels for cooking is scarce in many urban areas, and in such cases, it is not an effective gauge of energy poverty. In contrast, in rural areas, especially in developing countries, this indicator can be used effectively to identify energy poverty.

Traditional identification of energy poverty has relied extensively on population survey data. In recent years, with the development of information technology, GIS and machine learning methods have been increasingly applied to the identification

Table 15.1 Energy poverty indicators

	Indicator description	Source
1	The percentage of household earnings dedicated to energy costs	Li et al. (2023)
2	A binary indicator that shows whether the percentage of income spent on energy is higher than 10%	Taylor (1993)
3	A restriction of the 10% threshold to low-income households	Kahouli (2020)
4	Households are considered to be in energy poverty if their ratio of energy spending to income is more than double the median ratio	Moore (2012)
5	Households experiencing energy poverty are those that spend more on energy compared to the median level, while their income falls below 50% of the median income	Hills (2011)
6	Energy expenses are below a reasonable level to meet basic living needs	Eisfeld and Seebauer (2022) and Betto et al. (2020)

of energy poverty. Nighttime lighting data can be used as an indicator of the level of economic activity and energy use, and analyzing nighttime lighting data can reveal energy use and help identify energy-poor areas. Machine learning techniques can analyze large amounts of data and identify patterns and can more accurately identify energy-poor areas by building predictive models.

Reading

Identifying Energy Poverty Based on Remote Sensing Data and Machine Learning

India has many challenges related to energy poverty such as a large population without access to electricity, depenedence on traditional biomass energy sources, etc. The research conducted by Wang et al. 2021 employed satellite remote sensing data associated with various geographic and environmental elements, including precipitation, temperature, soil moisture, vegetation, fine particulate matter (PM2.5), travel time to the nearest urban center, and nighttime lighting, among others. It utilized a random forest algorithm to evaluate and forecast energy poverty by classifying it into distinct dimensions, such as the capacity, duration, reliability, quality, and affordability of electricity access.

This particular machine learning algorithm, which integrates geographic and environmental remote sensing indicators, is capable of identifying 90.91% of regions with high energy poverty rates. It exhibits superior performance compared to algorithms that solely rely on socioeconomic metrics. The methodology is founded on satellite remote sensing data, which is relatively cost-effective and readily accessible, especially in regions where up-to-date household survey information is scarce. Consequently, it offers a novel and efficient approach to energy poverty management.

15.1.2 Global Distribution of Energy Poverty

Energy poverty is a worldwide concern that is particularly pronounced in developing nations and regions. The International Energy Agency reported that in 2021, 675 million people globally were deprived of electricity, with 567 million of them residing in sub-Saharan Africa, which represents more than 80% of the world's population lacking electrical access. In 2010, 84% of the world's population had electricity available.

This proportion climbed to 91% by 2021, signifying that over 1 billion people gained access to electrical services during this period. However, the rate of growth in electricity access has slackened from 2019 to 2021 compared to previous years. Rural electrification projects have contributed to the advancements, but significant gaps still exist in urban settings. Take Latin America and the Caribbean as an example, where energy poverty is characterized by unreliable and poor-quality electricity services. Although most countries in the region have achieved universal electricity access, problems like intermittent power supply and voltage fluctuations continue to afflict many households and businesses. In India, despite some improvement in electricity access, a large number of households still lack access to clean cooking fuels, thereby causing high levels of indoor air pollution and posing public health risks.

Reading

Electricity shortages in Africa

More than 600 million people in Africa lack access to electricity, and this has a continuing negative impact on the region's socio-economic development. At the same time, Africa has high levels of inequality in access to electricity. For example, Kenya has an electricity coverage rate of more than 75 percent, and cities in Ghana, Nigeria, and Senegal have electricity coverage rates of more than 80 percent. However, Burundi will have only 10% electricity coverage in 2021.

The development of renewable energy holds the promise of solving Africa's power shortages. About 81% of Kenya's electricity generation comes from renewable sources. Hydropower's share of Africa's electricity production averages 17% in 2019. Solar photovoltaics is expected to be the most affordable energy source in Africa by 2030. However, power generation from renewable sources such as hydropower and PV is affected by climatic conditions, and power instability is increasing as climate change intensifies.

There are still around 2.3 billion people who rely on polluting fuels and technologies for cooking, and this is mainly the case in sub-Saharan Africa and Asia (Fig. 15.1). When traditional biomass is used, families often have to spend up to 40 h every week on tasks such as collecting firewood and cooking, which restricts women from finding jobs or participating in local governance bodies and hinders children from going to school. About 3.2 million premature deaths each year can be associated with household air pollution caused by the use of polluting fuels and cooking technologies.

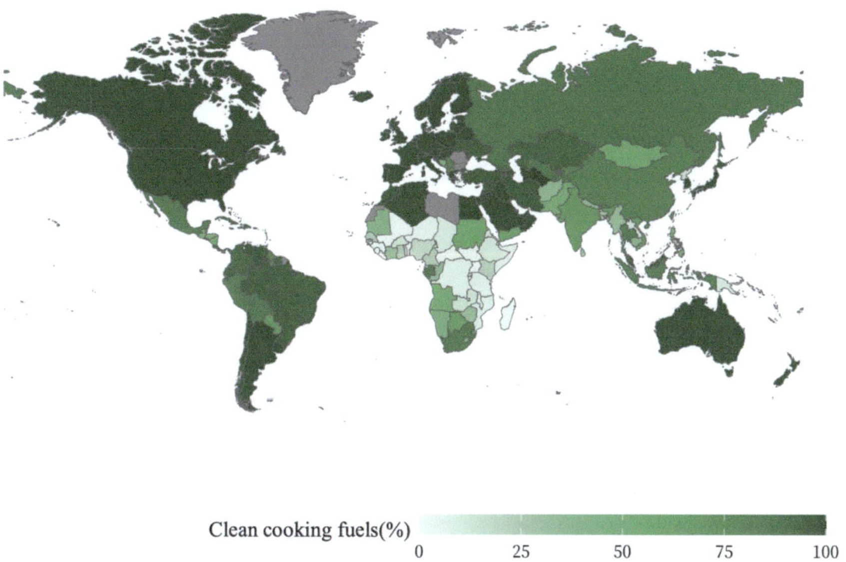

Fig. 15.1 Proportion of clean cooking fuels adoption in the world by country (2020). *Source* Li et al. (2025)

Energy poverty does not only occur in less developed regions, but is also widespread in developed countries. For example, more than 50 million households in the European Union live in energy poverty, with the highest rates of energy poverty in southern and south-eastern Europe. Especially in the aftermath of the Russian-Ukrainian conflict, the high cost of energy, such as natural gas, has made it impossible for many households to afford adequate energy consumption, leading to a poor quality of life (Fig. 15.2).

15.1.3 Practices to Address Energy Poverty

SDG7 of the United Nations aims to ensure universal access to affordable, reliable, sustainable, and modern energy. Key targets include increasing global access to affordable, reliable, and modern energy services by 2030, boosting the use of renewable energy in the global energy mix, doubling the rate of improvement in energy efficiency, strengthening international cooperation in clean energy research and technology, and expanding energy infrastructure and technology in developing countries. These targets are measured through indicators such as the proportion of the population with access to electricity, the share of renewable energy in the energy mix, energy intensity, international financial flows for clean energy research and development, and the installed renewable energy-generating capacity.

Reading

Russia-Ukraine Conflict and the Energy Crisis

The Russia-Ukraine conflict has had a significant impact on global energy prices, particularly in the oil and gas markets. As Russia is a globally important energy exporter or region, disruptions or reductions in its energy supply directly affect the stability and price volatility of the global energy market.

Since the outbreak of the Russia-Ukraine conflict, international oil prices have experienced dramatic fluctuations. Sanctions imposed by Western countries on Russia, including the ban on imports of Russian oil, have led to tensions in the global energy supply and further pushed up oil prices. In addition, the price of WTI crude oil in the United States had soared to US$130 per barrel during the conflict. Although oil prices eased after the release of the Strategic Petroleum Reserve in the United States to ease supply constraints, geopolitical factors continued to dominate oil price movements.

Fig. 15.2 The "Nord Stream" gas pipeline. *Note* The figure is retrieved from the European Space Agency (ESA) (2024)

The "Nord Stream" gas pipeline (Figure 15.2), which runs from Russia to Germany and delivers natural gas to Europe, began in February 2022 with the Russian-Ukrainian conflict, and the explosions of the Nord Stream I and

II pipelines on September 26th led to an energy crisis and panic in Europe. In order to make up for the Russian pipeline gas gap, Europe had to attract global liquefied natural gas resources to Europe at a premium, which broke the original market supply and demand balance and led to a significant increase in natural gas prices. In addition, as Russia is one of the key transit countries for natural gas in Europe, any military action could disrupt the energy supply chain throughout Europe and the wider region.

To achieve this goal, countries have adopted a variety of policy practices. In terms of energy infrastructure development, many developing countries have increased their investment in energy infrastructure such as electricity and natural gas. For example, China has upgraded its power grid facilities in rural areas on a large scale through the implementation of the "agricultural grid renovation" project, which has improved the reliability and stability of rural power supply and significantly improved rural residents' access to energy. In the process of "rural grid renovation," not only updated transmission lines, transformers and other power equipment, but also the use of smart grid technology, to improve the level of intelligent management of the power grid. India has launched a similar rural electrification program in an effort to expand grid coverage and provide electricity services to rural households. By providing subsidies and incentives, the Government of India encourages power companies to invest in the construction of power grids in rural areas.

In terms of energy subsidy policies, some countries and regions provide energy subsidies to low-income households or poor areas to increase the affordability of energy. For example, Iran subsidizes natural gas for residents to reduce the burden of energy expenditure. Hong Kong has implemented the Electricity Charges Subsidy Scheme since 2008. In 2023, for example, the Hong Kong Government will provide each eligible residential electricity account with a subsidy of HK$1000 for electricity charges, and the total amount of subsidy in that year will be more than HK$2.9 billion. However, energy subsidy policies also face a number of challenges, such as high fiscal pressure and the possibility of energy wastage.

Example

China's Policy Practices to Address Energy Poverty

As part of its national strategy to combat air pollution and reduce carbon emissions, China has implemented a series of measures to enhance access to clean cooking fuels for residents. The "West-East Gas Transmission" project, launched in 2000, addressed the uneven distribution of natural gas resources by laying a network of pipelines to alleviate natural gas shortages in eastern China. The Ministry of Agriculture's "National Rural Biogas Project Construction Plan" in 2006 significantly increased the construction of biogas digesters nationwide, promoting biogas as a key cooking fuel. By 2013, China had achieved universal electricity access through the "3-Year Plan to Solve the Electricity Problems of Underserved Populations", enabling electricity to be used for cooking in addition to lighting.

Furthermore, China has implemented policies to modernize traditional energy sources. The country's clean heating policy in 2017 aimed at reducing winter air pollution in northern China by transitioning from coal to cleaner energy sources like natural gas and electricity for heating. The "Home Appliances to the Countryside" policy, initiated in response to the 2008 economic crisis, and the "Energy-saving Product Benefit Project" in 2012 encouraged residents to purchase energy-efficient appliances through financial incentives, including electric rice cookers and clean stoves. These efforts have driven a swift energy transformation in China's residential sector.

With regard to the promotion of renewable energy, countries are actively encouraging the development of renewable energy sources, such as solar, wind and hydroelectric power, in order to improve the sustainability of energy supply. In Africa, many countries have utilized abundant solar energy resources to build solar photovoltaic power stations to provide electricity to remote areas. For example, Kenya's "off-grid solar" project has reduced reliance on traditional biomass energy sources by promoting small-scale solar power systems that provide clean electricity to a large number of households without electricity. The project has also promoted the sustainable development of solar technology in the region by establishing community solar service centers to provide residents with services such as solar equipment maintenance and technical training.

Despite significant efforts by countries worldwide to target energy poverty, the outlook for achieving the SDG7 target by 2030 remains bleak. In 2022, global electricity supply declines for the first time in a decade, largely as a result of COVID-19 and the conflict in Ukraine. The global electricity supply is expected to grow by 2030, with the global electricity supply declining for the first time in a decade. Despite improvements in energy intensity and renewable energy growth, international financial flows for clean energy in developing countries remain insufficient. At current rates, by 2030, 660 million people will still lack access to electricity and 1.8 billion people will be without access to clean cooking.

15.2 Energy and Health

15.2.1 Household Energy Use and Health

Household energy use has a significant impact on health, largely through indoor air pollution. Globally, over 2.1 billion people rely on polluting fuels and inefficient cooking practices, with solid fuels and kerosene being the primary sources. This disparity persists between urban and rural areas, with rural regions facing higher exposure rates.

The combustion of these fuels within homes releases harmful pollutants, including particulate matter that can penetrate the lungs and enter the bloodstream. In poorly ventilated spaces, indoor smoke can contain 100 times more particulate matter than permissible levels. Women and children are disproportionately affected due to their close proximity to cooking areas.

The health consequences of household air pollution are severe. The World Health Organization estimates that 3.2 million premature deaths occur annually due to diseases linked to this exposure. These diseases include:

Ischemic Heart Disease: Household air pollution accounts for 12% of all deaths related to ischemic heart disease.
Stroke: Household air pollution causes 12% of stroke deaths globally.
Lower Respiratory Tract Infections: Household air pollution increases the risk of LRTIs in children by nearly 50%, contributing to 44% of childhood pneumonia deaths and 22% of adult pneumonia deaths.
Chronic Obstructive Pulmonary Disease (COPD): In low- and middle-income countries, household air pollution is responsible for 23% of adult COPD deaths.
Lung Cancer: Exposure to household air pollution from solid fuels and kerosene increases the risk of lung cancer, accounting for 11% of adult lung cancer deaths.

In 2019, household air pollution resulted in the loss of an estimated 86 million healthy life years, primarily affecting women in low- and middle-income countries. It also contributes to health issues such as low birth weight, tuberculosis, cataracts, and certain types of cancers.

In addition, indoor air pollution may also have a negative impact on the mental health of residents. A poor indoor environment may lead to emotional problems such as anxiety and depression among residents, affecting the quality of family life and social harmony. Prolonged exposure to dim, smoky indoor environments will increase the psychological pressure on residents, making them prone to moods such as irritability and restlessness, which will in turn affect family relationships and social interactions.

Reading

Household Polluting Fuel Use and Child Mortality in India

The use of polluting fuels such as fuelwood and animal dung for cooking is common in Indian households. A study by Basu et al. (2024) analyzes the impact of indoor air pollution due to polluting cooking fuels on child mortality in India.

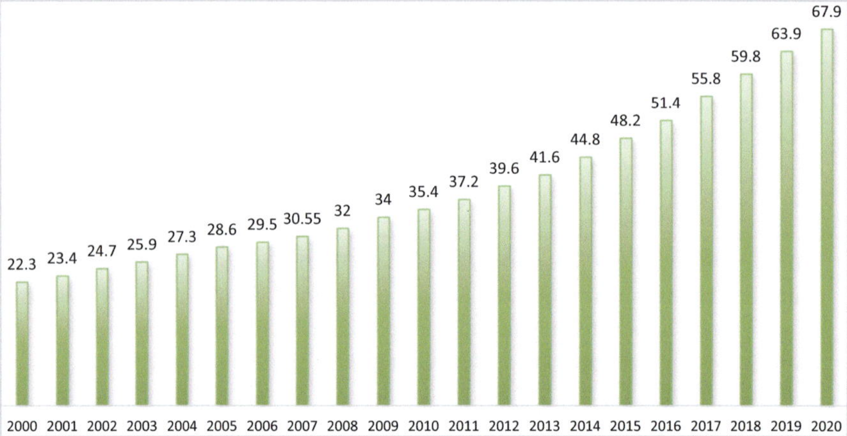

Fig. 15.3 Proportion of population using cleaner fuels in India. *Data source* World Bank (2024)

Figure 15.3 shows the proportion of the population using cleaner fuels in India. In the study,we found that cooking with polluted fuels significantly increased child mortality, especially neonatal mortality within the first 28 days of life. The mortality rate is higher for girls than for boys, and the effect is greater in relatively small households where the division of responsibility for childcare and cooking is limited in scope. If all households used cleaner fuels for cooking, 26.90 fewer children under five and 27.46 fewer infant deaths per 1,000 live births could be expected.

15.2.2 Fossil Fuel Use and Public Health

The widespread employment of fossil fuels has far-reaching implications not only for climate change but also for public health. During the various stages of fossil fuel handling, namely extraction, transportation, processing, and combustion, numerous pollutants such as sulfur dioxide, nitrogen oxides, particulate substances, and volatile organic compounds are emitted.

These pollutants partake in complex chemical interactions in the atmosphere, giving rise to phenomena like acid rain and haze, which are severe weather conditions. For example, in many industrial cities, prolonged hazy weather seriously affects the respiratory health of residents and increases the incidence of diseases such as asthma and lung cancer. At the same time, pollutants from fossil fuel combustion may also spread globally through atmospheric circulation, causing health impacts in areas far from the source of pollution. For example, although the Arctic region is far from industrial pollution sources, pollutants from fossil fuel combustion can settle in the Arctic region due to atmospheric circulation, affecting the local ecological environment. In addition, the use of fossil fuels is associated with the development of cardiovascular and neurological diseases. For example, long-term exposure to high concentrations of particulate matter may lead to an increased risk of cardiovascular diseases, such as coronary heart disease and myocardial infarction.

Historically, London's air pollution problems have been caused mainly by the burning of coal, a form of pollution known as "London smog" or "sulfide fog." This smog is produced mainly by the burning of coal with a high sulfur content, leading to a build-up of pollutants such as sulfur dioxide and particulate matter, which can be exacerbated, particularly in humid conditions. One of the worst cases was the London smog of December 1952, when a large concentration of pollutants from inversion phenomena and the burning of coal for winter heating led to the deaths of about 4000 people, and in the following months a further 8000 died of respiratory illnesses.

With the Industrial Revolution, the use of fossil fuels was not limited to coal, but also included other fuels such as gasoline. The combustion of these fuels produced new pollutants such as nitrogen oxides and volatile organic compounds, which reacted photochemically when exposed to sunlight to form photochemical smog. The Los Angeles photochemical smog event was one of the earliest such events observed, occurring in 1943, and was followed by several other occurrences in 1952, 1954, and 1955. Photochemical smog events resulted in the death of large pine forests and a reduction in citrus production in the mountains up to 2000 m above sea level, as far as 100 km away from the city. In 1950–1951 alone, the cost of air pollution in the United States amounted to 1.5 billion dollars; in 1955, more than 400 people over the age of 65 died of respiratory failure; and in 1970, more than 75% of the population suffered from pink eye.

15.3 Energy and Labor

15.3.1 Energy and Human Capital

The relationship between energy and labor is manifested in a number of ways, with energy having a particularly significant impact on human capital. Human capital is the sum of the elements of knowledge, skills and health that people accumulate through education, training and health. Energy, as an important basis for economic development, has a significant impact on the formation and development of human capital.

In energy-poor areas, the lack of a stable and reliable energy supply often makes it difficult for education and training facilities to function properly. For example, schools may not be able to use modern educational tools, such as multimedia teaching equipment and computers, due to insufficient electricity, which affects the quality of teaching and the learning outcomes of students. In rural schools in some African countries, the lack of electricity means that teachers can only rely on traditional blackboards and books for teaching, and are unable to present students with vivid multimedia teaching content, limiting their understanding and mastery of knowledge. In recent years, with the development of electronic information technology in education, the use of energy can also facilitate the development of new modes of education, such as distance education and online learning, and provide people with more convenient and efficient learning opportunities.

In addition, energy poverty may also affect the health status of the population, and, as discussed in Sect. 15.2, the poor health of workers due to energy-related health problems that prevent them from fully utilizing their labor capacity can also have a negative impact on human capital. For example, workers with chronic respiratory illnesses may take frequent time off from work due to ill health, resulting in low productivity, which affects their personal career development and the production and operation of their businesses.

To foster the harmonious advancement of energy and human potential, proactive interventions are imperative. Enhance energy governance by prioritizing clean energy adoption and renewable industry expansion while minimizing fossil fuel consumption and emissions. Invest in education and skill development programs to elevate the workforce's capabilities and ensure their employability in an evolving energy landscape. Simultaneously, bolster social protection and public services to safeguard vulnerable populations from energy disruptions and mitigate the repercussions on human capital growth.

Reading

Indoor Air Pollution and Academic Performance

Household energy use is a major causal factor of indoor air pollution. Xu et al. (2024) designed a randomized double-blind intervention experiment using indoor air purifiers in effect or not, and verified that indoor air pollution due to energy use could affect college students' performance in English exams. In recent years, with the increasing public awareness of air pollution health risk prevention, indoor air purifiers have been widely used, providing a database for conducting studies related to indoor air pollution exposure and health effects.

The study hypothesized that exposure to high levels of particulate matter would reduce cognitive functioning in healthy young adults, as manifested by a decrease in test scores, and thus a decrease in particulate matter concentrations would prompt a relative increase in test scores. To test the hypothesis, the research team used a randomized, double-blind, crossover design with separate interventions using ineffective and effective air purifiers to selectively filter indoor particulate matter. The results of the study showed that indoor air purifiers filtered out particulate matter and reduced indoor particulate matter concentrations, which were effective in improving college students' overall English test scores, with a significant improvement in reading scores.

Energy is not only responsible for influencing human capital, but in turn human capital contributes to the conservation of the resource environment. Based on Graff Zivin and Neidell (2013), we can establish a relationship between human capital and the environment as follows:

$$E = f(M(A), A(\mathrm{HC}, E_s), I_0) \tag{15.1}$$

where is E environmental pollution and M is pollution abatement which is a function of environmental awareness A. Environmental awareness A is a function of human capital HC and environmental signals E_s. I_0 is the reference value of pollution for a given level of production and consumption. The formula suggests that human capital is an important determinant of environmental awareness, which will affect resource and environmental quality by influencing the abatement behavior of households and firms.

To further determine the relationship between the impact of human capital on the environment, the above pollution function and HC can be derived:

$$\frac{\mathrm{d}E}{\mathrm{dIIC}} = \underbrace{\left(\frac{\partial E}{\partial M} \times \frac{\partial M}{\partial A} + \frac{\partial E}{\partial A}\right)}_{\frac{\mathrm{d}E}{\mathrm{d}A}} \times \underbrace{\left(\frac{\partial A}{\partial \mathrm{HC}} + \frac{\partial A}{\partial E_S} \times \frac{\partial E_S}{\partial \mathrm{HC}}\right)}_{\frac{\mathrm{d}A}{\mathrm{dHC}}} \tag{15.2}$$

where $\frac{\partial E}{\partial M} \times \frac{\partial M}{\partial A} + \frac{\partial E}{\partial A}$ can be denoted as $\frac{\mathrm{d}E}{\mathrm{d}A}$ and $\frac{\partial A}{\partial \mathrm{HC}} + \frac{\partial A}{\partial E_S} \times \frac{\partial E_S}{\partial \mathrm{HC}}$ can be denoted as $\frac{\mathrm{d}A}{\mathrm{dHC}}$. The sign of $\frac{\mathrm{d}E}{\mathrm{d}A}$ is negative and the sign of $\frac{\mathrm{d}A}{\mathrm{dHC}}$ is positive. To summarize, the combination of positive $\frac{\mathrm{d}A}{\mathrm{dHC}}$ and negative $\frac{\mathrm{d}E}{\mathrm{d}A}$ leads to the conclusion that $\frac{\mathrm{d}E}{\mathrm{dHC}}$ is negative, which proves the conclusion that the growth of human capital reduces environmental pollution.

15.3.2 Energy and Labor Supply

There is a close link between energy and labor supply. On the one hand, the state of energy supply affects the production and business activities of enterprises, which in turn affects employment opportunities and labor supply. In energy-deficient regions, enterprises may reduce the scale of production or stop production due to insufficient electricity and high energy costs, resulting in fewer jobs and unemployed or underemployed workers.

For example, in some regions that face frequent power interruptions, the productivity of manufacturing firms is substantially reduced, and in order to cut costs, firms may lay off workers or reduce the recruitment of new employees. In some garment manufacturing enterprises in some Southeast Asian countries, due to unstable power supply, machines and equipment on the production line are often down, resulting in delays in the production schedule, and the enterprises have to reduce the volume of orders, thus laying off employees to reduce costs.

On the other hand, the availability and affordability of energy also affects workers' willingness to work and labor participation rate. If household energy expenditures are too high, workers may need to spend more time and energy to obtain energy or to compensate for the financial stress caused by energy expenditures, thus reducing the time and energy devoted to labor supply. In some energy-poor areas, residents may need to spend a great deal of time collecting firewood or standing in line to purchase high-priced energy, which prevents them from fully engaging in their work.

Since 1967, Indonesia has been subsidizing the retail cost of kerosene. But following 2005, with the rise in global oil prices and the growth of oil consumption accompanying economic expansion, maintaining the kerosene subsidy became burdensome. Hence, in 2007, the Indonesian government introduced the Kerosene to LPG Conversion Program, with the key goal of reducing kerosene subsidy expenses and converting more than 70% of households not using LPG into LPG-using households within five years.

This energy transition policy has deeply influenced labor supply. A study by Verma and Imelda (2023) showed that women who were exposed to the program earlier had a 4% increase in lung capacity, equivalent to 11 L per minute, compared to those exposed later. The improvement in health may lead to labor supply changes. The study found that women got about 1 h of extra work time per day, which is a 20% increase, and men got about 0.9 h of extra work time per day, a 13% increase.

In energy-poor areas, residents often have to spend a great deal of time and effort collecting traditional biomass fuels, such as wood and animal dung, which takes up a great deal of work and study time (Fig. 15.4). For example, women and children are often tasked with fuel collection, which not only affects their work and studies, but also limits their development and ability to participate in economic activities. Energy poverty disproportionately affects women, who often have to spend a great deal of time collecting fuel, affecting their work and learning opportunities. For example, in some African countries, women spend hours each day collecting wood and animal dung, which not only takes time away from their work, but also affects their ability to learn and participate in social activities. Energy poverty also limits women's economic independence and self-development, making them more vulnerable to poverty and dependency.

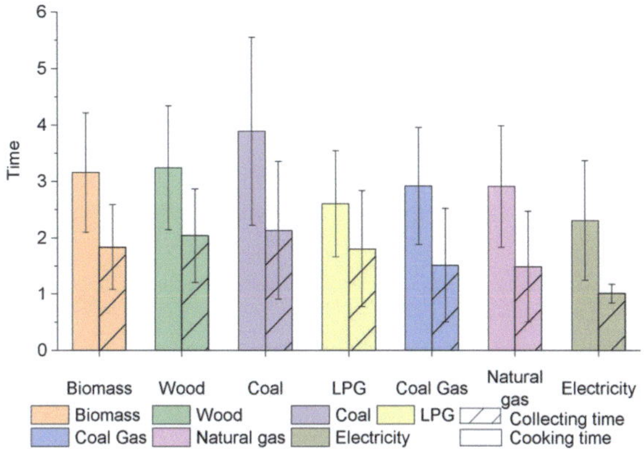

Fig. 15.4 Collecting time and cooking time with different fuels. *Note* for cooking time, 1 represents less than 15 min, 2 represents 15–30 min, 3 represents 30–45 min, 4 represents 45–60 min, 5 represents 1–2 h, and 6 represents 2–4 h. For collecting time, 1 represents no need to collect (zero-distance supply to household), 2 represents a 15–30 min walk (1–3 km), 3 represents a 30–60 min walk (3–5 km), 4 represents a 1–2 h walk (5–10 km), and 5 represents a walk of over 2 h (more than 10 km). *Source* Li et al. (2025)

15.4 Energy and Gender Equality

15.4.1 Energy and Women Development

In today's society, energy and women's development are inextricably linked, and the issue of energy has a profound impact on women's development process in many ways.

In energy-poor regions, women are often burdened with the task of collecting energy (Fig. 15.5). For example, in many rural areas of Africa and Asia, women spend a great deal of time collecting firewood or other traditional biomass fuels for household cooking and heating. This prevents them from investing their time in other more developmental activities, such as education, vocational skills training, or gainful employment. In some regions, women may spend as much as 2–3 h or more per day collecting fuel, which not only drains their physical strength but also severely limits their personal development.

The large amount of harmful pollutants, such as particulate matter and sulfur dioxide, produced by the combustion of traditional biomass fuels seriously jeopardizes women's health. Prolonged exposure to this poor indoor air pollution environment makes women more susceptible to respiratory diseases, cardiovascular diseases, and eye diseases. Clean energy programs significantly improve women's respiratory health. For example, Verma and Imelda (2023) showed that clean energy use led to a significant increase in women's lung capacity. Clean energy technologies such as electric stoves and improved cookware have reduced indoor air pollution, thereby improving the health of women and children. In Ethiopia, the large hydropower

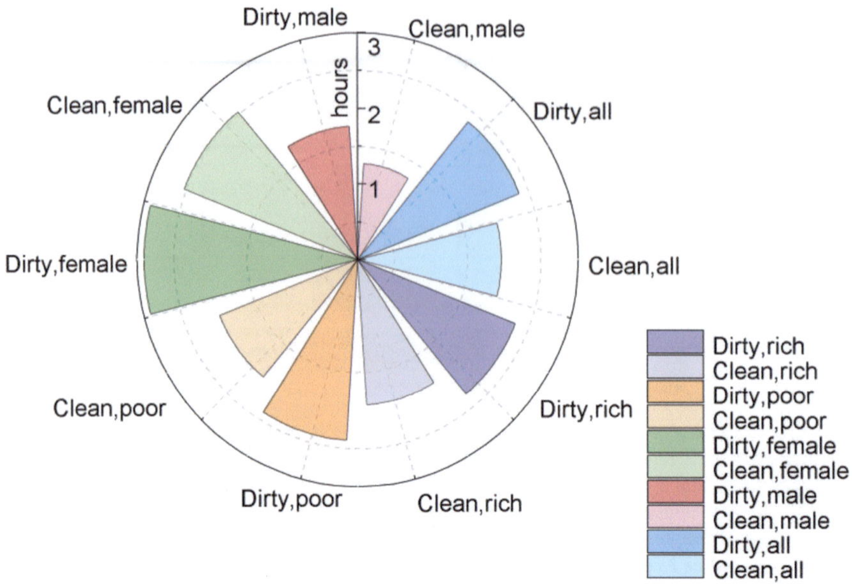

Fig. 15.5 Collecting time and cooking time with different fuels and gender. *Source* Li et al. (2025)

project Gedam has improved healthcare accessibility by providing clean energy, reducing the burden of firewood collection on women, and reducing health problems associated with the use of firewood or biomass.

Girls face many difficulties in the learning process due to inadequate household energy supply as a result of energy poverty. Lack of adequate lighting at night prevents them from studying effectively. Clean energy projects have allowed women to spend more time on educational activities. For example, in remote areas of Peru, the eMujer Clean Energy School keeps women informed by providing solar panels and training in clean energy technologies that enable them to work during the day and study at night. In Nepal, the Clean Energy Program reduces indoor air pollution and saves time by teaching clean cooking solutions, allowing women to spend more time on education.

Energy access is critical to women's daily lives, increasing their productivity, promoting employment and entrepreneurship, and improving economic conditions. For example, access to clean energy not only improves women's quality of life, but also creates new economic opportunities, supports livelihoods and generates new sources of income. In Yemen, for example, women have set up private solar micro-grids near the front lines of the conflict to provide clean electricity to their communities while earning an income. In India, women entrepreneurs use solar-powered electric sewing machines to produce face masks and sanitary napkins, providing much-needed hygiene products in rural areas.

Modern energy services have led to an improvement in women's socio-economic status and health. Nevertheless, the gender dimension in relation to women within energy policy still demands attention. Most existing energy policies often disregard gender differences and do not fully consider the needs of women and girls. Thus, it is essential to integrate a gender perspective into both the creation and implementation of energy policies, since women, just like men, play a vital role as catalysts for innovative and inclusive solutions.

As an example, the C3E initiative, led by the Clean Energy Ministerial and the International Energy Agency, aims to reduce the gender gap in energy and enhance the visibility of women leaders. The Solar Sahelis project in India, the Women in Energy project in Kenya, and the Solar Home Systems project in Bangladesh all aim to encourage women's participation in renewable energy through training, entrepreneurship, and mentorship programs. These projects generate economic opportunities, promote gender equality, and enhance the health and well-being of women and their families.

15.4.2 Energy and Fertility

The relationship between energy and fertility (Fig. 15.6) is a complex and multidimensional issue, involving social, economic, cultural, and gender roles.

At the broader societal level, an inverse relationship exists between the average energy usage per person and the number of children born per woman. Typically, as energy consumption per person rises, fertility rates drop. This trend is observable in less developed nations, where fertility rates remain high; for instance, Zambia records a fertility rate of 4.3, while in the Democratic Republic of Congo, it is even higher at 6.1.

Low energy availability can contribute to larger family sizes, which in turn may raise the birth rate and link to higher instances of both moderate and acute malnutrition. In regions like northern Tanzania, as much as half of the farming families do not own land and depend on firewood for energy. These families often suffer from inadequate access to essentials like clean water, sanitation, and electricity, problems that are exacerbated in larger households. Enhancing access to cleaner, modern energy sources could help moderate population growth, thereby potentially decreasing future demands for energy.

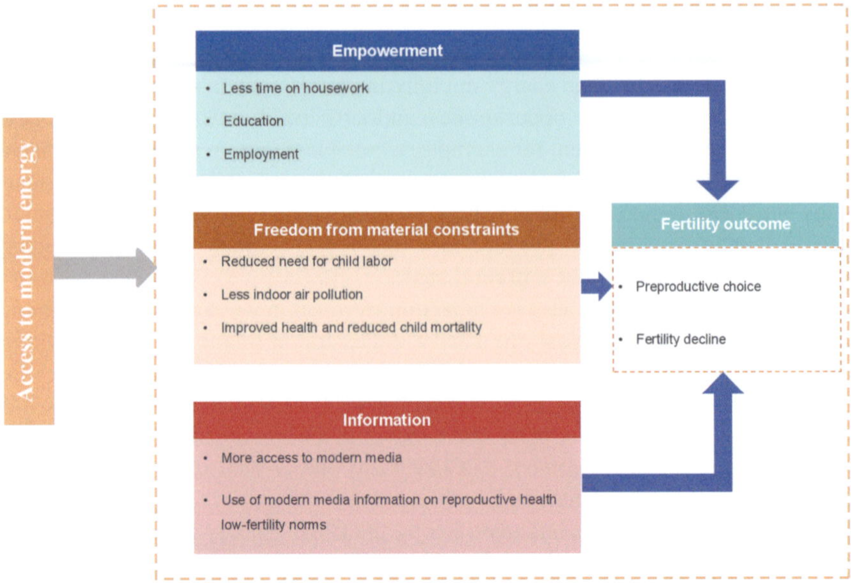

Fig. 15.6 The relationship between energy and fertility

Furthermore, the introduction of modern energy facilities significantly influences women's reproductive decisions (Belmin et al., 2022). Switching from traditional fuels like firewood or charcoal to modern energy options such as electricity and natural gas can fundamentally alter domestic tasks, specifically cooking. This not only improves health, but also reduces the burden on women who have to bear children due to heavy household chores such as collecting firewood or cooking over open fires. For example, the use of electric stoves saves a great deal of time and gives women greater access to information and education to realize their reproductive rights.

The availability of modern energy goes hand in hand with education, and both are pathways to greater female choice in the number of births. Electricity access is strongly associated with girls' education, promoting nighttime learning, reading and access to information, increasing education levels and preventing child marriage. Higher levels of education for women help to reduce fertility rates. In addition, energy use has a significant impact on women's health with measures such as improved cooking stoves and off-grid solar equipment, which will have a significant impact on future fertility rates.

The introduction of electricity has influenced fertility by providing households with new channels of information, such as family planning information, and by increasing employment opportunities. Universal access to electricity may also affect fertility through increased educational and employment opportunities. For example, projects to enhance the economic and social status of rural women in Ghana through capacity building, microfinance, and energy equipment may provide more job opportunities for rural females, thereby improving their socio-economic status. These increased opportunities may make women more inclined to delay or reduce childbearing in order to focus on professional development and personal growth. In addition, universal access to electricity may also affect fertility by improving access to medical services and public health. For example, universal access to electricity may improve public health conditions by improving access to rural health centers and reducing the use of carbon-based fuels such as kerosene or firewood.

Case

Influence of Power Interruptions on the Adoption of Electric Vehicles

Low-carbon electrification is crucial for deep decarbonization, with Electric Vehicles (EVs) playing a key role. However, a stable electricity infrastructure is essential for widespread EV adoption. China has made significant progress in the EV market but faces challenges like power interruptions due to extreme weather and energy restructuring. These interruptions affect electricity use and raise concerns about EV reliability, influencing consumer decisions.

Qiu et al. (2024) analyzed data from 310 Chinese cities (Nov 2019–Sep 2021) on power outages and EV sales. They found that power interruptions negatively impact EV adoption. An increase in power outages could reduce EV sales by approximately 4.01 units per month nationwide, leading to a significant loss in carbon reduction benefits. Doubling the annual outages per district could result in a $31.3 million annual loss in carbon reduction benefits, totaling $254 million over ten years. Hidden costs include reduced air quality and health benefits, which will worsen as the EV market grows.

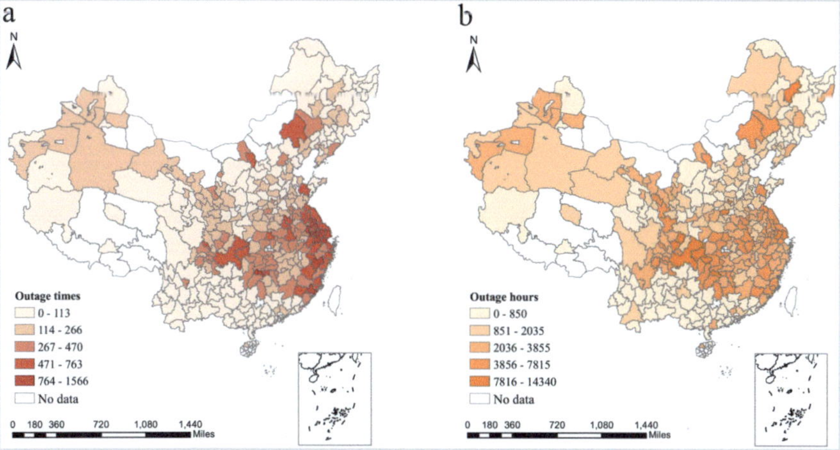

Fig. 15.7 The power disruptions in China. *Source* Qiu et al. (2024)

References

Belmin, C., Hoffmann, R., Pichler, P. P., & Weisz, H. (2022). Fertility transition powered by women's access to electricity and modern cooking fuels. *Nature Sustainability, 5*(3), 245–253.

Betto, F., Garengo, P., & Lorenzoni, A. (2020). A new measure of Italian hidden energy poverty. *Energy Policy, 138*, 111237.

Eisfeld, K., & Seebauer, S. (2022). The energy austerity pitfall: Linking hidden energy poverty with self-restriction in household use in Austria. *Energy Research & Social Science, 84*, 102427.

Epstein, A. (2014). *The moral case for fossil fuels*. Penguin.

Graff Zivin, J., & Neidell, M. (2013). Environment, health, and human capital. *Journal of Economic Literature, 51*(3), 689–730.

Hills, J. (2011). Fuel poverty: The problem and its measurement. *Interim report of the Fuel Poverty Review* (No. casereport69). Centre for Analysis of Social Exclusion, LSE.

Kahouli, S. (2020). An economic approach to the study of the relationship between housing hazards and health: The case of residential fuel poverty in France. *Energy Economics, 85*, 104592.

Li, C., Li, M., Zhang, L., Li, Q., Zheng, H., & Feldman, M. W. (2025). Energy–poverty–inequality SDGs: A large-scale household analysis and forecasting in China. *Proceedings of the National Academy of Sciences, 122*(1), e2408167121.

Li, X., Smyth, R., Xin, G., & Yao, Y. (2023). Warmer temperatures and energy poverty: Evidence from Chinese households. *Energy Economics, 120*, 106575.

Moore, R. (2012). Definitions of fuel poverty: Implications for policy. *Energy Policy, 49*, 19–26.

Pachauri, S., Poblete-Cazenave, M., Aktas, A., & Gidden, M. J. (2021). Access to clean cooking services in energy and emission scenarios after COVID-19. *Nature Energy, 6*(11), 1067–1076.

Qiu, Y., Deng, N., Wang, B., Shen, X., Wang, Z., Hultman, N., Shi, H., Liu, J., & Wang, Y. D. (2024). Power supply disruptions deter electric vehicle adoption in cities in China. *Nature Communications, 15*(1), 6041.

Taylor, L. (1993). Fuel poverty: From cold homes to affordable warmth. *Energy Policy, 21*(10), 1071–1072.

The European Space Agency. (ESA). (2024). *Nordstream pipeline map with shipping traffic*. https://www.esa.int/ESA_Multimedia/Images/2022/10/Nordstream_pipeline_map_with_shipping_traffic

Verma, A. P., & Imelda. (2023). Clean energy access: Gender disparity, health and labour supply. *The Economic Journal, 133*(650), 845–871.

Wang, H., Maruejols, L., & Yu, X. (2021). Predicting energy poverty with combinations of remote-sensing and socioeconomic survey data in India: Evidence from machine learning. *Energy Economics, 102*, 105510.

World Bank. (2024). *World development indicators*. https://databank.worldbank.org/source/world-development-indicators

Xu, J., Zhao, H., Zhang, Y., Yang, W., Wang, X., Geng, C., Li, Y., Guo, Y., Han, B., Bai, Z., Vedal, S., & Marshall, J. D. (2024). Reducing indoor particulate air pollution improves student test scores: A randomized double-blind crossover study. *Environmental Science & Technology, 58*(19), 8207–8214.

Chapter 16
Urban and City Sustainability

A city is not gauged by its length and width,
but by the broadness of its vision and the height of its dreams.
Herb Caen (1916–1997)

16.1 The Rise of Sustainability in Urban Studies

Regardless of whether one resides in a rural area, suburb, small town, or large city, built regions face increasing pressure to adopt sustainable practices. The primary reason for this is the uneven global distribution of resources. Natural resources including oil and coal, are finite, and essential urban elements, including housing and critical services, are also limited. Cities, in particular, are experiencing rapid growth that must be managed to meet the current needs of residents while anticipating future demands. Although present-day needs may be met, there is growing concern that future generations may not enjoy the same level of access to these vital resources. This underscores the importance of urban sustainability in addressing these challenges. This chapter explores various strategies for achieving sustainability in urban areas, examines methods for managing growth, and provides relevant examples to illustrate these approaches in practice.

Sustainable Development Goal 11 (SDG 11) aims to make cities and human settlements inclusive, safe, resilient, and sustainable. This goal is critical as urban areas continue to expand, necessitating strategies that ensure sustainable development. SDG 11 encompasses targets such as improving access to adequate, safe, and affordable housing, providing sustainable transport systems, enhancing inclusive and sustainable urbanization, and reducing the adverse environmental impact of cities. These targets are designed to mitigate the multifaceted challenges within urban areas, including those related to resource distribution, infrastructure, and environmental sustainability.

L. Zhang, *Sustainability Economics*, https://doi.org/10.1007/978-3-031-87829-9_16

16.1.1 The Evolution of Urban Sustainability Paradigms

The evolution of urban sustainability paradigms reflects a dynamic response to multifaceted challenges, such as resource depletion, rapid urbanization, and climate change. From the early environmental movement of the 1960s and 1970s to the adoption of international frameworks like Agenda 21 and the SDGs, the concept of urban sustainability has continually evolved. This progression has given rise to innovative paradigms such as smart cities, green cities, resilient cities, and inclusive cities, each addressing different aspects of sustainability. The integration of technology, natural elements, resilience planning, and social equity into urban development underscores the necessity for cities to balance economic growth with environmental stewardship and social justice. As urban areas continue to face complex challenges, the principles of sustainability will remain central to urban planning and policy, guiding efforts to create more resilient, inclusive, and sustainable urban environments. This ongoing evolution highlights the critical need for comprehensive and adaptive approaches to urban sustainability, ensuring that cities can thrive in the face of future uncertainties.

Early Environmental Awareness

Urban sustainability has emerged as a vital area of inquiry in response to the complex challenges associated with resource depletion, rapid urbanization, and climate change. The notion of sustainability within urban studies has undergone significant evolution, shaped by the broader environmental movement of the 1960s and 1970s. This movement highlighted the imperative for development practices that reconcile economic growth with environmental stewardship and social equity.

The rise of urban sustainability can be traced through pivotal international agreements and frameworks, such as the United Nations' Agenda 21 and the SDGs. These frameworks have established a global blueprint for sustainable urban development, encouraging cities to implement policies and practices that foster sustainability. Concurrently, academic research and urban planning methodologies have increasingly emphasized the integration of sustainability principles into urban development, giving rise to new paradigms such as smart cities, green cities, resilient cities, and inclusive cities.

The historical context of urban sustainability underscores the evolution of environmental awareness and the formulation of key frameworks and paradigms that have shaped the discipline. From the early environmental movement to the adoption of the SDGs and the New Urban Agenda, the incorporation of sustainability into urban studies reflects a growing acknowledgment of the necessity for cities to evolve in ways that balance economic growth, environmental protection, and social equity. As urban areas continue to confront multifaceted challenges, the principles of sustainability will remain central to urban planning and policy, guiding efforts to cultivate more resilient, inclusive, and sustainable urban environments. Table 16.1 shows the overview of the different urban sustainability paradigms.

Table 16.1 The overview of the different urban sustainability paradigms

City type	Key features	Examples
Smart cities	Utilizes technology and data to enhance urban management and services	Smart grids for energy distribution, intelligent transportation systems for traffic management
Green cities	Incorporates natural elements into urban design	Parks, green roofs, urban forests, sustainable transportation options, energy-efficient buildings
Resilient cities	Designed to endure and recover from environmental, economic, and social shocks	Flood defenses, stormwater management systems, flexible and adaptive design principles
Inclusive cities	Ensures access to essential services, affordable housing, and economic opportunities for all residents	Affordable housing initiatives, inclusive transportation policies, participatory urban planning

Smart Cities

The smart city paradigm harnesses technology and data to enhance urban management and services. In this model, cities employ sensors, Internet of Things (IoT) devices, and data analytics to optimize energy consumption, traffic management, and waste disposal. This approach aims to improve operational efficiency, reduce costs, and elevate the overall quality of urban life. The integration of digital technologies into urban infrastructure promotes real-time monitoring and management, resulting in more responsive and adaptive urban systems. For example, smart grids can optimize energy distribution, minimizing waste and enhancing the reliability of power supplies. Similarly, intelligent transportation systems can mitigate traffic congestion and lower emissions by optimizing traffic flow and encouraging the use of public transportation.

Green Cities

Green cities prioritize the incorporation of natural elements into urban design. This includes the creation of parks, green roofs, and urban forests, which enhance biodiversity, improve air quality, and mitigate the urban heat island effect. Green cities also advocate for sustainable transportation options and energy-efficient buildings. The integration of green infrastructure not only provides ecological benefits but also enhances the aesthetic and recreational value of urban spaces. For example, urban parks and greenways offer residents opportunities for recreation and relaxation, contributing to both physical and mental well-being. Furthermore, green roofs and walls have an edge in reducing energy consumption in buildings by offering natural insulation and cooling.

Resilient Cities

Resilient cities are designed to endure and recover from environmental, economic, and social shocks. This involves proactive planning for climate change impacts, such as rising sea levels and extreme weather events, ensuring that urban infrastructure and communities can adapt to these changes. Resilience planning encompasses the development of robust infrastructure, emergency response systems, and community preparedness initiatives. For instance, cities vulnerable to flooding may invest in flood defenses, such as levees and stormwater management systems, to safeguard at-risk areas. Additionally, resilient cities often incorporate flexible and adaptive design principles, enabling them to respond effectively to evolving conditions and emerging threats.

Inclusive Cities

Inclusive cities emphasize the significance of ensuring that all residents, irrespective of socioeconomic status, have access to essential services, affordable housing, and economic opportunities. This paradigm prioritizes social equity and the necessity of addressing social inequalities within urban development. Inclusive urban planning involves participatory processes that engage diverse stakeholders in decision-making, ensuring that the priorities and needs of marginalized communities are adequately represented. For example, affordable housing initiatives aim to provide safe and decent housing for low-income residents, while inclusive transportation policies seek to enhance access to public transit for underserved areas.

16.1.2 Examples of Urban Sustainability

Numerous cities worldwide have emerged as leaders in urban sustainability, implementing innovative practices and policies that serve as exemplary models for others to follow.

Singapore: Singapore's approach to urban sustainability encompasses innovative water management systems, abundant green spaces, and the integration of smart city technologies. The city-state's policies prioritize efficiency and resilience, positioning it as a model for sustainable urban development in densely populated environments. Singapore's water management strategy, referred to as the "Four National Taps" includes local catchment water, imported water, NEWater (recycled water), and desalinated water, thereby ensuring a reliable and sustainable water supply. Additionally, the city incorporates green spaces into its urban landscape through initiatives such as the Park Connector Network and vertical greenery in high-rise buildings.

Copenhagen: Copenhagen is widely recognized for its unwavering commitment to sustainability. The city has developed an extensive cycling infrastructure, integrated green roofs, and initiated various renewable energy projects. With the ambitious goal of achieving carbon neutrality by 2025, Copenhagen exemplifies the potential of comprehensive urban sustainability strategies. The city's emphasis on sustainable transportation, characterized by a robust network of bike lanes and pedestrian-friendly streets, has significantly diminished car dependency and reduced greenhouse gas emissions. Furthermore, Copenhagen's investments in wind energy and district heating systems have played a crucial role in advancing its carbon neutrality objectives.

16.2 Key Components of City Sustainability

City sustainability is a complex and multifaceted concept that encompasses various interrelated components, each having a crucial impact on the overall sustainability of urban areas. Essential elements such as sustainable urban planning, green infrastructure, energy efficiency, water management, waste management, and social equity collectively work to make livable, resilient, and equitable cities. By adding these components to urban planning and policy frameworks, cities can effectively tackle the intricate challenges associated with urbanization, climate change, and resource depletion, thereby ensuring a sustainable future for all residents.

16.2.1 Sustainable Urban Planning

Effective urban planning is essential for managing urban growth and development in a sustainable manner (Wei et al., 2015). This involves the creation of compact, mixed-use developments that minimize the need for car travel, promote public transportation, and preserve green spaces. The objective of sustainable urban planning is to balance the demands of the current population with those of future generations, ensuring that urban areas can accommodate growth without compromising environmental integrity or quality of life.

Compact, mixed-use developments are designed to bring residential, commercial, and recreational spaces closer together, thereby reducing reliance on private vehicles while encouraging walking, cycling, and the use of public transport. This approach not only alleviates traffic congestion and air pollution but also fosters vibrant and diverse communities. Moreover, the preservation of green spaces within urban areas provides residents with access to nature, which has been demonstrated to improve mental health and overall well-being.

Urban planners also prioritize the development of resilient infrastructure capable of withstanding environmental challenges such as climate change and natural disasters. This includes designing buildings and public spaces that are adaptable to changing conditions and implementing zoning regulations that promote sustainable land use.

16.2.2 Green Infrastructure

The integration of natural elements into urban design, including urban forests, green roofs, and parks, constitutes a vital component of city sustainability. Green infrastructure enhances biodiversity, enhances air quality, and mitigates the urban heat island effect (Tian et al., 2021). Additionally, it provides recreational spaces and contributes positively to residents' mental health and overall well-being.

Parks and urban forests serve as vital "green lungs" for cities, effectively absorbing carbon dioxide and releasing oxygen while providing habitats for wildlife, thus contributing to urban biodiversity. Green roofs and walls, which are covered with vegetation, offer further benefits by managing stormwater runoff, reducing energy consumption, and insulating buildings.

Green infrastructure projects often involve the restoration of natural ecosystems within urban settings. For instance, the establishment of wetlands and riparian buffers can enhance water quality and provide flood protection. These initiatives not only improve the ecological health of urban environments but also offer recreational and educational opportunities for residents.

16.2.3 Energy Management

The adoption of renewable energy sources and improving energy efficiency are critical for achieving city sustainability. This effort includes the implementation of energy-efficient building codes, retrofitting existing structures, and investing in infrastructures of renewable energy (Song et al., 2017). Energy efficiency measures enable cities to reduce their carbon footprint, lower energy costs, and enhance energy security.

Energy-efficient building codes establish standards for new construction and significant renovations, ensuring that buildings are designed to minimize energy consumption. These codes typically include requirements for insulation, windows, heating and cooling systems, and lighting. Retrofitting existing buildings with energy-efficient technologies, such as LED lighting, high-efficiency HVAC systems, and smart thermostats, can lead to significant reductions in energy use.

Investing in renewable energy infrastructure, including solar panels, wind turbines, and geothermal systems, allows cities to generate clean, sustainable energy. Many cities are also exploring district energy systems, which provide heating and cooling to multiple buildings from a central plant, thereby improving efficiency and reducing emissions.

16.2.4 Water Management

Sustainable water management practices are critical for the effective management of water resources (An et al., 2022a) and reducing the risk of flooding. These practices include rainwater harvesting, wastewater recycling, and the use of permeable surfaces. By implementing these strategies, cities can ensure a reliable water supply, protect water quality, and enhance resilience to climate change.

Rainwater harvesting entails the collection and storage of rainwater for subsequent use, thereby alleviating pressure on municipal water supplies. This harvested water can be utilized for irrigation, toilet flushing, and other non-potable applications. Wastewater recycling, or water reclamation, treats wastewater to a level suitable for reuse in various applications, such as landscape irrigation and industrial processes.

The use of permeable surfaces, including permeable pavements and green roofs, aids in stormwater management by allowing water to infiltrate the ground rather than flowing into storm drains. This approach reduces the risk of flooding and improves water quality by filtering pollutants. Additionally, cities are having investment projects of green infrastructure, such as bioswales and rain gardens, which utilize vegetation and soil to absorb and filter stormwater.

16.2.5 Waste Management

Effective waste management systems that promote recycling, composting, and waste-to-energy technologies are essential for minimizing the environmental impact of urban waste. These systems enable cities to reduce landfill use, lower greenhouse gas emissions, and recover valuable resources (Kumar & Samadder, 2017).

Recycling programs collect and process materials, including paper, plastic, glass, and metals, diverting them from landfills and reintegrating them into the manufacturing process. Composting initiatives convert organic waste into nutrient-rich compost, including food scraps and yard trimmings, which can enhance soil health and support urban agriculture.

Waste-to-energy technologies, such as anaerobic digestion and incineration with energy recovery, convert waste materials into energy. Anaerobic digestion breaks down organic waste in the absence of oxygen, producing biogas that can be utilized for electricity generation or heating. Incineration with energy recovery burns waste to generate steam, which can be used for electricity production or district heating.

16.2.6 Social Equity

Ensuring that all residents have access to basic services, affordable housing, and economic opportunities is a key component of city sustainability. This involves addressing social inequalities and promoting inclusive urban development. Social equity ensures that the benefits of sustainability initiatives are shared by all members of the community, regardless of their socioeconomic status (Fiack et al., 2021).

Inclusive urban development involves creating affordable housing options, improving access to quality education and healthcare, and providing economic opportunities for all residents. Affordable housing initiatives aim to ensure that low- and moderate-income households can find safe, decent, and affordable places to live. This may involve the construction of new affordable housing units, the preservation of existing affordable housing, and the provision of rental assistance programs.

Enhancing access to high-quality education and healthcare is essential for promoting social equity. Cities can invest in public schools, libraries, and community centers to provide educational resources and support services. Additionally, expanding access to healthcare services, including preventive care and mental health services, helps ensure that all residents can lead healthy lives.

Economic opportunities are also crucial for social equity. Cities can support job creation and workforce development through initiatives such as job training programs, small business support, and economic development incentives. By fostering a diverse and inclusive economy, cities can create pathways to economic mobility and reduce income inequality.

16.3 Indicators and Measurements of City Sustainability

Measuring the sustainability of cities involves the use of various indicators and metrics that assess different aspects of urban sustainability (Verma & Raghubanshi, 2018). These indicators are vital tools for policymakers and urban planners, enabling them to monitor the progress of sustainability initiatives and pinpoint areas for improvement. This section explores the primary indicators used to measure the sustainability of cities, emphasizing their significance and the methodologies employed to assess them.

16.3.1 Environmental Indicators

Environmental indicators are essential for evaluating the ecological health of cities and the environmental impact of urban activities. These indicators play a crucial role in assessing the success of sustainability strategies aimed at reducing environmental degradation and fostering ecological balance. Key environmental indicators include:

Greenhouse Gas Emissions: A measurement of a city's carbon footprint involves calculating the total greenhouse gas emissions across various sectors, including transportation, industry, and residential areas. This indicator provides valuable insights into the city's progress in addressing climate change problems by decreasing emissions and transitioning to low-carbon economies.

Air Quality: Air quality is assessed through indicators like concentrations of particulate matter ($PM_{2.5}$) and nitrogen dioxide (NO_2). These pollutants significantly impact public health, making it crucial for cities to regularly monitor and enhance air quality to ensure the well-being of residents.

Water Quality: Water quality indicators, including the concentration of contaminants and the presence of harmful microorganisms, assess the safety and purity of water sources. Access to clean water is a cornerstone of urban sustainability and public health.

Biodiversity: Biodiversity indicators measure the variety and abundance of species within urban environments. These metrics help to understand the impact of urbanization on local ecosystems and the success of conservation efforts in preserving urban biodiversity.

Green Spaces: The quantity and accessibility of green spaces, such as parks, urban forests, and community gardens, are key indicators of environmental sustainability. Green spaces offer ecological advantages like carbon sequestration, temperature regulation, and biodiversity support while promoting the overall life quality for urban residents.

16.3.2 Economic Indicators

Economic sustainability indicators assess the financial health and resilience of cities, providing insights into their capacity to foster economic growth, generate employment, and ensure the equitable distribution of resources. Key economic indicators include:

Economic Growth: Indicators such as Gross Domestic Product (GDP) and Gross Value Added (GVA) reflect the overall economic performance of a city. Sustainable economic growth is fundamental to creating job opportunities and improving living standards, enabling cities to meet the needs of both present and future populations.

Employment Rates: Employment indicators, including the unemployment rate and labor force participation rate, gauge the availability of jobs and the inclusiveness of the labor market. High employment rates are closely linked to economic stability, social well-being, and poverty reduction.

Income Distribution: The assessment of income inequality through indicators like the Gini coefficient and income quintile share ratio measures the degree of economic disparity within a city. Reducing income inequality is critical for promoting social equity, reducing poverty, and fostering social cohesion.

Affordable Housing: The availability and affordability of housing are central to economic sustainability. Housing affordability is often measured by the ratio of housing costs to household income. Ensuring access to affordable housing helps prevent homelessness and promotes long-term social stability.

16.3.3 Social Indicators

Social sustainability indicators focus on the well-being and quality of life of urban residents, reflecting the social aspects of sustainability such as access to essential services, social inclusion, and community engagement. Important social indicators include:

Access to Education: Indicators related to educational access, including school enrollment rates and literacy rates, measure the availability and quality of educational opportunities within the city. Education is a key driver of social mobility, economic development, and community resilience.

Healthcare Access: Healthcare indicators, such as the number of healthcare facilities per capita and the density of healthcare professionals, assess the availability and quality of healthcare services. Ensuring equitable access to healthcare is vital for promoting public health and fostering social well-being.

Public Transportation: Indicators related to public transportation, such as the coverage, frequency, and affordability of transit services, assess the accessibility and efficiency of urban mobility systems. Efficient public transport networks reduce dependence on private vehicles, mitigate traffic congestion, and contribute to the overall sustainability of cities.

Social Services: Access to social services, such as childcare, eldercare, and social welfare programs, is a critical aspect of social sustainability. These services support vulnerable populations, enhance social inclusion, and help create equitable urban environments.

Social Inclusion and Community Engagement: Social inclusion indicators, including community participation rates and the representation of marginalized groups in decision-making, assess the level of social cohesion and civic engagement within a city. Promoting social inclusion is essential for building resilient and equitable communities.

16.3.4 Integrated Indicators

Some frameworks use integrated indicators that combine environmental, economic, and social dimensions to offer a holistic assessment of city sustainability. These indicators enable cities to evaluate synergies and potential trade-offs between different sustainability goals. Table 16.2 shows the overview of the different forms of indicators. Notable integrated indicators include:

Global City Indicators Facility (GCIF): The GCIF provides a standardized set of indicators designed to assess city performance across various dimensions of sustainability, including governance, economy, environment, and quality of life. These indicators allow cities to benchmark their performance against global standards, promoting continuous improvement and knowledge exchange.

Table 16.2 The overview of the different forms of indicators

Indictor	Dimensions	Evaluation
Global City Indicators Facility	Governance, economy, environment, quality of life	Various indicators across governance, economic performance, environmental sustainability, and quality of life
European Green City Index	Energy consumption, greenhouse gas emissions, waste management	Energy efficiency, carbon footprint, waste recycling rates, air quality
Sustainable Cities Index	People, planet, profit	Social well-being, ecological sustainability, financial stability

European Green City Index: This index evaluates the environmental performance of European cities by examining indicators such as energy consumption, greenhouse gas emissions, and waste management practices. The index offers a comprehensive overview of urban sustainability, highlighting best practices and areas in need of improvement.

Sustainable Cities Index: The Sustainable Cities Index ranks cities worldwide based on indicators related to people, planet, and profit. It provides a balanced perspective on urban sustainability, incorporating social equity, environmental impact, and economic resilience to offer a comprehensive assessment of cities' overall sustainability.

16.4 City Sustainability Initiatives

Urban sustainability initiatives are vital for tackling the multifaceted challenges faced by climate change, rapid urbanization, and the depletion of natural resources. These initiatives generally focus on several core elements of urban sustainability, including reducing carbon emissions, enhancing resilience to environmental changes, and enhancing the overall quality of urban life. Notable approaches such as sponge cities, smart cities, low-carbon cities, eco-cities, and resilient cities each offer distinct strategies that contribute to sustainability while enhancing the livability of urban areas. By adopting these approaches, cities can foster environments that are more sustainable, resilient, and conducive to a high quality of life. The following section discusses these initiatives in greater detail, exploring their goals, methodologies, and impacts.

16.4.1 Sponge Cities

The concept of sponge cities is centered on the idea of designing urban environments that can effectively absorb and manage rainwater, thereby reducing the risks associated with flooding and improving water quality. This is achieved through the incorporation of permeable surfaces, green roofs, and rain gardens. The primary aim of sponge cities is to mimic natural hydrological cycles by allowing rainwater to infiltrate, be stored, and reused within urban areas, thereby enhancing both water management and urban resilience.

Permeable Surfaces: Materials such as permeable pavements and porous asphalt allow rainwater to seep into the ground rather than flow into storm drains. This reduces surface runoff, mitigates flooding, and helps recharge groundwater supplies.

Green Roofs: Vegetated roofs, which are covered with soil and plant life, absorb rainwater, decrease runoff, and provide insulation. In addition to their hydrological benefits, green roofs can reduce the urban heat island effect and create habitats for wildlife.

Rain Gardens: Shallow, landscaped basins designed to capture and filter stormwater runoff, rain gardens help mitigate flooding and improve water quality by naturally filtering pollutants before they enter stormwater systems.

Example

Freiburg as an Eco-City

Freiburg, Germany, stands as a beacon of sustainable urban development, showcasing how cities can harmoniously integrate environmental stewardship with urban living. The city's extensive green spaces not only enhance biodiversity but also provide residents with recreational areas that promote well-being. Freiburg's comprehensive public transportation network reduces reliance on private vehicles, thereby lowering carbon emissions and traffic congestion. Additionally, the city's forwardthinking policies on renewable energy and energy efficiency have led to significant reductions in greenhouse gas emissions. By prioritizing solar energy, Freiburg has become an renewable energy adoption leader, with numerous solar panels adorning rooftops across the city. This holistic approach to sustainability has not only improved the life quality for Freiburg's residents but also set a benchmark for other cities aiming to achieve environmental responsibility and resilience.

Case

Sponge City Initiative in Shenzhen

The concept of sponge cities, as exemplified by Shenzhen, China, represents a groundbreaking approach to urban water management and sustainable development (see Fig. 16.1). Shenzhen, a rapidly expanding metropolis, has encountered significant challenges related to waterlogging, water pollution, and inadequate drainage infrastructure. To address these issues, the Shenzhen government introduced the "Interim Measures for the Construction Management of Shenzhen Sponge City" in 2018. This policy integrates various aspects of urban planning, construction, and management to enhance the city's capacity to absorb, store, and utilize rainwater. Key initiatives include the development of permeable pavements, green roofs, sunken rain gardens, and comprehensive drainage systems. These measures have significantly improved the urban water environment, reduced waterlogging incidents, and promoted the sustainable use of water resources.

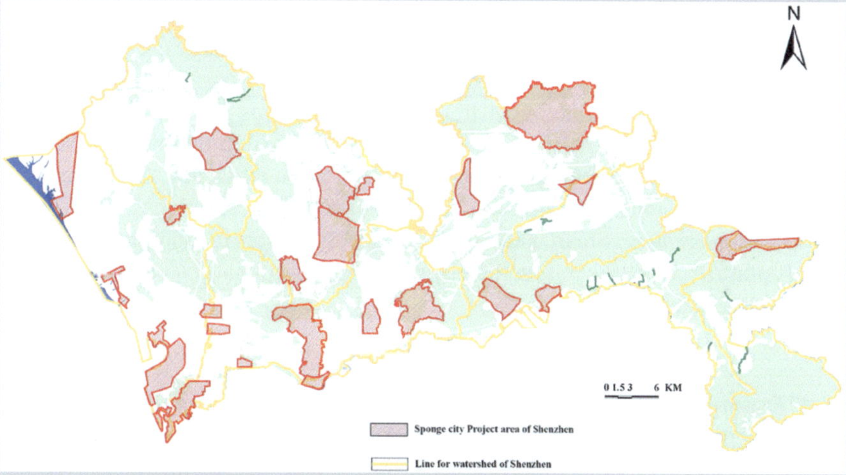

Fig. 16.1 Sponge city project areas within Shenzhen watersheds. *Source* Wang et al. (2022)

Shenzhen's sponge city initiative has achieved remarkable results, with over 1361 projects completed and 276 km^2 of sponge city area established, exceeding the national target. Notable projects include the Baishida Elementary School, which features a permeable playground to prevent water accumulation, and the Futian Mangrove Ecological Park, which effectively utilizes 92% of annual rainfall. Additionally, the Vanke Cloud City Project demonstrates the successful integration of construction and ecological restoration. Despite these achievements, the policy still faces challenges, such as the need for greater public participation and the development of a comprehensive rainwater management database. Overall, Shenzhen's experience offers valuable insights into the potential of sponge cities to improve urban resilience and sustainability in the face of climate-related risks.

16.4.2 Smart Cities

Smart cities integrate advanced technologies and data-driven solutions to optimize urban management and services. By employing sensors and IoT devices, these cities can monitor and optimize systems such as energy consumption, traffic flow, and waste management. The goal is to increase operational efficiency, reduce costs, and enhance the quality of urban life.

Energy Management: Energy management systems and smart grids optimize energy distribution and consumption, reducing waste and improving system reliability. For example, smart meters offer real-time data on energy use, empowering consumers to make more informed decisions.

Traffic Management: Intelligent transportation systems utilize advanced technology, such as data analytics and sensor technologies, to optimize traffic flow, reduce congestion, and enhance public transportation efficiency. Adaptive traffic signal systems adjust timings based on real-time traffic conditions.

Waste Management: Smart waste systems employ sensors to monitor waste levels and optimize collection routes, thereby reducing operational costs and minimizing environmental impacts.

Example

Rotterdam as a Resilient City

Rotterdam, Netherlands, exemplifies resilience through its cutting-edge flood protection strategies and comprehensive climate adaptation measures. The Maeslantkering storm surge barrier, a marvel of engineering, is a pivotal component of the city's defense against rising sea levels and storm surges. Additionally, Rotterdam has embraced innovative solutions such as floating buildings, which not only provide housing but also adapt to fluctuating water levels. The city's approach is holistic, integrating emergency preparedness and active community engagement to ensure that residents are well-informed and involved in resilience efforts. By cultivating a culture of preparedness and innovation, Rotterdam has established itself as a leader role of urban resilience globally, demonstrating how cities can effectively respond to the challenges posed by climate change and extreme weather events. Reading Smart Cities Mitigate Climate Risks Smart cities play an essential role in mitigating climate risks by harnessing advanced technologies and innovative urban planning strategies. The integration of information and communication technologies (ICTs) within these urban environments enhances their resilience to climate-related challenges.

Reading

Smart Cities Mitigate Climate Risks

Smart cities play an essential role in mitigating climate risks by harnessing advanced technologies and innovative urban planning strategies. The integration of information and communication technologies (ICTs) within these urban environments enhances their resilience to climate-related challenges. For instance, smart cities can utilize real-time data generated by sensors and IoT devices to monitor environmental conditions, predict extreme weather events, and implement adaptive measures promptly. This proactive approach reduces the vulnerability of urban infrastructure and communities to climate risks such as floods, heatwaves, and storms. Additionally, smart city initiatives, such as energy-efficient buildings, smart grids, and sustainable transportation systems, not only contribute to a decrease in greenhouse gas emissions but also promote overall environmental sustainability (see Fig. 16.2). By fostering a more adaptive and responsive urban environment, smart cities enhance the overall life quality for their residents while ensuring long-term resilience against the negative impacts of climate change.

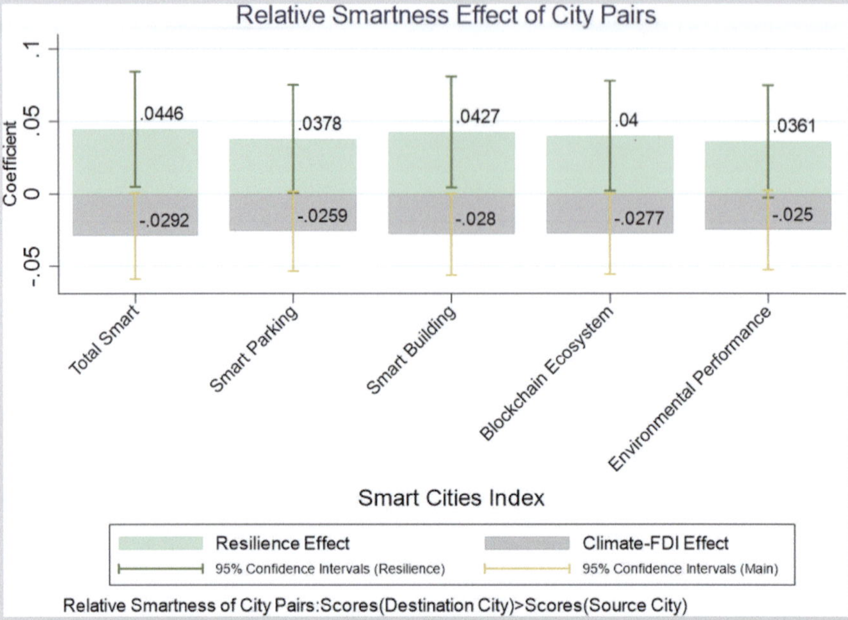

Fig. 16.2 Histogram of climate risk difference between city pairs. *Source* An et al. (2022b)

16.4.3 Low-Carbon Cities

Low-carbon cities aim to significantly reduce greenhouse gas emissions by promoting renewable energy, enhancing energy efficiency, and encouraging sustainable transportation. These cities are focused on transitioning to low-carbon economies in order to mitigate the adverse effects of climate change.

Renewable Energy: Low-carbon cities prioritize the adoption of renewable energy sources, including solar, wind, and geothermal power. Urban installations, like rooftop solar panels and wind turbines, play a significant role in generating clean energy while decreasing reliance on fossil fuels.

Energy Efficiency: Implementing stringent energy-efficient building codes, retrofitting older buildings, and promoting energy-saving technologies are key strategies for reducing overall energy consumption. Energy-efficient buildings, for example, utilize less energy for heating, cooling, and lighting, which directly lowers emissions.

Sustainable Transportation: Cities promote the use of electric vehicles (EVs), invest in public transportation systems, and create infrastructure for cycling and walking. These measures reduce transportation-related emissions and promote more sustainable urban mobility.

China's low-carbon city policy represents a pioneering effort to address the environmental challenges posed by rapid urbanization and industrialization. Initiated in 2010, the policy framework aims to reduce carbon emissions, promote energy efficiency, and enhance sustainable urban development. The central government, in collaboration with local authorities, has implemented a series of regulations and initiatives to achieve these goals. Key measures include improving industrial energy efficiency, reducing building energy consumption, and enhancing transportation fuel economy. Additionally, the policy encourages the development of clean energy sources and the adoption of green technologies. The low-carbon city pilot program, which includes 42 pilot projects across various provinces and cities, serves as a testing ground for innovative strategies and practices. These pilots have demonstrated significant progress in reducing carbon footprints and improving urban ecological efficiency. For instance, cities like Shenzhen and Hangzhou have implemented comprehensive plans that integrate renewable energy, smart grids, and sustainable transportation systems. The success of these pilot projects has provided valuable insights and best practices that can be scaled up and replicated across the country. Overall, China's low-carbon city policy underscores the nation's commitment to balancing economic growth with environmental protection, paving a more sustainable way in the future.

16.4.4 Eco-cities

Eco-cities are urban areas designed with the primary goal of minimizing environmental impact while promoting sustainable living practices. These cities integrate green infrastructure, encourage sustainable transportation modes, and foster eco-friendly behaviors among residents.

Green Infrastructure: Eco-cities incorporate a variety of green spaces, urban forests, and green roofs to enhance air quality, improve biodiversity, and offer recreational opportunities. These green elements also mitigate the urban heat island effect and assist in managing stormwater.

Sustainable Transportation: Prioritizing public transit, cycling, and walking over private car use is a hallmark of eco-cities. This reduces traffic congestion, lowers emissions, and promotes healthier lifestyles. Investments in bike lanes, pedestrian-friendly streets, and efficient public transportation systems are common strategies.

Eco-Friendly Behaviors: Encouraging residents to engage in sustainable practices, including recycling, composting, and conserving resources like water and energy, is integral to the success of eco-cities. Public campaigns and incentives can help raise awareness and foster responsible behaviors.

Csae

Barcelona's Smart City Projects

Barcelona is widely recognized for its pioneering role in the development of smart city projects that focus on sustainability. The city's smart initiatives aim to enhance energy efficiency,transportation systems, and public services while promoting environmental sustainability and digital inclusion.

Smart Grids: Barcelona has deployed smart grids across the city to optimize the distribution and consumption of energy. By utilizing real-time data, these grids balance supply and demand, decreasing energy waste and enhancing the reliability of the electricity network. Additionally, Barcelona encourages the adoption of renewable energy sources, including solar power, to lower carbon emissions.

Traffic Management: The city's smart traffic management system uses sensors and data analytics to monitor traffic flow and optimize signal timings. This system has proven effective in reducing congestion, lowering emissions, and improving the overall efficiency of the city's transportation infrastructure.

Waste Management: Barcelona has also implemented a smart waste management system that uses sensors to monitor the fill levels of waste bins. This technology helps to optimize collection routes, reduce operational costs, and minimize the environmental impact of waste collection while improving service efficiency.

Digital Inclusion: In recognition of the importance of digital inclusion, Barcelona has invested in initiatives aimed at ensuring equitable access to digital technologies. These include digital literacy programs and the provision of free Wi-Fi in public spaces, designed to reduce the digital divide and empower all residents to participate in the digital economy.

Environmental Sustainability: Barcelona's smart city projects are rooted in a commitment to environmental sustainability. The city has adopted green building standards, promoted energyefficient technologies, and encouraged the use of sustainable transportation options to minimize its ecological footprint.

Example

Freiburg as an Eco-City

Freiburg, Germany, stands as a beacon of sustainable urban development, showcasing how cities can harmoniously integrate environmental stewardship with urban living. The city's extensive green spaces not only enhance biodiversity but also provide residents with recreational areas that promote well-being. Freiburg's comprehensive public transportation network reduces reliance on private vehicles, thereby lowering carbon emissions and traffic congestion. Additionally, the city's forward-thinking policies on renewable energy and energy efficiency have led to significant reductions in greenhouse gas emissions. By prioritizing solar energy, Freiburg has become an renewable energy adoption leader, with numerous solar panels adorning rooftops across the city. This holistic approach to sustainability has not only improved the life quality for Freiburg's residents but also set a benchmark for other cities aiming to achieve environmental responsibility and resilience.

16.4.5 Resilient Cities

Resilient cities are those that are built to withstand and recover from a variety of environmental, economic, and social shocks. These cities focus on preparing for and adapting to the impacts of climate change, such as rising sea levels and extreme weather events, while ensuring that infrastructure and communities remain adaptable in facing such challenges.

Climate Adaptation: Resilient cities develop and implement comprehensive climate adaptation strategies, which may include flood defenses, enhanced stormwater management systems, and the protection of critical infrastructure.

Emergency Preparedness: Effective disaster response systems and community preparedness programs are vital for resilient cities. These measures ensure that urban areas can respond quickly and recover effectively in the event of a disaster.

Community Resilience: Building social cohesion and fostering strong community networks are essential components of resilience. Community-based initiatives, such as local resilience hubs or neighborhood support programs, play a crucial role in ensuring that residents can cope with and recover from adverse events.

Example

Rotterdam as a Resilient City

Rotterdam, Netherlands, exemplifies resilience through its cutting-edge flood protection strategies and comprehensive climate adaptation measures. The Maeslantkering storm surge barrier, a marvel of engineering, is a pivotal component of the city's defense against rising sea levels and storm surges. Additionally, Rotterdam has embraced innovative solutions such as floating buildings, which not only provide housing but also adapt to fluctuating water levels. The city's approach is holistic, integrating emergency preparedness and active community engagement to ensure that residents are well-informed and involved in resilience efforts. By cultivating a culture of preparedness and innovation, Rotterdam has established itself as a leader role of urban resilience globally, demonstrating how cities can effectively respond to the challenges posed by climate change and extreme weather events.

16.5 Sustainability Challenges in the Age of Digitalization

The digitalization of cities offers significant opportunities for enhancing urban sustainability, but it also presents several challenges that must be addressed. Digital twins, data privacy and security, the digital divide, the environmental impact of digital technologies, and the integration with existing infrastructure are critical areas that require careful consideration. By addressing these challenges, cities can harness the potential of digital technologies to make more efficient, resilient, and sustainable urban environments. This requires a collaborative approach, involving policymakers, urban planners, technology providers, and residents. Through thoughtful planning and implementation, cities can navigate the complexities of digitalization and achieve their sustainability goals.

16.5.1 Digital Twins

Digital twins are virtual replicas of physical urban environments that can be used to simulate and optimize urban systems. They enable real-time monitoring and analysis of urban infrastructure, helping cities to improve efficiency and resilience. However, the implementation of digital twins requires significant investment in technology and data management.

Digital twins can model various urban systems, including transportation networks, energy grids, and water management systems. By simulating different scenarios, cities can identify potential issues and optimize their operations. For example, a digital twin of a city's transportation network can help planners understand traffic patterns and develop strategies to reduce congestion and emissions.

Despite their potential benefits, digital twins pose several challenges. The creation and maintenance of digital twins require substantial financial resources and technical expertise. Additionally, the integration of digital twins with existing urban infrastructure can be complex, necessitating careful planning and coordination.

16.5.2 Digital Divide

The digital divide refers to the gap between those who have access to digital technologies and those who do not. Addressing this divide is significant for all residents; they can benefit from the opportunities presented by digitalization.

The digital divide can manifest in various ways, including disparities in access to high-speed internet, digital literacy, and the availability of digital devices. These disparities can exacerbate existing social inequalities, limiting access to employment, education, and other important services.

To bridge the digital divide, cities must invest in digital infrastructure and provide affordable internet access to all residents. Additionally, digital literacy programs can help individuals develop the needed skills to navigate the digital landscape effectively. By promoting digital inclusion, cities can ensure that all residents can participate in and benefit from digitalization.

16.5.3 Challenges of Digital Technology

Data Privacy and Security

The increasing use of digital technologies in cities raises concerns about data privacy and security. Ensuring that residents' data is protected and used responsibly is essential for maintaining public trust and promoting sustainable digitalization.

Smart city technologies collect vast amounts of data from various sources, including sensors, cameras, and mobile devices. This data can be used to improve urban services, but it also raises privacy concerns. Unauthorized access to personal data can lead to identity theft, surveillance, and other privacy violations.

To address these problems, cities must implement robust data protection measures. This includes encrypting data, establishing clear data governance policies, and ensuring transparency in data collection and use. Additionally, cities should engage with residents to build trust and ensure that their privacy concerns are addressed.

Environmental Impact of Digital Technologies

The production and use of digital technologies have environmental impacts, such as energy consumption and electronic waste. Sustainable digitalization involves minimizing these impacts through the usage of energy-efficient technologies and responsible e-waste management.

Digital technologies, including data centers, IoT devices, and smart grids, consume significant amounts of energy. To reduce their environmental footprint, cities can adopt energy-efficient technologies and practices. For example, data centers can implement energy-saving measures, including optimizing cooling systems and using renewable energy sources.

Electronic waste, or called e-waste, is considered as another significant environmental concern. The rapid turnover of digital devices leads to the accumulation of e-waste, which can contain hazardous materials. Cities must develop effective e-waste management strategies, including recycling programs and regulations to ensure the safe disposal of electronic devices.

Integration with Existing Infrastructure

Integrating digital technologies with existing urban infrastructure can be challenging. Cities need to ensure that new technologies are compatible with existing systems and that they enhance, rather than disrupt, urban sustainability efforts.

Digital technologies require careful planning and coordination. For example, the implementation of smart grids must be aligned with existing energy infrastructure to ensure seamless operation. Similarly, the deployment of IoT devices for traffic management must be integrated with existing transportation systems.

Cities must also consider the sustainability of digital technologies for the long term. This includes assessing the lifecycle impacts of digital devices and ensuring that they can be maintained and upgraded over time. By adopting a holistic approach to integration, cities can maximize the benefits of digital technologies while minimizing potential disruptions.

16.5.4 Integrating Digital Technologies for Sustainable City Construction

In recent years, cities across the globe have increasingly turned to digital technologies to tackle sustainability challenges, making significant strides toward enhancing efficiency, resilience, and environmental sustainability. Through initiatives like digital twins, smart grids, and green infrastructure, cities such as Rotterdam, Singapore, and Barcelona are integrating advanced technologies to address urban sustainability issues. These cases underscore the significance of a comprehensive and coordinated approach to urban development, ensuring that technological advancements contribute not only to the optimization of urban systems but also to the well-being of all residents and the preservation of the environment (Fig. 16.3).

Case

Barcelona's Smart City Projects

Barcelona is widely recognized for its pioneering role in the development of smart city projects that focus on sustainability. The city's smart initiatives aim to enhance energy efficiency, transportation systems, and public services while promoting environmental sustainability and digital inclusion.

Smart Grids: Barcelona has deployed smart grids across the city to optimize the distribution and consumption of energy. By utilizing real-time data, these grids balance supply and demand, decreasing energy waste and enhancing the reliability of the electricity network. Additionally, Barcelona encourages the adoption of renewable energy sources, including solar power, to lower carbon emissions.

Traffic Management: The city's smart traffic management system uses sensors and data analytics to monitor traffic flow and optimize signal timings.

Fig. 16.3 Smart shower device in Singapore. *Note* Smart water technologies can help consumers track their water usage for specific household activities. An example is the smart shower device which provides users with real-time feedback on their water consumption. This allows users to save water by setting water conservation goals and adjusting their habits

This system has proven effective in reducing congestion, lowering emissions, and improving the overall efficiency of the city's transportation infrastructure.

Waste Management: Barcelona has also implemented a smart waste management system that uses sensors to monitor the fill levels of waste bins. This technology helps to optimize collection routes, reduce operational costs, and minimize the environmental impact of waste collection while improving service efficiency.

Digital Inclusion: In recognition of the importance of digital inclusion, Barcelona has invested in initiatives aimed at ensuring equitable access to digital technologies. These include digital literacy programs and the provision of free Wi-Fi in public spaces, designed to reduce the digital divide and empower all residents to participate in the digital economy.

Environmental Sustainability: Barcelona's smart city projects are rooted in a commitment to environmental sustainability. The city has adopted green building standards, promoted energy-efficient technologies, and encouraged the use of sustainable transportation options to minimize its ecological footprint.

References

An, Y., Li, Q., & Zhang, L. (2022). Managing agricultural water use in a changing climate in China. *Sustainable Production and Consumption, 33*, 978–990.

An, Y., Liu, N., Zhang, L., & Zheng, H. (2022). Adapting to climate risks through cross-border investments: Industrial vulnerability and smart city resilience. *Climatic Change, 174*, 10.

Fiack, D., Cumberbatch, J., Sutherland, M., & Zerphey, N. (2021). Sustainable adaptation: Social equity and local climate adaptation planning in U.S. cities. *Cities, 115*, 103235.

Kumar, A., & Samadder, S. R. (2017). A review on technological options of waste to energy for effective management of municipal solid waste. *Waste Management, 69*, 407–422.

Song, Q., Li, J., Duan, H., Yu, D., & Wang, Z. (2017). Towards to sustainable energy-efficient city: A case study of Macau. *Renewable and Sustainable Energy Reviews, 75*, 504–514.

Tian, P., Li, J., Cao, L., Pu, R., Wang, Z., Zhang, H., Chen, H., & Gong, H. (2021). Assessing spatiotemporal characteristics of urban heat islands from the perspective of an urban expansion and green infrastructure. *Sustainable Cities and Society, 74*, 103208.

Verma, P., & Raghubanshi, A. S. (2018). Urban sustainability indicators: Challenges and opportunities. *Ecological Indicators, 93*, 282–291.

Wang, Y., Jiang, Z., & Zhang, L. (2022). Sponge city policy and sustainable city development: The case of Shenzhen. *Frontiers in Environmental Science, 9*, 772490.

Wei, Y., Huang, C., Lam, P. T. I., & Yuan, Z. (2015). Sustainable urban development: A review on urban carrying capacity assessment. *Habitat International, 46*, 64–71.

Chapter 17
Human–Environment Interactions

Each person, human or no, is bound to every other in a reciprocal relationship. Just as all beings have a duty to me, I have a duty to them. If an animal gives its life to feed me, I am in turn bound to support its life. If I receive a stream's gift of pure water, then I am responsible for returning a gift in kind. An integral part of a human's education is to know those duties and how to perform them.
"Braiding Sweetgrass", Robin Wall Kimmerer (2013)

17.1 Population and Environment

17.1.1 Malthusian Theory of Population

Thomas Robert Malthus was a famous British demographer and political economist. He systematically elaborated his ideas on population theory in his book "The Principles of Population" published in 1798, and this theory had an extremely wide and far-reaching influence at that time and for a long time afterward.

The core ideas of Malthusian population theory are based on two basic assumptions. The first assumption is that food is necessary for human survival, and the second assumption is that lust between the sexes is inevitable and will remain almost as it is. Based on these two assumptions, he believed that population, if unrestricted, would grow geometrically, while the means of subsistence would only grow arithmetically. For example, the population might grow in a geometric progression of 1, 2, 4, 8, 16, … while the food supply might grow only in an arithmetic progression of 1, 2, 3, 4, 5, …. This huge difference in growth rates inevitably leads to an imbalance between population growth and the availability of means of subsistence.

Over time, the rate of population growth far exceeds the rate of growth of the means of subsistence, which inevitably leads to a series of serious consequences. When the size of the population exceeds what the means of subsistence can carry,

L. Zhang, *Sustainability Economics*, https://doi.org/10.1007/978-3-031-87829-9_17

phenomena such as poverty, famine, disease and war occur. These so-called "positive disincentives" bring the population back to a level appropriate to the means of subsistence. For example, many times in history, when a region was overpopulated and food production could not meet demand, it was prone to large-scale famines that resulted in large numbers of deaths, thus causing a sharp decline in population numbers.

At the same time, Malthus also mentioned "preventive inhibitions", which included moral restraints, such as late marriage, celibacy and other behaviors. He believed that people could control the rate of population growth through these preventive measures in order to avoid various crises caused by excessive population growth. However, he also pointed out that in real society, the effect of preventive inhibition is often limited because human eroticism makes it difficult for people to completely restrain the desire to procreate.

Malthusian demography had a significant impact on classical economics. It prompted economists to think more deeply about the relationship between the finiteness of resources and population growth, and provided an important dimension of thinking for the subsequent development of economic theory. For example, the pressures and constraints imposed on the economic system by demographic factors had to be considered in the study of economic growth and resource allocation.

The theory also triggered extensive social discussion and research. It has drawn people's attention to the root causes of the problem of poverty in society, recognizing that excessive population growth may be an important factor contributing to the increase in poverty. It has also influenced the formulation of social policies to a certain extent, for example, some countries have begun to pay attention to the study and implementation of population control policies to alleviate the tension between population and resources.

The ideas of Malthus had an important role in inspiring Darwin's theory of evolution. In his study of biological evolution, Darwin drew on Malthus's view that the growth of biological populations was limited by environmental resources, and put forward the core idea of the theory of evolution, which is "the survival of the fittest". Biological populations are bound to compete for survival in an environment of limited resources, and only those individuals and species that are adapted to the environment and able to acquire sufficient resources can survive and reproduce.

Malthusian population theory has certain limitations. First, his population growth model is too simplistic and absolute. In reality, population growth is not entirely geometric and continuous; with social and economic development, scientific and technological progress and the change of people's concept of reproduction, the rate of population growth will change greatly. For example, the population growth of many developed countries has been very slow or even negative in modern times. Secondly, he has not fully taken into account the enormous role of scientific and technological progress in increasing the production capacity of the means of living. The rapid development of modern agricultural technology, industrial production technology and so on has greatly enhanced the supply capacity of the means of subsistence, which to a certain extent can satisfy the demand of the ever-increasing population.

Reading

Unlocking Adolescents' Potential for Sustainable Engagement

Thomaes et al. (2023) explore adolescents' sustainable engagement in the context of climate change. Adolescents are uniquely positioned as they are concerned about the planet, impacted by climate change, and motivated to contribute, yet often show low sustainable behavior rates. The sustainability motive-alignment hypothesis is proposed. It suggests that when adolescents see sustainable behavior as fulfilling their motives for autonomy and status, they are internally motivated. For example, if an adolescent views advocating for environmental protection as a way to assert independence from adults' perceived narrow views and gain respect among peers, they are more likely to engage.

Barriers such as cognitive limitations, where adolescents may doubt their ability to make an impact, normative influences like the lack of parental or peer modeling of sustainable behaviors, and emotional responses including climate anxiety or unfounded hope, impede their sustainable engagement. Thomaes et al. (2023) discover that by reframing sustainability to resonate with adolescents' desires for autonomy and status, policies can become more effective. For instance, a school project that positions recycling as a trendy and responsible choice among peers can lead to increased participation. Media campaigns can also play a crucial role by depicting sustainable actions as a means of expressing individuality and social consciousness.

Despite its limitations, the Malthusian theory of population also has positive significance that cannot be ignored. For the first time, it systematically raised the issue of the relationship between population and means of subsistence, which aroused great attention to the population issue on a global scale and prompted governments, scholars and all walks of life to conduct in-depth studies on population laws and policies. It also laid the foundation for subsequent demographic research and promoted the continuous development and improvement of demography as a discipline.

In conclusion, Malthusian population theory is an important milestone in the history of demography; although its theory has certain defects, the thinking and discussion it triggered have had a lasting and profound impact in many fields such as economics, sociology, biology, and so on, and in today's world, the relationship between population and resources and environment is still one of the important challenges faced by the world (Fig. 17.1), and some of the ideas in Malthusian population theory are still worthy of our in-depth study and reflection. We should study and reflect on them in depth.

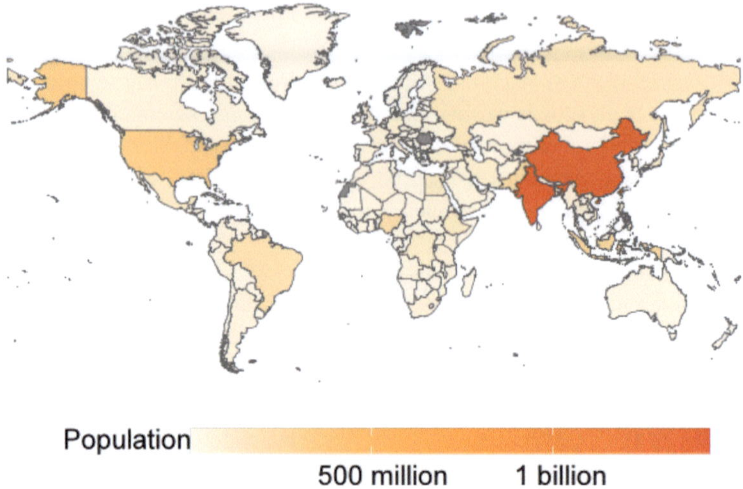

Fig. 17.1 World population distribution (2023). *Data source* World Bank (2024)

17.1.2 Population Prediction Method

Population prediction is of great significance in the fields of demographic research, social and economic planning, public policy-making, etc. Accurate population prediction can provide a key basis for governmental decision-making, urban planning, and resource allocation.

Trend extrapolation

Trend extrapolation makes projections of future populations based on the assumption that a certain trend exhibited by historical population data will continue in the future. It assumes that population growth or change follows certain patterns and that future conditions can be inferred by analyzing past data. The most commonly used extrapolation formula for linear trends is:

$$P_t = P_0 + kt \tag{17.1}$$

where P_t is the number of people in the projection period, P_0 is the number of people in the base period, k is the slope of population growth (i.e., the amount of population growth per year), and t is the time interval between the projection period and the base period.

For example, if it is known that the population of a region was ten million ($P_0 =$ 1,000,000) in the year 2000 (base period), and that the population has been growing at an average rate of twenty thousand (k = 20,000) per year over the past few decades, the population is projected to be fifteen million (15,000,000) in the year 2025 ($P_{2025} = 10 + 0.2 \times 25 = 15$ million).

Nonlinear trend extrapolation may take the form of a quadratic:

$$P_t = P_0 + kt + at^2 \tag{17.2}$$

or exponential function:

$$P_t = P_0 e^{rt} \tag{17.3}$$

where r is the population growth rate, and the appropriate form of the function is selected for fitting and forecasting based on the characteristics of the data.

Cohort Element Method

The cohort element method groups the population by age and sex (cohort) and considers the effects of the elements of births, deaths, and migration on the size and structure of the population for each cohort separately. It is assumed that the elements of population dynamics in each cohort will follow certain patterns in the future and that these patterns can be estimated from historical data. The basic formula for projecting the future population size of an age cohort is:

$$P_{x,t+1} = P_{x,t} - D_{x,t} + B_{x,t} - M_{x,t} + I_{x,t} \tag{17.4}$$

where $P_{x,t}$ indicates the number of people aged x in the period t, $D_{x,t}$ is the number of people aged x in the period who died, $B_{x,t}$ is the number of babies born to women aged x in the period (births), $M_{x,t}$ is the number of people aged x in the period who moved out, and $I_{x,t}$ is the number of people aged x in the period who moved in.

For example, to project the number of females in the 25–29 age group in a region for the next year ($t + 1$), it is necessary to know the number of females in that age group in the current year t ($P_{25-29,t}$), the number of deaths ($D_{25-29,t}$), the number of births ($B_{25-29,t}$), the number of out-migrants ($M_{25-29,t}$) and the number of in-migrants ($I_{25-29,t}$) of females in that age group in the current year, and then substituting them into the formula. For the number of births $B_{x,t}$, it is usually also necessary to consider the fertility rate $f_{x,t}$ and the number of women of childbearing age in the population $W_{x,t}$:

$$B_{x,t} = f_{x,t} \times W_{x,t} \tag{17.5}$$

Gray Prediction Model

The gray prediction model is suitable for the situation where the amount of data is small and the information is not completely clear. It is based on the gray system theory that the population system is a gray system with part of the information known and part of the information unknown, and through the cumulative generation and other processing of the original population data, it mines the intrinsic laws of the data and establishes a differential equation model to predict population. The basic form of the most commonly used GM (1, 1) model is:

$$\frac{dx^{(1)}(k)}{dk} + ax^{(1)}(k) = u \tag{17.6}$$

where $x^{(1)}(k)$ is a sequence of one cumulative generation of original population data $x^{(0)}(k)$ and a and u are the parameters to be estimated. The original population data are first cumulatively generated, and then the least squares method is utilized to estimate the parameters a and u, thus, building a prediction model. For example, the population data for the past few years of a region are known, and after cumulative generation and parameter estimation, a forecasting model is obtained, which can be used to forecast the population size in the coming years.

Reading

Machine Learning Methods in Population Prediction

Machine learning, as a powerful data-driven technology, brings new methods and ideas for population prediction, and has significant advantages in handling complex population data and improving prediction accuracy. Machine learning methods are based on a data-driven approach that builds predictive models by learning from large amounts of historical demographic data and data on related factors, automatically discovering patterns and relationships in the data. Assuming that patterns of demographic change are embedded in the historical data, the models are able to learn and generalize to future projections.

Taking neural network as an example, its model structure is more complex, containing an input layer, hidden layer and output layer. The input layer receives data on various factors affecting the population (e.g., age structure, economic development indicators, environmental factors, etc.), and the output layer outputs the predicted population size or demographic indicators. The connection weights between neurons are adjusted by training a large amount of data to minimize the model prediction error. For example, a neural network model is trained using years of population data and related economic, social, and environmental data, and then hypothetical data on future related factors are input to predict the future population.

Different population projection methods have their own advantages, disadvantages and scope of application. The trend extrapolation method is simple and intuitive, but the reliability of long-term forecasting is low; the cohort element method is comprehensive in its consideration of factors, but the data requirements are high; the regression analysis method can reveal the relationship between the population and other factors, but assumes that the relationship between the factors is stable; the grey forecasting model is suitable for the situation of limited data; and the machine-learning method can deal with the relationship between the complex data, but the model is poorly interpreted. In practical application, it is usually necessary to consider multiple methods comprehensively, choose a suitable prediction model in the light of the specific situation, and continuously adjust and optimize it according to new data and the actual situation, so as to improve the accuracy and reliability of population prediction.

17.1.3 Role of Population on Environment

The IPAT methodology is a crucial analytical tool in sustainable development economics, enhancing our understanding of how human activities impact the environment. Developed in the early 1970s by ecologists Paul R. Ehrlich and John P. Holdren, IPAT stands for Impact = Population × Affluence × Technology (refer to Chap. 5). This formula highlights the combined effect of population size, economic development, and technological applications on environmental impact. Over time, scholars have refined IPAT by incorporating elements like elasticity coefficients and decomposition analysis, transforming it into dynamic models that better handle real-world data and long-term trends. Additionally, the model has evolved to include factors such as consumption patterns and trade, improving its explanatory power. Recent integration with other methodologies, like Life Cycle Assessment and Input–Output Analysis, and concepts such as the Ecological Footprint, has broadened IPAT's application. This integration enables a more comprehensive analysis of environmental issues, making IPAT an increasingly valuable tool in environmental research and sustainable development assessments.

Table 17.1 illustrates the application of IPAT in carbon emissions analysis. The first row presents the 2023 world GDP per capita in 2023 PPP US$ (A) and the 2023 global carbon dioxide emissions from UNPD (I). The value for T is derived by dividing I by P multiplied by A to determine the amount of carbon dioxide emitted per unit of GDP. In the second row, the implications of a 50% increase in world population on I are calculated, assuming constant affluence and technology. This increase in population is chosen as a conservative estimate for the expected growth by 2100. The third row reflects the effects of the population increase as well as a doubling of per capita GDP on I. The doubling of GDP is a rough estimate of what would be required to eradicate poverty. Lastly, the fourth row solves IPAT for T when I is set at the level from the first row, with P and A values from the third row. It demonstrates that in order to maintain total emissions at the current level with a 50% population increase and a doubling of affluence, carbon dioxide emissions per unit GDP would need to be reduced to one-third of their current level.

Although the IPAT methodology offers a simplified framework to measure environmental impacts, it faces several challenges in practical applications.

Factor Simplification and Practical Complexity: The IPAT methodology reduces environmental impacts to the three main factors of population, affluence, and technology, and while it helps to capture key factors for analysis, in practice,

Table 17.1 Global carbon dioxide scenarios

	P (billions)	A (PPP US \$)	T (tonnes per \$)	I (billions of tonnes)
Current (2023)	8.06	13,169.6	0.0003560	37.79
$P \times 1.5$	12.09	13,169.6	0.0003560	56.68
$P \times 1.5$ and $A \times 2$	12.09	26,339.2	0.0003560	113.36
$P \times 1.5$ and $A \times 2$ with I at current	12.09	26,339.2	0.0001187	37.79

Source World Bank (2024)

environmental issues are influenced by the interaction of many complex factors, such as culture, institutions, and social structure, which are not adequately represented in the IPAT model and may result in less than these factors are not fully reflected in the IPAT model, which may lead to less comprehensive and accurate analysis results.

Data and quality issues: The application of the IPAT methodology requires a large amount of accurate data to quantify demographic, economic, and technological factors and their relationship to environmental impacts. However, there are often difficulties in obtaining quality data, especially in developing countries or regions where the statistical data systems may be inadequate and the accuracy, completeness and timeliness of the data are difficult to ensure, which may affect the reliability of the model analysis.

Assumptions of causality: The IPAT method assumes a simple multiplicative relationship between factors, but in reality, the relationship between demographic, economic, and technological and environmental impacts is not completely linear and independent, and there may be complex non-linear relationships, feedback mechanisms, and threshold effects. For example, technological advances may in some cases create new environmental problems rather than simply reduce environmental impacts, a complex causal relationship that is more difficult to accurately portray in the IPAT model.

17.2 Environment Footprint

17.2.1 Concepts and Categories of Environmental Footprints

In recent years, with the growing popularity of sustainability economics, the term "footprint" has been applied to a wide range of environmentally related factors, and has become a broad concept of how human production and consumption behaviors affect the Earth's environment, which can be applied to the whole of humankind, countries, cities, firms, communities, and individuals, among others, and which is collectively referred to as the "environmental footprint". Thus, the vast majority of environmental footprints can be viewed as indicators that assess and quantify the total human demand for and impact on global natural resources, and can be used to guide scientific analysis and decision-making. However, unlike the assessment of direct resource consumption, which is visible to the naked eye, the environmental footprint tends to emphasize the virtual, hidden and invisible consumption of resources, i.e., the total resource consumption of a product or service from the production of raw materials all the way to the end of its life cycle (Fig. 17.2).

For instance, the ecological footprint, developed over 30 years ago, measures the environmental impact of resource consumption and waste production in terms of productive land area needed for human activities, expressed in "global hectares" (gha). Its counterpart, biocapacity, refers to an area's ability to regenerate resources and absorb waste. For instance, Qatar's ecological footprint in 2018 was 14.3 gha per capita, while its biocapacity was only 0.9 gha per capita, indicating an ecological deficit. The carbon footprint, introduced in 1999, calculates the total carbon emissions produced directly or indirectly by an entity over its life cycle. Oil-producing countries with small populations, like Qatar, have the highest per capita carbon emissions, with Qatar emitting 35.59 tons/year in 2021. The biodiversity footprint measures the loss of biodiversity due to land-use change, resource exploitation, and invasive species. International trade has been shown to impact biodiversity more in developing countries, while developed countries have reduced their biodiversity footprints through cleaner production technologies.

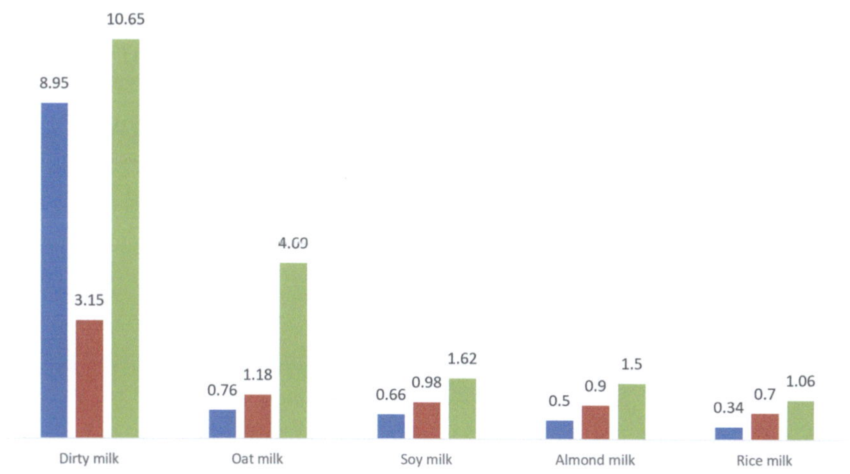

Fig. 17.2 Environmental footprints of dairy and plant-based milks. *Data source* Poore and Nemecek (2018)

17.2.2 Methods for Calculating Carbon Footprint

The concept of "carbon footprint" is derived from the "ecological footprint," which expresses the total amount of Greenhouse Gases (GHGs) emitted during human production and consumption activities in terms of carbon dioxide equivalents (CO2equivalent, abbreviated as CO2eq). Accurate calculation of carbon footprints is important for assessing the impact of human activities on climate change, formulating emission reduction strategies and promoting sustainable development. Currently, the main methods for calculating carbon footprints include the inventory factor method and input–output analysis.

Inventory Factor Method (IFM)

This method uses national GHG inventories compiled by the IPCC and the corresponding emission factors to calculate the emissions of various GHGs (Fig. 17.3). The key to this method is to determine the extent to which various GHGs contribute to climate change, which is recommended by the IPCC to be quantified using the Global Warming Potential (GWP). For example, the GWP100 calculated in the IPCC Fifth Assessment Report is 25, indicating that the greenhouse effect over a 100a time period is 25 times that of the same mass.

The approach offers several advantages, including a comprehensive examination of greenhouse gas emissions from the burning of diverse fossil fuels, convenient access to data, a more straightforward calculation procedure, and applicability to energy carbon footprint accounting at all scales. However, it also has disadvantages, such as the inability to cover implied indirect carbon emissions and the need for

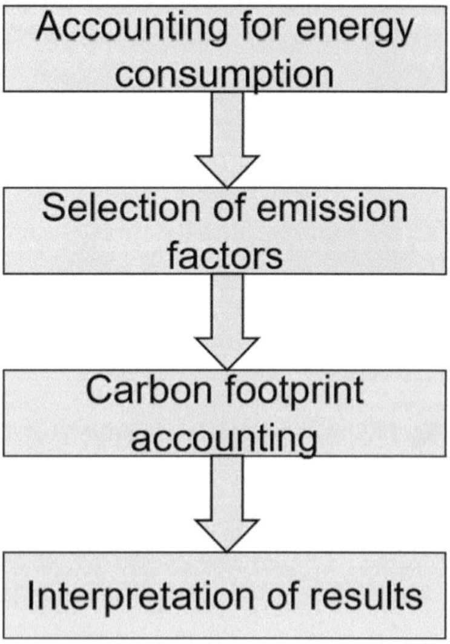

Fig. 17.3 Inventory factor method

correction with process analysis methods when applied at micro and meso levels. Additionally, regional differences in production technology and energy quality make it challenging to select appropriate regional emission factors, potentially leading to high calculation results.

Example

Calculating Carbon Footprint

Assuming that a factory consumes 1,000 tons of coal in a year, and based on the emission factor for coal (assumed to be 2.5 tons/ton of coal), the carbon dioxide emissions from coal combustion are 1000 × 2.5 = 2,500 tons.

If the plant also emits a certain amount of methane (CH_4), assuming that the emissions are 10 tons and the GWP100 is 25, the converted carbon dioxide equivalent is 10 × 25 = 250 tons CO_2-eq. Adding the CO_2 emissions from coal combustion to the converted CO_2-eq. Methane gives the carbon footprint of this part of the plant's activities (other GHGs and emission sources are not considered yet).

Input-Output Analysis (IOA)

IOA is a "top-down" analytical method, which demonstrates the connections among initial inputs, intermediate inputs, total inputs as well as intermediate outputs, final outputs, and total outputs within each sector. Input–output models can be categorized into 3 forms in studying resource and environmental issues: value-based, in-kind, and hybrid. At their core, all of these aim to transform the economic ties between production sectors or regions into the physical relationship regarding greenhouse gas emissions. In this way, the exchange process of emissions can be clearly presented and allocated to each sector or region, so as to make the relationship between direct and indirect carbon emissions clear.

It can be summed up in four steps: choosing and processing the input–output table (IOT), building the carbon footprint model, performing carbon footprint accounting, and analyzing the results (see Fig. 17.4). One must pick IOTs that have an appropriate scale and time frame in accordance with the accounting subject, and pre-process them using sectoral aggregation to curtail the errors resulting from discrepancies in production technology and scale among sectors. The central components of the model construction mechanism include the carbon intensity factor matrix, the Leontief inverse matrix, and the final demand matrix, among others. Specifically, the carbon intensity factor matrix can be derived from the energy consumption of each sector, the energy and carbon emission factors, and the sectoral output value listed in the statistical yearbook, while the data for the Leontief inverse matrix and the final demand matrix are generally computed using the IOT.

It has become the main method for carbon footprint accounting at the medium and macro level, which can comprehensively reflect the direct and indirect carbon emission relationship of each sector in the economic system, overcome the problem

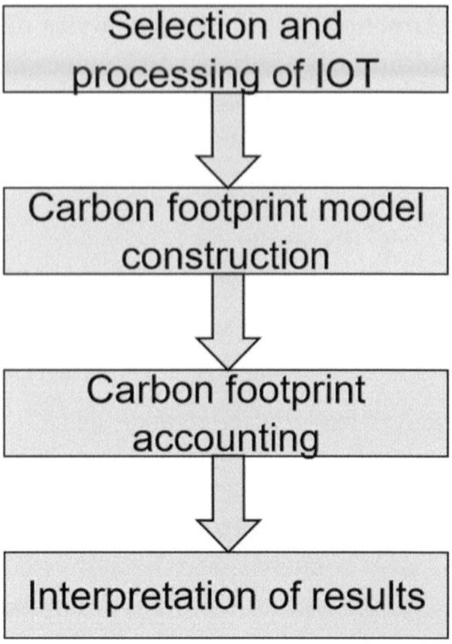

Fig. 17.4 Input–output analysis

of duplicated or omitted calculations due to the complexity of inter-sectoral production relationships, and reduce the uncertainty brought by the delineation of system boundaries, which has a higher economic advantage compared with LCA.

However, IOA takes a lot of time and effort to compile IOT, so there is an obvious lag in using IOT to calculate the carbon footprint. The energy consumption data may differ from the sectoral aggregation in IOT, which leads to certain errors in the carbon footprinting results, and the sectoral aggregation also makes the accuracy of the corresponding carbon emission factors insufficient. Carbon footprinting based on IOA can only be applied to the sectoral or regional level, and there is a big obstacle at the micro level.

17.3 Ecology and Economics

Natural capital and ecosystem services are important concepts at the intersection of ecology and economics, which profoundly reveal the close connection between nature and human society and are of key significance in understanding the value of ecosystems and promoting sustainable development.

Natural capital is the source of ecosystem services, and ecosystem services are the functional expression of natural capital. The state of natural capital (e.g., quantity, quality and structure) directly affects the ability to provide ecosystem services. For example, the abundance of forest resources (natural capital) determines the level of timber supply services, carbon sink regulation services, and ecotourism and cultural services that it can provide. At the same time, over-utilization or irrational management of ecosystem services by humans may lead to the degradation of natural capital, which in turn affects the sustainable provision of ecosystem services.

17.3.1 Natural Capital

Natural capital refers to the assets of the natural environment, including land, forests, water, minerals, biodiversity, etc., that can provide useful goods and services to human beings. These natural assets are valuable assets endowed by nature to human beings and constitute the material basis for human survival and development. Natural capital presents several distinctive characteristics:

Scarcity: With the intensification of human activities, the quantity and quality of natural capital are declining and becoming more and more scarce. For example, clean water resources have become extremely precious in some areas, and high-quality arable land resources are under pressure to diminish.

Renewable and non-renewable: Some natural capital, such as forests and fishery resources, can be renewed under rational management, but some mineral resources are non-renewable, and their reserves will gradually diminish once they are exploited.

Multifunctionality: Natural capital often has multiple functions, for example, forests not only provide timber and other material products, but also have a variety of ecological functions such as water conservation, climate regulation, and protection of biodiversity.

The valuation of natural capital is a complex but important task. Traditionally, people mainly focus on the economic value of natural capital, such as the value of forest timber and the development value of land through market prices. However, natural capital also has significant ecological and social values, such as the value of ecosystem services (which will be elaborated in detail below), and cultural and aesthetic values. In recent years, it has become a research hotspot to synthesize a variety of valuation methods, including the market value method, the alternative cost method, and the conditional value method, in order to assess the value of natural capital in a more comprehensive way.

17.3.2 Ecosystem Services

Ecosystem services refer to the environmental conditions and utility that ecosystems form and maintain for human survival and development, and mainly include provisioning services, regulating services, supporting services and cultural services.

Provisioning services: Ecosystems provide humans with a variety of material products, such as food (crops, fishery products, etc.), raw materials (wood, fiber, etc.), energy (e.g., hydroelectricity, biofuels, etc.) and water resources. For example, food produced by agro-ecosystems is the basic material security for human survival.

Regulating services: regulating processes such as climate, hydrology, soil, etc., within ecosystems, thereby providing benefits to humans. For example, wetland ecosystems can regulate flood flow and reduce the risk of flooding; forest ecosystems can absorb carbon dioxide and mitigate climate change; and the root system of vegetation can hold soil and prevent erosion.

Support services: maintain the basic functions of the Earth's life-support systems, such as soil formation, nutrient cycling, and primary production. These services are the basis for the production of other ecosystem services and, although not directly consumed by humans, are essential for ecosystem stability and sustainability. For example, microorganisms in the soil are involved in the nutrient cycling process, enabling plants to obtain the nutrients they need and thus maintain the productivity of the entire ecosystem.

Cultural services: Non-material benefits that fulfill human spiritual and cultural needs, including eco-tourism, cultural and historical sites, aesthetic values, educational and scientific research values. For example, beautiful natural landscapes attract tourists and promote local tourism; ancient forests or mountain ranges carry cultural and historical significance and inspire artistic creation and cultural inheritance; and natural ecosystems are also natural laboratories for scientific research, providing rich material for human understanding of the laws of nature.

While the concept of ecosystem services captures the importance and value of ecosystems to humans, it does not directly reflect the extent to which human activities require and depend on ecosystems. Ecosystem service dependence refers to the extent to which human society relies on the various services provided by ecosystems for economic, social and livelihood purposes. Ecosystem services include provisioning services (e.g., provisioning of food, water, raw materials, etc.), regulating services (e.g., climate regulation, flood regulation, water purification, etc.), supporting services (e.g., soil formation, nutrient cycling, etc.) and cultural services (e.g., eco-tourism, cultural landscapes, spiritual values, etc.). Human survival and development cannot be achieved without these ecosystem services, and ecosystem service dependence quantifies the strength of this dependence.

According to the definition of the Ecosystem Service Dependency Index, the larger the value of the IDES total index and its sub-indexes, the higher the human dependency on the corresponding ecosystem service. The IDES total index and sub-indexes are calculated as follows:

$$\mathrm{IDES} = \sum_{i=1}^{3} \mathrm{IDES}_i \tag{17.7}$$

$$\mathrm{IDES}_i = \frac{\mathrm{ENB}_i}{\left|\sum_{i=1}^{S} \mathrm{ENB}_i + \mathrm{SNB}\right|} \tag{17.8}$$

where i denotes the type of ecosystem service, IDES is a total index measuring human dependence on ecosystem services, IDES_i is a sub-index of human dependence on ecosystem services of type i, ENB_i is the total net benefit to humans from ecosystem services of type i, and SNB is the total net benefit to humans from socio-economic activities.

The ecosystem service is crucial in several aspects: First, sustaining life-support systems: Ecosystem services directly or indirectly guarantee human survival and health. For example, clean air and water are essential for human survival, and ecosystems provide these necessary resources through their own purification functions. Second, promote economic development: Many industries depend on ecosystem services, such as agriculture, forestry, fisheries, and tourism. Healthy ecosystems provide stable material products and a favorable ecological environment, creating favorable conditions for economic activities while reducing the risk of economic loss due to ecological damage. Third, enhancing human well-being: Rich ecosystem services enrich human spiritual and cultural life, provide opportunities for recreation and leisure, enhance people's sense of identity and belonging to nature, and help to improve their quality of life and sense of well-being.

17.3.3 Gross Ecosystem Product

GEP (Gross Ecosystem Product) represents the aggregated economic worth of all the ultimate products and services that ecosystems supply to humanity within a particular region during a defined time frame. It plays a crucial role in precisely gauging the value that ecosystems bring to humans and is highly instrumental in driving ecological preservation and sustainable development efforts.

Globally, the stockpile of ecosystem assets such as forests, grasslands, wetlands, fertile soils, and biodiversity, along with the stream of services they offer, are facing substantial stress and disruption. Against this backdrop, the GEP has been devised to assess the overall economic value of the end products and services related to ecosystems within a specific region and during a particular accounting period. GEP accounting zeroes in on the utilization of ecosystem services throughout the economic production chain and appraises them based on their societal advantages. Essentially, it converts the input of ecosystems to human beings into economic and monetary figures. Comparable to GDP, yet with a heightened emphasis on nature's role in the economy, it serves as a valuable addition to GDP by spotlighting the contribution of nature that is typically ignored in GDP computations.

GEP accounting consists of four key steps:

Accounting for the stock of ecosystem assets: The continuous provision of sustainable ecosystem services hinges upon the preservation of a reserve of ecosystem assets. These assets can be classified into natural ones like forests, scrublands, grasslands, rivers, and lakes, as well as improved ones such as agricultural land and reservoirs. To conduct an evaluation, a holistic approach is required, considering factors such as the scope (area), spatial configuration, and condition (quality) of the ecosystems. Data collection typically involves the utilization of field investigations and remote sensing techniques.

Accounting for ecosystem service provision: Placing emphasis on the biophysical production function of ecosystem services, various biophysical models and software, like InVEST and SWAT, are utilized. By integrating ecosystem asset data with other biophysical details like soil, slope, and climate data, the provision of ecosystem services can be precisely quantified. For accurate accounting, it is crucial to conduct ecosystem surveys that maintain consistent spatial and temporal scales.

Valuation of ecosystem services: The determination of the quantity (actual utilization) and price of services involves integrating the supply and demand aspects of ecosystem services. The demand side is manifested as the inclination to pay for these services, thereby quantifying their contribution to human welfare in monetary terms. The value of services is obtained by multiplying the amounts and prices. In the accounting process, monetary valuation relies on exchange prices (either market prices or approximated exchange prices for non-market services). This ensures the aggregation and comparability of different services and facilitates the application of GEP in diverse economic analyses.

Aggregation to GEP: The last step involves combining the value of the diverse ecosystem services into GEP, calculated as follows:

$$\mathrm{GEP} = \sum_{i \in I} r_i p_i q_i \tag{17.9}$$

In this equation, I represents the collection of ecosystem services. Here, r_i is the ratio that indicates the portion of the accounting value of a particular ecosystem service originating from nature. The symbol p_i refers to the accounting price of the service in this equation, and q_i is the quantity of that service. For regulating services $r_i = 1$, and $r_i < 1$ for other services (e.g., many physical services).

17.3.4 GEP Application in China

China has been actively engaged in GEP projects and has so far implemented a considerable number of GEP projects across various scales, with 16 at the provincial level, 29 at the municipal level, and 151 at the county level, amounting to a total of 196. Additionally, 16 technical specifications/guidelines for GEP accounting have been issued, among which 1 is at the national scope, and 15 are at the local scope.

Many governments have incorporated GEP into their administrative evaluation systems, clarifying the responsibilities of governments at all levels, and making GEP and GDP joint assessment indicators, so as to encourage officials to emphasize ecological protection and restoration, and at the same time, set up incentives and disincentive mechanisms to encourage investment in natural capital, and promote green development, for example, in Lishui, Shenzhen (Fig. 17.5), Pu'er, and Ezhou, among other places.

Based on GEP accounting, ecological compensation policies are more operational in terms of determining compensation standards, evaluating benefits and rationally distributing funds. For example, the ecological compensation policy in Beijing's Miyun Reservoir Basin uses GEP to account for the cost-effectiveness of service flows, and Ezhou uses GEP to optimize the allocation of funds and improve the supply of ecosystem services.

Fig. 17.5 GEP of Shenzhen in 2021. *Note* The data unit is billion CNY. *Data source* Shenzhen government (2024)

Example

The ecological compensation in Beijing's Miyun Reservoir Basin

The ecological protection compensation system is a public institutional arrangement that provides reasonable compensation to ecological protectors, clearly defines the rights and obligations of ecological protectors and beneficiaries, and internalizes the economic externalities of ecological protection. Beijing's largest surface water source, Miyun Reservoir, has 2/3 of its catchment area in Hebei Province. 2018–2020 and 2021–2025, Beijing and Hebei Province have signed two rounds of horizontal ecological protection compensation agreements for the Chaobai River Basin Water Catchment Area upstream of the Miyun Reservoir to support the water environment management of the Chaobai River Basin Water Catchment Area upstream of the Miyun Reservoir. In 2023, Beijing established a linkage mechanism between the value of ecosystem regulating services (GEP-R) and comprehensive ecological protection compensation policy fund allocation for watershed conservation zones to guide governments at all levels to strengthen their responsibilities for ecological protection and green development using GEP as a green policy indicator.

Thanks to the implementation of the eco-protection compensation mechanism, all 77 sub-watersheds upstream of Miyun Reservoir have reached the eco-cleanliness standard.The value of GEP-R in Miyun District in 2023 will be 64.58 billion CNY. From 2021 to 2023, Miyun District receives 2.1 billion yuan of ecological protection compensation funds. The Miyun Reservoir Strategic Reserve Water Beneficiary Area provides horizontal compensation to the Water Conservation Contributing Area, with annual funding of about 1 billion CNY. In addition, Beijing has made cross-provincial compensation to mobilize the upstream areas of the reservoir for water conservation, and has accumulated 1.955 billion CNY of horizontal subsidy funds to the Chaobai River Basin in Hebei Province since 2018.

17.3.5 Limitations and Challenges

As GEP applications become more widespread, it is necessary to regularize the methodology to guarantee compatibility and comparability, and this should be based on the work of the United Nations, while factors such as the location of beneficiaries need to be taken into account when integrating multiple biophysical models, in order to fill the gaps between ecological modeling and the valuation of services.

GEP accounting relies on a large amount of data, and while high-quality, spatially explicit data can improve credibility, it faces many challenges, such as difficulties in selecting indicators due to ecosystem diversity, and problems with data compatibility and sharing, etc. In the future, it will be necessary to improve the ecosystem monitoring network in order to enhance data availability.

Although GEP accounting and its policy application are still in the nascent stages of development, they have already undergone extensive piloting and practical implementation in China and other locations. Looking ahead, it will be necessary to standardize the methodology, utilize more spatially detailed data, integrate additional services (such as the health benefits derived from nature experiences), and draw from a wider range of policy applications to more effectively support global green and inclusive development.

References

Kimmerer, R. W. (2013). *Braiding sweetgrass: Indigenous wisdom, scientific knowledge and the teachings of plants*. Milkweed editions.

Poore, J., & Nemecek, T. (2018). Reducing food's environmental impacts through producers and consumers. *Science, 360*(6392), 987–992.

Shenzhen government. (2024). *Shenzhen releases Gross Ecological Product (GEP) accounting results for 2021*. https://www.sz.gov.cn/szzt2010/wgkzl/jcgk/jcygk/zdzcjc/content/post_10359192.html

Thomaes, S., Grapsas, S., van de Wetering, J., Spitzer, J., & Poorthuis, A. (2023). Green teens: Understanding and promoting adolescents' sustainable engagement. *One Earth, 6*(4), 352–361.

World Bank. (2024). *World Development Indicators*. https://databank.worldbank.org/source/world-development-indicators

Part V
Policy-and Practice-Related Sustainability Issues

Chapter 18
Economic Policy Instruments

I think I shall never see a billboard lovely as a tree.
Perhaps, unless the billboards fall, I'll never see a tree at all.
"Trees," Ogden Nash (1938)

18.1 History of Environmental Policy

The emergence of modern environmental policy is deeply rooted in a growing awareness of environmental challenges. The 1960s marked a watershed moment in this awareness, epitomized by Rachel Carson's groundbreaking work, "Silent Spring." Opening with the poetic yet haunting question, "What has silenced the voices of spring in our backyards?" Carson systematically exposed the environmental crisis stemming from the widespread misuse of pesticides during World War II. This work sparked unprecedented public concern about synthetic pesticides in American society, ultimately leading to a nationwide ban on DDT and facilitating the establishment of the United States Environmental Protection Agency (EPA).

Public awareness of environmental issues grew from a series of disasters: the 1952 London Killer Fog, the 1959 Minamata disease in Japan, the 1967 Torrey Canyon oil spill, the 1978 Love Canal toxic waste incident in the U.S., and the 1984 Bhopal gas leak in India. These events underscored the economic problem of polluters not bearing the full social costs of their actions.

Environmental policies evolved through three phases (see Fig. 18.1):

1. *Command-and-Control Policies* (1970–1987): This phase focused on direct regulation, marked by the U.S. Clean Air Act of 1970, the Clean Water Act of 1972, and the Resource Conservation and Recovery Act of 1976. These laws aimed at immediate pollution control and public health protection.

L. Zhang, *Sustainability Economics*, https://doi.org/10.1007/978-3-031-87829-9_18

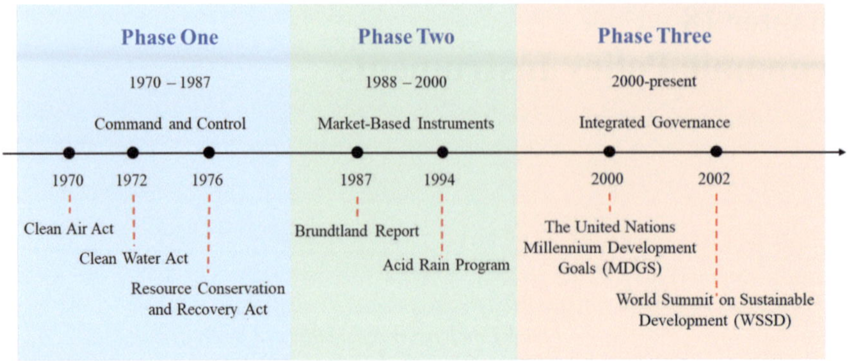

Fig. 18.1 Timeline of environmental policy evolution

2. *Market-Based Instruments* (1987–2000): Influenced by the 1987 Brundtland Report, this phase introduced market-based solutions for environmental protection. The U.S. implemented emissions trading programs like the Leaded Gasoline Trading Program, the Regional Clean Air Markets Program, and the NO_x Trading Program. The 1994 Acid Rain Program successfully reduced SO_2 emissions and compliance costs.
3. *Integrated Environmental Governance* (2000–present): This phase emphasizes comprehensive environmental, economic, and social objectives. The 2000 Millennium Development Goals and the 2002 World Summit on Sustainable Development established the "three pillars" of sustainable development: environmental protection, economic development, and social equity. This approach combines regulation with market mechanisms and voluntary initiatives.

From initial focuses on direct pollution control, through the integration of economic efficiency considerations, to today's comprehensive approach to sustainable development, environmental policy objectives have become increasingly sophisticated and interconnected.

The transformation of these objectives parallels developments in economic thinking about environmental issues. The root causes of environmental degradation lie in the externality of environmental resources: polluters often don't bear the full social costs of their actions, while environmental protection benefits typically have public good properties, making efficient resource allocation through market mechanisms alone difficult. Early policies focused primarily on correcting clear market failures through direct regulation. As understanding grew about the economic dimensions of environmental problems, objectives expanded to include cost-effective solutions through market-based instruments and then further intertwined with broader patterns of economic and social development.

18.2 The Economics of Pollution Control

The economic analysis of environmental pollution control starts with understanding pollution as a complex economic phenomenon involving resource flows, environmental carrying capacity, and economic activities. Net pollution occurs when emissions exceed the environment's ability to absorb them, highlighting the importance of both emission scale and absorption capacity. Environmental systems can naturally purify and restore to some extent, but their capacity is limited. When emissions surpass this capacity, pollutants accumulate, causing environmental degradation.

The economic framework for pollution control centers on two key concepts: the Marginal Damage Cost (*MDC*) and the Marginal Abatement Cost (*MAC*).

The Marginal Damage Cost function represents the additional environmental damage caused by each incremental unit of pollution. This relationship typically exhibits an upward slope, indicating that as pollution levels increase, the damage caused by each additional unit becomes progressively more severe. The total environmental damage at any given pollution level can be visualized as the area under the *MDC* curve, representing the cumulative impact of all pollution up to that point.

The Marginal Abatement Cost function, on the other hand, represents the additional cost required to reduce pollution by one more unit. These costs encompass various components, including labor, capital investment, energy requirements, and opportunity costs from reduced production. The *MAC* curve typically slopes downward with respect to emissions, reflecting the fact that initial pollution reductions are relatively inexpensive, but costs increase substantially as firms attempt to achieve higher levels of pollution control. The total abatement cost for any given reduction in emissions can be calculated as the area under the *MAC* curve.

Combining these two concepts, as shown in Fig. 18.2, we can identify the economically efficient level of pollution control. At the intersection point Q^*, the marginal cost of reducing one more unit of pollution exactly equals the marginal benefit from avoiding that unit's environmental damage. At emission levels above Q^*, the damage from additional pollution exceeds the cost savings from reduced control measures, suggesting that more stringent controls would benefit society. Conversely, at levels below Q^*, the cost of additional control exceeds the environmental benefits, indicating that resources might be better allocated elsewhere. This represents the optimal balance between environmental protection and economic costs, where neither increasing nor decreasing pollution control would improve overall social welfare.

Having established the fundamental concepts of pollution control and the economic framework of marginal costs, we now turn to the practical challenges of implementing effective environmental policies. The theoretical intersection of MDC and MAC provides an elegant solution for determining optimal pollution levels. However, translating this theoretical framework into practical policy faces several significant challenges.

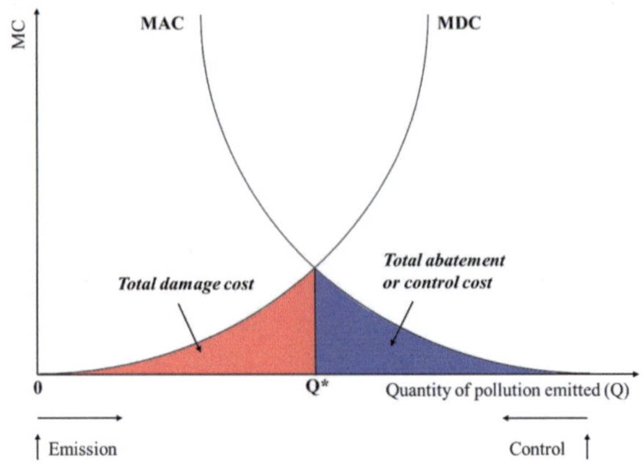

Fig. 18.2 The relationship between MAC and MDC

First, it is challenging to accurately measure both environmental damage and abatement costs. Environmental damage often manifests in complex, long-term ways that are difficult to quantify. It is hard to precisely value the health impacts, such as air pollution, the loss of biodiversity, the degradation of ecosystem services, or even subjective well-being. Similarly, abatement costs can be challenging to estimate, particularly when considering indirect costs such as reduced economic competitiveness or technological adaptation requirements.

Geographic variation presents another significant challenge. Both environmental damage and abatement costs can vary substantially across different regions. A given level of pollution might cause more severe damage in densely populated areas compared to sparsely populated regions. Similarly, abatement costs might be higher in regions with older industrial infrastructure or limited access to clean technologies. The spillover effects of pollution further complicate these circumstances. This spatial variation makes it difficult to implement one-size-fits-all policy solutions.

Most importantly, both MDC and MAC curves are dynamic rather than static. Technological progress can significantly shift the MAC curve by making pollution control more cost-effective. For instance, advances in renewable energy technologies have dramatically reduced the cost of reducing carbon emissions. Meanwhile, new scientific understanding of environmental impacts can alter our assessment of the MDC curve, as seen with evolving research on the health effects of various pollutants.

These challenges have led to the development of different policy approaches, each attempting to address these complexities in distinct ways. Market-based instruments, such as emissions trading systems and environmental taxes, harness market mechanisms to achieve cost-effective pollution reduction while adapting to changing conditions. Command-and-control regulations provide direct oversight of pollution sources, offering greater certainty in environmental outcomes but potentially at higher economic costs. Hybrid approaches combine elements of both systems, attempting to balance their respective advantages and limitations.

18.3 Market-Based Instruments

Market-based policies leverage economic incentives to drive businesses and individuals toward reducing carbon emissions and adopting sustainable practices. Usually, market-based policies create financial incentives or penalties, allowing participants to make cost-effective decisions.

18.3.1 Environmental and Carbon Tax

Environmental taxes, also known as Pigouvian taxes, were first proposed by the economist Arthur Pigou. The fundamental idea behind these taxes is to internalize the external costs of pollution by making polluters pay for the negative impact of emissions. Figure 18.3 illustrates how environmental tax works: If the government imposes a tax (*tax*) equal to the external costs, firms must pay both their private marginal cost and the tax, where their private marginal cost (MPC) curve shifts to align with the marginal external cost (MEC) curve. This adjustment leads to market equilibrium, where the higher price ($P \rightarrow P^*$) reduces demand ($Q \rightarrow Q^*$), resulting in lower emissions and increased government revenue.

Carbon and energy taxes are designed based on Pigouvian taxes. These taxes are imposed on fossil fuels depending on their carbon content, calorific value, or the amount of carbon emissions. The goal is to adjust the price of fossil fuels to reflect the environmental costs associated with their consumption and encourage cleaner energy alternatives. Globally, 37 carbon tax programs have been implemented as of 2024, some of which are shown in Fig. 18.4. For example, British Columbia introduced a carbon tax in 2008, South Africa implemented one in 2019, and Boulder, Colorado, became the first U.S. city to pass a voter-approved carbon tax in 2006. Countries that have implemented independent carbon taxes also include Finland, the Netherlands,

Fig. 18.3 An example of Pigouvian tax

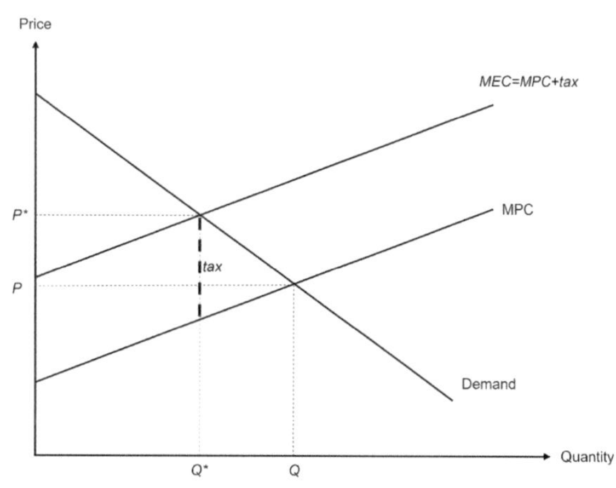

Sweden, Norway, and Denmark. These nations primarily focus on taxing fossil fuels, gradually expanding to include final products. Although the tax rates differ across countries, they all adhere to the principle of tax neutrality, meaning that the goal is not to increase fiscal revenue but to address environmental concerns.

Example

Finland and UK's Climate Tax

Finland introduced the world's first carbon tax in 1990 to reduce CO_2 emissions by targeting fossil fuels based on their carbon content. Initially modest, the tax was reformed in 1997 and 2011, becoming a combined carbon and energy tax with current rates of 18.05 per ton of CO_2. While it effectively shifted energy consumption patterns, exemptions for peat and the wood industry raised environmental concerns. To ensure fairness, Finland adopted a revenue-neutral approach, redistributing tax revenues through income tax cuts, benefiting lower-income groups and pensioners.

The UK's Energy Profits Levy (EPL), introduced in May 2022, targets the extraordinary profits of oil and gas companies with a 35% tax rate on upstream activities, bringing the total tax rate to 75%. The levy includes allowances for investment and decarbonisation and is set to expire in 2029, but may end earlier if oil and gas prices fall below certain thresholds. Starting in November 2024, the levy will increase to 38%, raising the total tax t0 78%. The EPL aims to boost public revenue while promoting investment in green energy projects through targeted allowances, balancing fiscal needs with energy transition goals.

18.3.2 Cap-and-Trade System

Carbon emissions have become a new commodity. Figure 18.5 shows the trading system, where an overall cap is set on emissions and allocates permits or allowances to firms, each representing a unit of permissible emissions. The government distributes these permits through free allocation or auction. Firms must either hold enough allowances to cover their emissions or buy additional allowances from other companies. This creates a market price for permits that fluctuates based on supply and demand. When the economy is strong or emissions are high, the price of allowance rises. Conversely, if renewable energy reduces emissions from traditional sources, the price may fall. This flexibility encourages cost-effective reductions, as firms with lower reduction costs sell permits to those facing higher costs, ensuring the overall emissions cap is met at the lowest cost.

Figures 18.6 and 18.7 further explain the trading activities of the two firms. Imagine two curves: the Marginal Social Cost (MSC) curve and the Marginal External Cost (MEC) curve. As the MSC curve gently slopes downward, it interacts with MEC

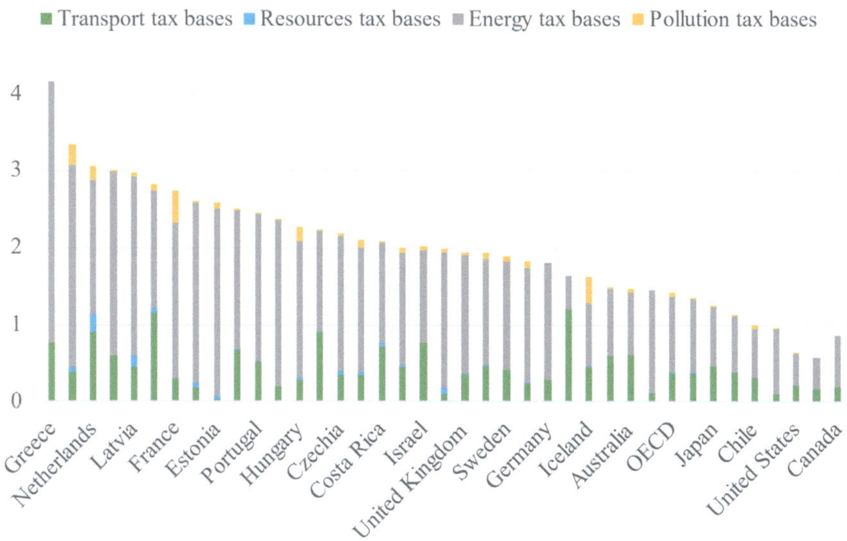

Fig. 18.4 The environmental tax (% of GDP) around the world, 2022. *Data source* OECD (2022)

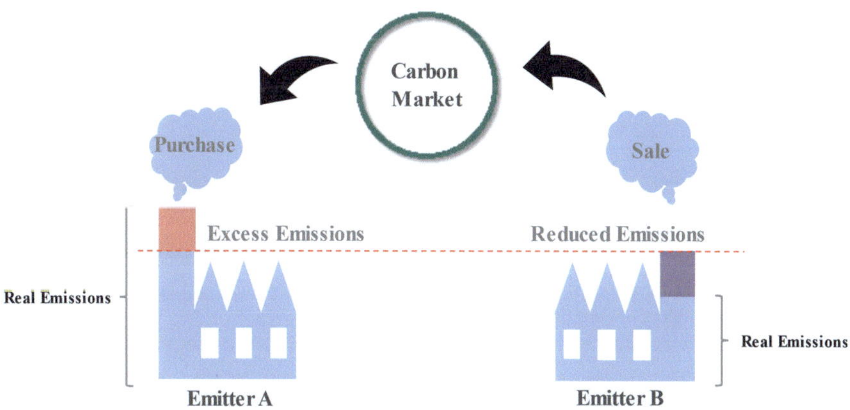

Fig. 18.5 How carbon emission trading works

at (Q^*, P_e^*), which marks the optimal emission level (Q^*) and the optimal market price of allowance (Q^*). Here is a detailed example to show this process. If:

- Firm 1's marginal abatement cost: $\text{MAC}_1 = 100 - E_1$
- Firm 2's marginal abatement cost: $\text{MAC}_2 = 100 - E_2$
- The government cap for total emissions is 80 units, and 40 units are allocated to each firm:

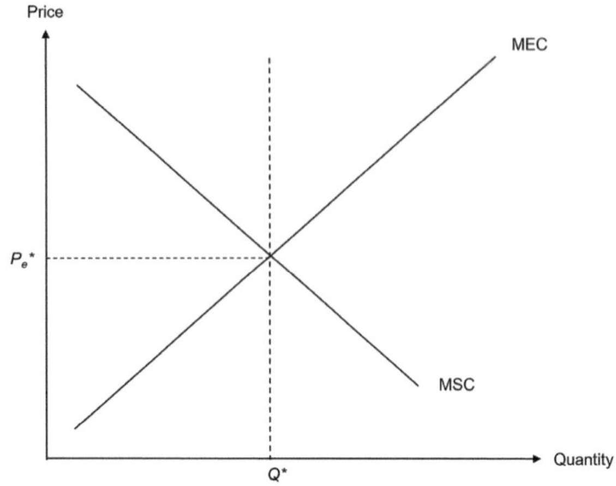

Fig. 18.6 The price of allowance under perfect information

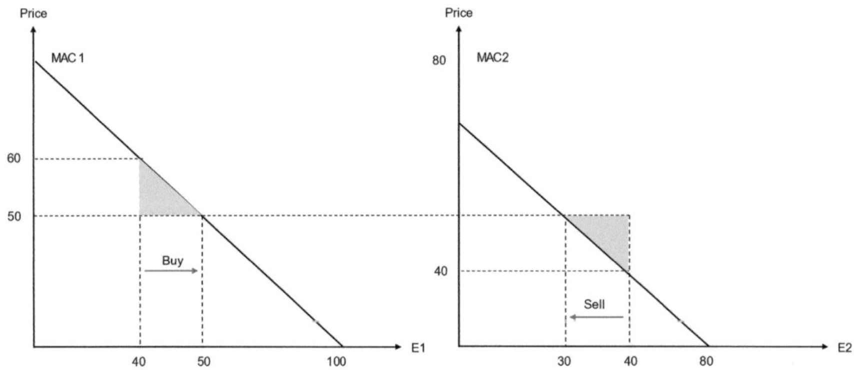

Fig. 18.7 An example showing the benefits of emission trading

Firm 1 needs to reduce 60 (=100 − 40) units of emissions, and Firm 2 needs to reduce 40 (=80 − 40) units of emissions.

Since $\text{MAC}_1 > \text{MAC}_2$ at the initial allocation, Firm 1 is willing to buy allowances, while Firm 2 is willing to sell allowances. The trade will continue until $\text{MAC}_1 = \text{MAC}_2$ at emissions levels such that $E_1 + E_2 = 80$. At this point, $E_1 = 50$, $E_2 = 30$, $MAC_1 = \text{MAC}_2 = 50$. Therefore, only if $P_p = 50$, Firm 1's demand for allowance equals Firm 2's supply of allowance.

- Firm 1's cost changes from 1800 to 1750, with a reduction of 50.
- Firm 2's cost changes from 800 to 750, with a reduction of 50.

Through emission trading, both firms reduce their abatement costs, and the total emissions remain at the government cap of 80 units. This illustrates the core benefit of emission trading—reducing the overall cost of achieving emission reduction goals by allowing firms with lower abatement costs to sell allowances to firms with higher abatement costs.

The effectiveness of a carbon market depends on several key elements. Together, these factors ensure that carbon markets can drive real emissions reductions while fostering innovation and providing cost-effective pathways to sustainability.

- Emissions Cap: Sets the upper limit for emissions from a group of sources and defines the emissions reduction target.
- Allowances: A permit to release a unit of emissions (e.g., one ton) within a specific period.
- Offset and credits: Allows entities to compensate for their greenhouse gas emissions by investing in projects that reduce or remove emissions elsewhere. In return, they receive carbon credits, each representing a reduction of one metric ton of CO_2 or its equivalent. These credits, after certification by a government or independent body, can be traded between entities. Eligible offset projects include reforestation, renewable energy, energy efficiency, and carbon capture utilization and storage (CCUS).
- Monitoring, Reporting, and Verification (MRV): A process used to quantify and report greenhouse gas emission reductions from activities like deforestation prevention, which are then verified by third parties. These verified reductions are converted into carbon credits, with one credit representing one ton of CO_2 equivalent emissions avoided.
- Price or Supply Adjustment Mechanisms (PSAM): Regulate the supply of allowances based on specific criteria, such as price or quantity thresholds. They help maintain a stable and effective market by ensuring consistent prices for long-term decarbonization while avoiding excessive costs. PSAMs can manage both high and low-price scenarios through methods like auction reserve prices, price floors, or cost-containment reserves (CCR).

Allocation: Grandfathering or Auction

Emissions permits can be allocated to firms through several methods. One common approach is free allocation, often called "grandfathering". Each firm receives a set number of permits based on its historical emission levels, helping avoid upfront costs, particularly for industries with high initial emissions. However, this method can create perverse incentives, as firms receiving free permits may lack motivation to invest in cleaner technologies. Moreover, it might lead to market inefficiencies, as firms that reduce emissions early are not rewarded for their efforts.

Alternatively, permits could be allocated through auctioning, where businesses must purchase their emissions allowances from the government. This method generates revenue for the government, which can be reinvested in low-carbon initiatives or used to reduce taxes. Auctioning also promotes fairness and transparency, as permits are distributed based on market demand rather than historical emissions. Unlike grandfathering, auctioning provides a stronger financial incentive for companies to reduce emissions, as they will have to pay for any additional permits they need. This approach encourages innovation and the adoption of cleaner technologies, as firms strive to minimize their costs.

18.3.3 Other Examples of Carbon Pricing

Carbon pricing assigns a cost to greenhouse gas emissions, reflecting their external impacts like health and environmental damage. There are currently 75 carbon pricing instruments worldwide, covering 24% of global emissions, up from 7% two decades ago. In 2023, global carbon pricing generated a record $104 billion, with emissions trading schemes contributing most of this revenue (World Bank, 2024). Over half of the funds were directed to climate and nature programs. In addition to carbon taxes and cap-and-trade systems, there are other forms of carbon pricing:

- **Crediting mechanism**: A system that tracks and certifies GHG emission reductions from specific projects or programs. These reductions, measured according to an established protocol, generate carbon credits that could be traded domestically or internationally.
- **Results-based climate finance (RBCF)**: A funding model where payments are made after verified climate outcomes are achieved. This approach encourages projects that not only reduce emissions but also provide additional benefits, including poverty alleviation, improved access to clean energy, and enhanced community well-being.
- **Subsidies and incentives**: Providing financial support for renewable energy projects, electric vehicle adoption, or energy efficiency upgrades. According to the World Bank's new Green Subsidies Database, the number of subsidies with environmental goals is rising (Signoret & Jose, 2024).

18.4 Climate Change and Climate Policy

Climate change is a critical environmental challenge characterized by long-term shifts in the Earth's temperature and weather patterns. Since the late nineteenth century, scientists have been tracking global surface temperatures. It can be observed from Fig. 18.8 that the planet is warming. By 2023, Earth's temperature was approximately 2.45°F (1.36 °C) higher than the preindustrial average of 1850–1900. This warming is largely driven by an increase in greenhouse gases, such as carbon dioxide (CO_2), methane (CH_4), and water vapor (H_2O). This phenomenon, known as the greenhouse effect, leads to rising global temperatures. Notably, the past decade has been the warmest on record, underscoring the growing urgency of addressing climate change.

The consequences of climate change are far-reaching and multifaceted. Rapid changes in climate can disrupt ecosystems, as species do not have enough time to adapt to new conditions, potentially leading to extinction. Additionally, climate

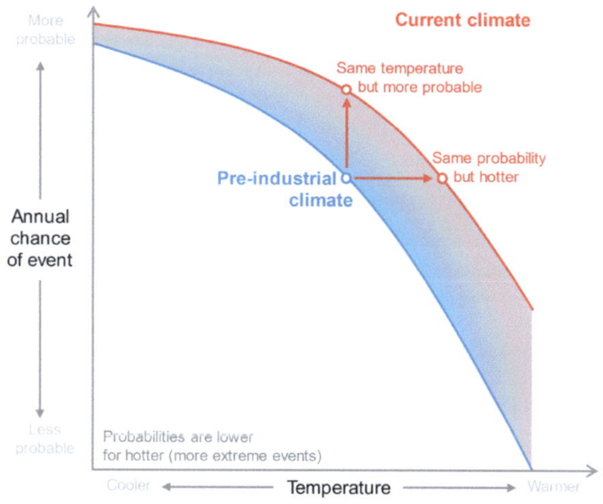

Fig. 18.8 Climate change and extreme event. *Data source* IPCC (2021)

change disrupts weather patterns, making extreme weather events such as hurricanes, droughts, and floods more frequent and severe. Rising global temperatures also cause sea levels to increase, primarily due to the melting of ice sheets and glaciers, as well as the thermal expansion of ocean water. This rise in sea levels threatens coastal areas, potentially leading to habitat destruction and displacement of populations. The loss of sea ice also contributes to a feedback loop, where less sunlight is reflected back into space, leading to further ocean warming and a further reduction in sea ice. This feedback mechanism accelerates the rate of climate change. Despite efforts to mitigate climate change, the global temperature is expected to continue rising due to the greenhouse gases already present in the atmosphere.

The problems we are currently experiencing will likely worsen. Addressing climate change requires global cooperation to reduce greenhouse gas emissions and transition to more sustainable practices. Over the past 25 years, thousands of climate policies have been introduced worldwide to combat climate change. According to Stechemesser et al. (2024), scaling up the 63 most successful policies could close 26% to 41% of the current emissions gap needed to align with the Paris Agreement goals.

18.4.1 Global Climate Negotiations

Despite several benefits of climate governance, it involves substantial investment, long timelines, and relatively low immediate returns, making it unattractive to private investors. Therefore, it is crucial to promote the establishment of a fair, reasonable, and cooperative global climate governance system that ensures mutual benefits and win–win outcomes for all. Over decades, governments have collectively pledged to slow global warming. Among various international collaborations (see

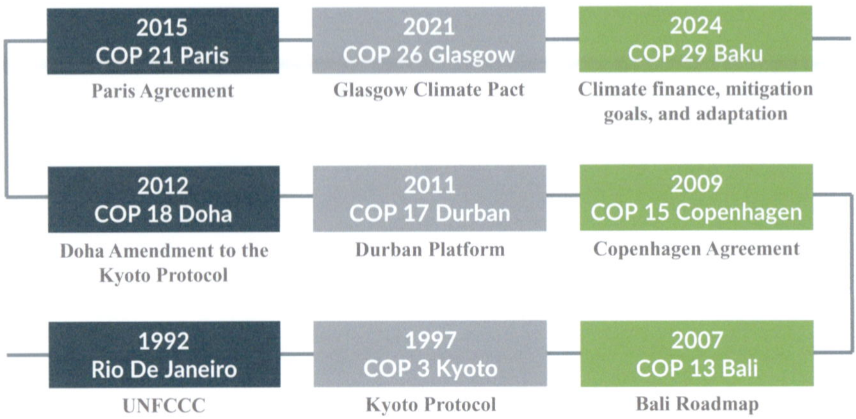

Fig. 18.9 The main climate summits and achievements

Fig. 18.9), three key milestones stand out in this journey: the United Nations Framework Convention on Climate Change (UNFCCC), the Kyoto Protocol, and the Paris Agreement. Together, they form the backbone of international climate diplomacy, highlighting humanity's commitment to securing a livable planet.

Framework: UNFCCC, 1992

Adopted at the Earth Summit in Rio de Janeiro in 1992, the UNFCCC was the first global effort to address climate change. Initially signed by 166 countries, it came into force on March 21, 1994, and has since been approved by 197 nations. The main goal is to stabilize greenhouse gas concentrations in the atmosphere to prevent dangerous interference with the climate system. The UNFCCC established a framework for action based on equity and common but differentiated responsibilities (CBDR), calling on industrialized nations, historically the largest emitters, to lead mitigation efforts while supporting developing countries through financial and technological means. The Convention is governed by the Conference of the Parties (COP), which meets annually to track progress. It also promotes cooperation through initiatives like Action for Climate Empowerment, focusing on education and public participation to strengthen global climate efforts.

Commitments: The Kyoto Protocol (1997)

Building on the UNFCCC, the Kyoto Protocol, adopted in 1997 and entering into force in 2005, was a significant step in global climate negotiations. It was the first treaty to set legally binding emission reduction targets for developed nations, acknowledging their historical role in climate change. It introduced flexible mechanisms such as emissions trading (ET), joint implementation (JI), and the Clean Development Mechanism (CDM), allowing countries to meet their targets through international cooperation:

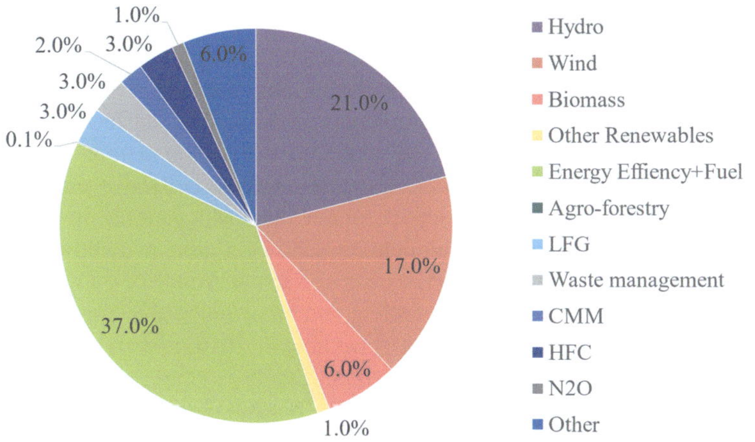

Fig. 18.10 CDM projects types, 2008. *Data source* World Bank (2009)

- ET allows countries with excess emissions reduction capacity to sell their surplus allowances to countries that are struggling to meet their targets. This creates a financial incentive for countries to reduce emissions more efficiently and cost-effectively.
- JI enables developed countries to invest in emission-reduction projects in other developed countries. This allows for cost-effective reductions by implementing projects where they are most affordable, fostering international collaboration.
- CDM permits developed countries to invest in emission-reduction projects in developing countries, earning credits toward their own targets. This not only helps reduce emissions but also promotes sustainable development in developing nations by transferring technology and financial resources. Figure 18.10 summarizes the global registered CDM projects in 2009.

The Protocol followed the principle of common but differentiated responsibilities, with developed nations taking the lead in emission reductions. Despite some successes, such as emissions cuts in certain countries, global emissions continued to rise, revealing the limitations of the Protocol. Key emitters like the United States did not participate, and developing nations like China and India had no binding targets. The first commitment period (2008–2012) saw some compliance, but global emissions grew by 32% between 1990 and 2010. A second commitment period, the Doha Amendment, extended the Protocol to 2020, but many countries, including Japan and Russia, did not take on new targets. These challenges paved the way for the 2015 Paris Agreement, which introduced a more inclusive and flexible approach to climate action.

Ambition: The Paris Agreement (2015)

The adoption of the Paris Agreement in 2015 marked a break in global climate governance. Unlike previous agreements, it embraced a universal approach, requiring all countries—developed and developing alike—to contribute to climate change mitigation. The central goal of the Agreement is to limit global temperature rise to well below 2 °C above pre-industrial levels, with an aspirational target of 1.5 °C.

The Paris Agreement introduced Nationally Determined Contributions (NDCs), allowing countries to tailor their climate action plans based on their unique capabilities. This approach promotes flexibility while ensuring collective progress. The Agreement emphasizes transparency, accountability, and periodic reviews, with countries expected to strengthen their commitments over time. In addition, the Paris Agreement formalized market-based mechanisms under Article 6, enabling international collaboration through carbon trading and other innovative approaches. The Paris Agreement entered into force on November 4, 2016, and continues to drive global efforts toward a sustainable and resilient future.

Since the Paris Agreement was signed in 2015, progress has been made in reducing projected greenhouse gas emissions. At the time, emissions in 2030 were expected to increase by 16%, but now that projection has been reduced to 3%. However, to stay on track with the Paris Agreement's 2 °C pathway, emissions must still be reduced by 28%, and by 42% for the 1.5 °C pathway. Currently, fully implementing unconditional Nationally Determined Contributions (NDCs) would result in a temperature rise of 2.9 °C, while fully implementing conditional NDCs could limit it to 2.5 °C (UNEP, 2023).

18.4.2 Global Carbon Market

The global carbon market is expanding rapidly as countries intensify efforts to address climate change and meet carbon neutrality targets. Figure 18.11 demonstrates the trend of global carbon pricing, including both ETS and carbon tax implementation. Major economies like the European Union (EU), the United States (US), the United Kingdom (UK), and countries in Asia such as Japan and South Korea have committed to achieving net-zero emissions by 2050 or earlier. Emissions trading systems (ETS) are central to these efforts, with many countries and regions adopting them to reduce greenhouse gas emissions. The EU led the way by establishing the first multinational ETS in 2005, and since then, numerous national and regional systems have been launched globally. Key examples include national ETSs in South Korea, New Zealand, and the UK, as well as regional systems like California's Cap-and-Trade Program and the Regional Greenhouse Gas Initiative (RGGI) in the US. In 2021, China launched its own ETS, marking a significant step in global carbon pricing. These systems primarily target high-emission sectors, such as electricity, industry, transportation, and aviation. They operate by setting a cap on total emissions, distributing tradable carbon allowances to companies, and allowing firms to buy and sell allowances in the market. This market-based approach provides flexibility, enabling businesses to reduce emissions at the lowest cost.

As of 2022, global ETSs cover approximately 8.9 gigatons of CO_2 equivalent (CO_2eq), up from 2.1 gigatons in 2005, demonstrating significant growth in the scale

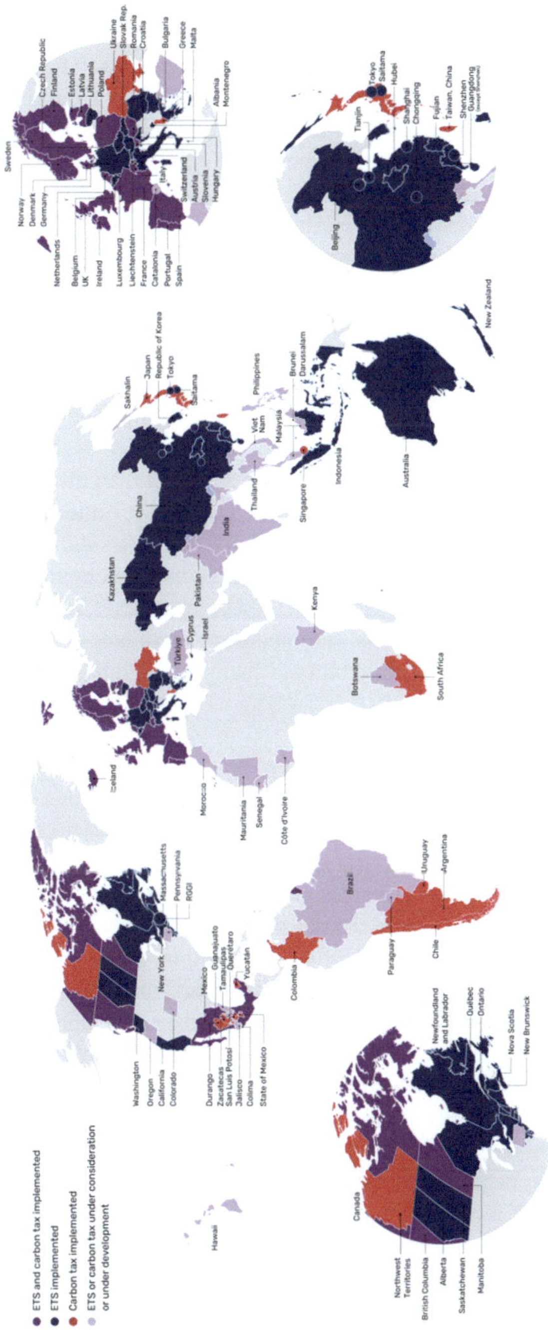

Fig. 18.11 Map of carbon pricing. *Data source* State and Trends of Carbon Pricing 2024 (World Bank, 2024)

of carbon pricing. Currently, 36 ETSs are operational worldwide, and an additional 22 are under development or consideration. These systems now cover 58% of global GDP, with emerging economies such as Argentina, Brazil, and India increasingly adopting carbon markets to address local climate challenges. The global carbon market reached a milestone in 2023, generating over USD 74 billion in revenue, underscoring the growing importance of carbon pricing as a tool for reducing emissions. As more countries and regions implement ETSs, the global carbon market remains essential in achieving climate goals, offering flexibility and economic efficiency in reducing greenhouse gas emissions.

On the positive side, carbon trading systems provide a cost-effective approach to combat climate change by incentivizing companies to adopt cleaner technologies. These systems also generate revenue for governments, which can be invested in public projects, including those aimed at sustainability. In addition, carbon markets promote transparency by holding companies accountable for their emissions. However, carbon trading is not without its challenges. It does not always encourage businesses to significantly reduce their emissions, as the allocated emission caps may be too lenient, and purchasing emission credits can sometimes be cheaper than investing in cleaner technologies. The effectiveness of carbon markets can also be limited by weak monitoring systems, allowing some businesses to falsify emissions data. Moreover, the lack of uniform participation and differing standards between countries can cause carbon leakage. Finally, the additional costs faced by businesses are often passed on to consumers. Despite these criticisms, carbon trading continues to be a key element in many strategies aimed at mitigating climate change and global warming.

18.4.3 EU ETS

The EU Emissions Trading System (EU ETS), launched in 2005, is the world's first and one of the largest carbon markets designed to reduce greenhouse gas emissions through a cap-and-trade mechanism. Under this system, a cap is set on total emissions from sectors like electricity, industry, and aviation, while companies receive allowances that permit limited emissions. Over the years, the ETS has evolved into a cornerstone of the EU's climate policy, influencing global carbon market frameworks and covering around 40% of the EU's total emissions. In 2023, the EU ETS achieved a significant 16.5% reduction in emissions from stationary installations, driven by increased renewable energy generation, particularly wind and solar, and a continued shift from coal to natural gas in power production (European Commission, 2024). Emissions from these installations are now 47.6% below 2005 levels, aligning well with the 2030 target of a 62% reduction. However, emissions from intra-European aviation increased by 10%, reflecting growing sector activity. The system's effectiveness was supported by a robust carbon price signal and the MSR's role in managing surplus allowances, ensuring market stability. Meanwhile, ETS revenues reached EUR 43.6 billion in 2023, contributing to climate initiatives across Member States, innovation, and modernization funds.

The development of the EU Emissions Trading System (EU ETS) can be divided into four key phases, each representing significant steps toward strengthening and expanding the system, as summarized in Table 18.1.

- **Phase 1 (2005–2007): Pilot Phase**. This phase primarily served as a pilot stage to explore the foundational framework. The sectors mainly included power generation and energy-intensive industries. The allocation of carbon allowances was largely based on historical emissions (accounting for around 95%), with a relatively lenient overall cap. As a result, there was a surplus of allowances, resulting in low carbon prices—close to zero—and providing limited incentives for firms to reduce emissions.
- **Phase 2 (2008–2012): Alignment with the Kyoto Protocol**. This phase introduced a more stringent emissions cap and began the transition toward auctioning allowances, reducing the reliance on free allocation. The system expanded to cover more countries and sectors, and more ambitious emission reduction goals were set. However, like Phase 1, there were challenges, including price volatility and concerns about the impact on industries exposed to international competition.
- **Phase 3 (2013–2020): Structural Reforms**. This phase marked a significant overhaul of the EU ETS. The EU implemented stricter emission caps and introduced a single EU-wide carbon market, replacing the fragmented national markets from previous phases. The system moved toward greater auctioning of allowances, with free allocations becoming more limited. A key reform was the establishment of the Market Stability Reserve (MSR), designed to address price volatility by adjusting the supply of allowances. The cap was reduced more aggressively, aiming for a 20% emissions reduction by 2020 compared to 1990 levels.
- **Phase 4 (2021–2030): Expansion**. The current phase focuses on further tightening the cap and aligning the EU ETS with the European Green Deal and the EU's 2050 climate neutrality goal. The system now covers more sectors and includes stricter emissions reduction targets. The cap continues to decline at a faster rate, and the use of auctioning is further increased. The introduction of new mechanisms, such as the Carbon Border Adjustment Mechanism (CBAM), aims to protect EU industries from carbon leakage and ensure global competitiveness while meeting climate goals.

18.4.4 Carbon Border Adjustment Mechanism

When considering international trade, the imposition of a carbon tax may weaken the competitiveness of domestic industries. To address this challenge, carbon border tax has emerged as a solution. The government imposes taxes on carbon-intensive imports, ensuring that goods produced in regions without stringent emissions regulations are taxed at the border. The rationale behind carbon tariffs is to level the playing field for domestic industries in countries that have implemented carbon pricing mechanisms, while discouraging the relocation of industries to regions with weaker climate

Table 18.1 The development of EU ETS

	Phase 1 (2005–2007)	Phase 2 (2008–2012)	Phase 3 (2013–2020)	Phase 4 (2021–2030)
Goals	A learning-by-doing exercise	8% reduction relative to 1990 levels	20% reduction relative to 1990 levels	55% reduction relative to 1990 levels
Geographic scope	15 EU member states	30 countries (27 EU members plus Norway, Iceland, Lichtenstein	Same as Phase 2	Same as Phase 2
GHG	CO_2	CO_2, N_2O	CO_2, N_2O, PFCs	Same as Phase 3
Sectors	Combustion installations with more than 20 MW, oil refineries, coke ovens, iron and steel works, mineral, pulp and paper industry, electricity generators included	Same as Phase 2 plus aviation	Same as Phase 3 plus ammonia, and aluminum industries	Same as Phase 3
Allowance	Allocated for free based on historic emissions through grandfathering approach	Auction up to 10% of the allowances Allowed to use credits from the CDM and JI	Shifted to auctioning as the main principle, with some remaining free allocation based on benchmarks	Decline annually at a rate of 2.2%, up from 1.74% Up to 57% of the general allowances will be auctioned, with the remaining allowances allocated for free
Price	Excessive EUA allocation to businesses led to oversupply, causing prices to fall to zero by the end of the period	The additional credits and the 2008 economic crisis, created a large EUA surplus, driving the price down from €30 to less than €7	The large surplus of EUAs carried over from the second period, resulting in an EUA price of only €3–7	The price averaged €83 in 2023, with a significant rise from €25/tCO_2 in 2020
Penalty	40 Euro per ton of CO_2	100 Euro per ton of CO_2	Same as Phase 2	Same as Phase 2

policies. The EU's Carbon Border Adjustment Mechanism (CBAM) is a tax that aims to address carbon leakage and ensure fairness by taxing carbon-intensive imports at the EU border. Figure 18.12 visualizes the concept and development of CBAM.

By aligning the carbon price of imports with that of domestic production under the EU Emissions Trading System (ETS), CBAM encourages cleaner production globally and supports the EU's climate goals. The system is designed to comply with WTO rules and will be implemented in two phases: the transitional phase (2023–2025) and the definitive regime beginning in 2026. During the transitional phase, importers of goods such as cement, iron and steel, aluminum, fertilizers, electricity, and hydrogen must report the greenhouse gas (GHG) emissions embedded in their imports without purchasing certificates. Reporting flexibility is allowed initially, with companies able to use EU methods, equivalent methods, or default reference values

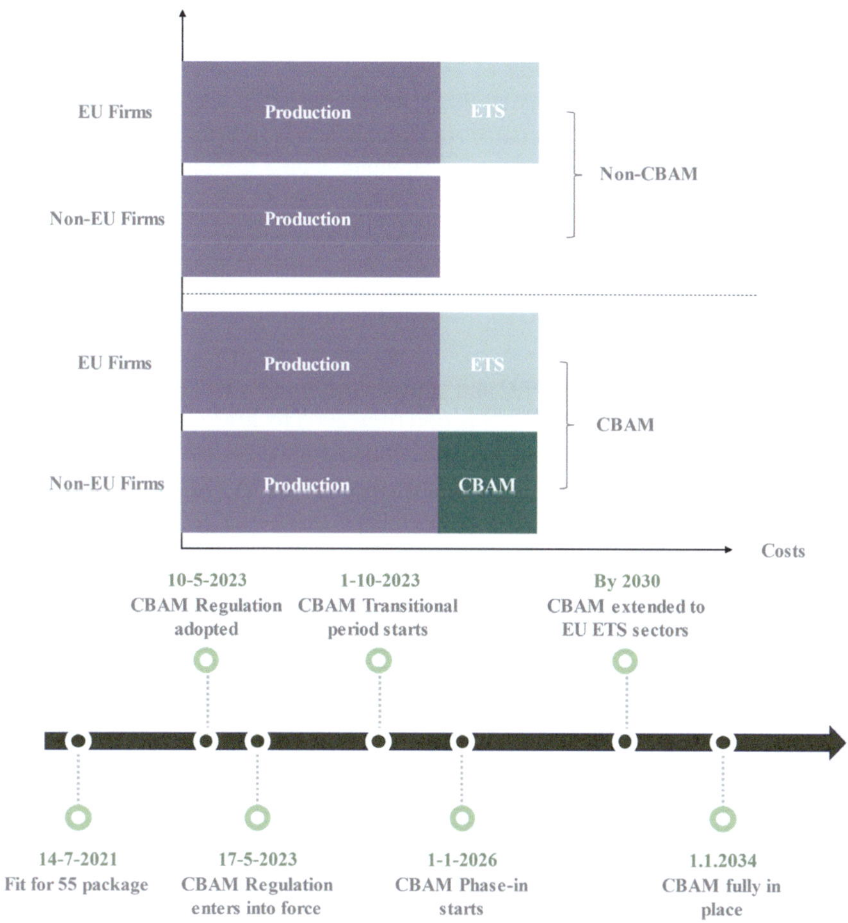

Fig. 18.12 The mechanism and development of CBAM

until 2024. From 2025, stricter reporting rules will apply, and a dedicated CBAM portal will facilitate emissions data submissions for installation operators outside the EU, streamlining compliance for EU importers.

Starting in 2026, CBAM will require EU importers to purchase and surrender CBAM certificates annually, priced according to the weekly average ETS auction price. Importers can deduct carbon prices already paid in the exporting country. The product scope will be reviewed before the definitive phase, with the potential to include additional goods by 2030. CBAM's gradual introduction aligns with the phase-out of free allowances under the ETS, ensuring a balanced transition for businesses and authorities while advancing the EU's decarbonization agenda.

18.5 ETS in China

China, the world's largest emitter of greenhouse gases, has set ambitious climate goals: to peak carbon emissions by 2030 and achieve carbon neutrality by 2060. To support this green economic transformation, China launched its carbon markets, progressing through three key stages: CDM, regional pilot markets, and the national carbon market.

18.5.1 CDM in China

China signed the UNFCCC in 1992 and ratified the Kyoto Protocol in 2002, enabling participation in the CDM. The CDM period from 2005 to 2012 was significant, as it allowed Chinese companies to participate in international carbon trading by generating Certified Emission Reductions (CERs) through CDM projects. From just 18 projects in 2005, the market expanded significantly, reaching over 4400 approved projects by 2012. However, projects are unevenly distributed across the country, with a significant concentration in five central and western provinces that have abundant hydropower and wind resources. These regions account for over 38% of approved projects, while more developed areas like Beijing, Shanghai, and Tianjin have fewer projects. The distribution aligns with the CDM's goal of promoting sustainable development and facilitating technology transfer to less economically developed regions (Hong et al., 2013).

18.5.2 Regional Pilot Carbon Markets

China's journey towards a national carbon market began with local pilot markets. Starting in 2013, pilot carbon markets were launched in Beijing, Shanghai, Tianjin, Chongqing, Hubei, Guangdong, and Shenzhen, with Fujian joining in 2016.

Table 18.2 indicates some key characteristics of these pilots. By the end of December 2021, these pilots involved over 2900 enterprises and allocated approximately 8 billion tons of carbon allowances. Cumulative trading volumes across the pilots reached 36.26 million tons, with a total transaction value of 1.167 billion yuan. Among the seven carbon trading pilots, Guangdong had the highest cumulative trading volume and value, surpassing 100 million tons, followed by Hubei, which showed strong trading activity despite a later launch. Beijing had the highest carbon price at 56.42 yuan/ton, nearly double that of other markets. The pilots displayed significant institutional differences: Beijing, Shanghai, and Shenzhen focused more on the tertiary sector with lower inclusion thresholds, while Hubei, Guangdong, Tianjin, and Chongqing prioritized heavy industries with higher thresholds. It was found that the cost efficiency of power plants improved following the pilot's implementation, establishing a strong foundation for the inclusion of the power sector in the national carbon market (Zhang et al., 2024).

18.5.3 National Carbon Market

The National ETS officially launched trading on July 16, 2021 (see Fig. 18.13), initially covering the power sector, with additional sectors like steel, petrochemicals, and construction materials expected to join in future, positioning it as the largest carbon market by volume worldwide. By the end of 2023, the total transaction volume reached 442 million tons, with a total value of 24.92 billion yuan. However, several challenges persist. First, unclear policy expectations have led to market hesitation and a "hold-back" mentality among companies, resulting in low turnover rates—around 2–3%, much lower than the 5% average turnover rate of the pilot markets and far below the EU carbon market's turnover rate of approximately 500%. Second, the market experiences significant "tide-like" fluctuations, with most trading volume occurring at the end of the compliance period. Third, large block trades dominate the market, distorting price signals, as these trades often occur at prices lower than market values and do not reflect the true cost of emissions reductions. As of mid-2023, over 80% of transactions were large block trades, with prices averaging about 9% lower than the spot market price. These issues hinder the transparency and efficiency of the market, increasing transaction costs. In January 2024, the State Council issued the Interim Regulations on Carbon Emission Trading, laying a legal foundation for the market's continued growth.

Table 18.2 Overview of China's regional ETS

Region	Industry scope	Qualified enterprises	Carbon-based products
Beijing	Industrial: power, cement, petrochemical, manufacturing Non-industrial: services, public transport, aviation	Enterprises with emissions ≥ 5000 tons CO_2/year (direct + indirect)	Beijing emission allowances, CCERs, forestry carbon sequestration (FCERs), PCERs
Shanghai	Industrial: power, water supply, cement, petrochemical, iron and steel Non-industrial: transport, buildings, aviation	Power/industry ≥ 20,000 tons CO_2/year or ≥ 10,000 tons coal equivalent/year	Shanghai emission allowances, CCERs, Shanghai emission allowances forwards (SHEAFs)
Tianjin	Industrial: power, iron and steel, cement, petrochemical, construction	Enterprises with emissions ≥ 20,000 tons CO_2/year	Tianjin emission allowances, CCERs, Tianjin regional forestry offsets (TFCER)
Shenzhen	Industrial: power, gas, manufacturing Non-industrial: public buildings, transport	Enterprises with emissions ≥ 3000 tons CO_2/year or large public buildings	Shenzhen emission allowances, CCERs
Guangdong	Industrial: power, cement, iron and steel, petrochemical Non-industrial: transport, buildings	Enterprises with emissions ≥ 10,000 tons CO_2/year or energy consumption ≥ 5000 tons coal equivalent/year	Guangdong emission allowances, CCERs, PHCERs, eco-compensation CERs
Chongqing	Industrial: power, iron and steel, electrolytic aluminum	Enterprises with emissions ≥ 20,000 tons CO_2/year or energy consumption ≥ 10,000 tons coal equivalent/year	Chongqing emission allowances, CCERs, CQCERs
Hubei	Industrial: power, cement, steel, chemicals, construction	Enterprises with energy consumption ≥ 10,000 tons coal equivalent/year (2016–2018)	Hubei emission allowances, CCERs
Fujian	Industrial: petrochemical, chemical, construction	Enterprises with energy consumption ≥ 5000 tons coal equivalent/year (2013–2020)	Fujian emission allowances, forestry CERs, CCERs

Source HKEX (2023)

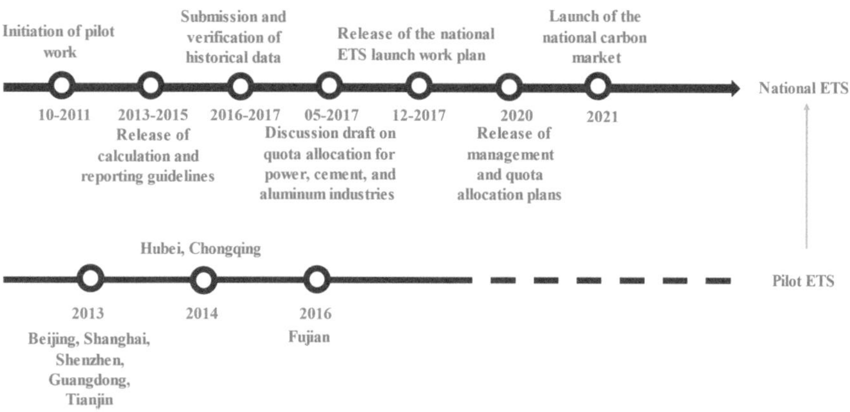

Fig. 18.13 The development of China's ETS

References

European Commission. (2024). *2024 Carbon market report*. https://climate.ec.europa.eu/document/download/92ec0ab3-24cf-4814-ad59-81c15e310bea_en?filename=2024_carbon_market_report_en.pdf

Intergovernmental Panel on Climate Change (IPCC). (2021). *Climate change 2021: The physical science basis. Contribution of working group I to the sixth assessment report of the intergovernmental Panel on Climate Change*. https://www.ipcc.ch/report/ar6/wg1

Hong Kong Exchanges and Clearing Limited (HKEX). (2023). *The growth of global carbon markets and opportunities for Hong Kong*. https://www.hkex.com.hk/-/media/HKEX-Market/News/Research-Reports/HKEx-Research-Papers/2023/CCEO_CarbonMkt_202301_e.pdf

Hong, J., Guo, X., Marinova, D., Yang, F., & Yu, W. (2013). Clean development mechanism in China: Regional distribution and prospects. *Mathematics and Computers in Simulation, 93*, 151–163.

Organization for Economic Co-operation and Development (OECD). (2022). *Environmental tax*. https://www.oecd.org/en/data/indicators/environmental-tax.html?oecdcontrol-8027380c62-var3=2022

Signoret, J., & Jose, E. (2024). *Green subsidies database: Data for unfair advantage—Distortive subsidies and their effects on global trade*. http://documents.worldbank.org/curated/en/099644006032423503/IDU1b117b6921f0b1ae4c1dd6fbd4c89d9

Stechemesser, A., Koch, N., Mark, E., Dilger, E., Klösel, P., Menicacci, L., Nachtigall, D., Pretis, F., Ritter, N., Schwarz, M., Vossen, H., & Wenzel, A. (2024). Climate policies that achieved major emission reductions: Global evidence from two decades. *Science, 385*(6711), 884–892.

World Bank. (2024). *State and trends of carbon pricing 2024*. http://hdl.handle.net/10986/41544

World Bank. (2009). *State and trends of the carbon market 2009*. https://documents.worldbank.org/en/publication/documents-reports/documentdetail/281731468336837686/state-and-trends-of-the-carbon-market-2009

Zhang, H., Zhang, L., & Zhang, N. (2024). When and under what conditions does an emission trading scheme become cost effective? *The Energy Journal, 45*(2), 261–294.

Chapter 19
Business Practice and Green Finance

When sustainability is viewed as being a matter of survival for your business, I believe you can create massive change.
Cameron Sinclair

19.1 Climate Change and Business

19.1.1 Climate Change Risks

Climate change risks that a company confronts include physical risks and transition risks. The physical risks refer to the direct damage to asset values and business activities, which can be either acute or chronic in nature. Acute physical risks are driven by extreme weather events such as floods, droughts, and storms. These risks cause immediate damage to infrastructure, supply chains, and even overall business operations. For example, severe storms may disrupt power grids, leading to energy supply interruptions, while floods can destroy production facilities and the availability of natural resources. Chronic physical risks, on the other hand, are related to longer-term shifts in climate patterns, including global warming, rising sea levels, and changing precipitation patterns. These gradual changes create persistent challenges, such as rising temperatures affecting agriculture, or the slow but steady encroachment of seawater on coastal assets. The cumulative effect of these physical risks could be significant as they lead to higher operating costs, reduced production capacity, and lower efficiency in business operations. As a result, companies face greater financial pressures and have to invest more in adapting to these climate-induced challenges.

The transition risks relate to the indirect impact during sustainable development, due to strengthening regulations, technological innovation, and shifts in consumer preference, imposing extra costs and pressures on climate-sensitive firms. To combat climate change, the government has introduced different types of regulations, such as emissions trading, efficiency standards, and carbon taxes, which constrain corporate operations and production. Since the Paris Agreement, climate change risks have

L. Zhang, *Sustainability Economics*, https://doi.org/10.1007/978-3-031-87829-9_19

Table 19.1 The category and description of climate risks

Category		Description
Physical risk	Acute	Event-driven risks from extreme weather events, such as cyclones, hurricanes, heat waves, or floods
	Chronic	Long-term climate evolution, like rising temperatures, sea level rise, and changing precipitation patterns
Transition risk	Policy and legal	Risks arise from regulation changes, increasing litigation due to failure in mitigating or adapting to climate change, and insufficient disclosures
	Technology	As new technologies replace outdated systems, they disrupt existing industries, creating both winners and losers. However, the timing of technology development and implementation remains a significant uncertainty in evaluating technology risk
	Market	Climate change affects market dynamics through shifts in supply and demand for certain goods and services as climate risks and opportunities are integrated
	Reputation	Changes in customer and community perceptions about an organization's role in addressing climate change can pose reputational risks

Data sources TCFD (2017)

been priced into loans and thus change firms' ratings, making it more challenging to access lending and yields. Some climate-related non-price terms of loan contracts, like collateral and covenants, impose additional financial costs and limit corporate capital management. The energy transition has also decreased the efficiency of non-renewable energy, and firms are exposed to the risk of stranded assets. Climate-sensitive industries are suffering from significant uncertainties with the advancement in climate awareness and changes in customer and investor preferences. Table 19.1 shows an overview of firms' climate risks.

19.1.2 Climate Change Impacts on Business

Businesses are navigating a period of transition, where the economic impacts of climate change are still unfolding. Key challenges include disruptions to operations, resource availability, legal compliance, and shifts in insurance costs and consumer behaviors. These factors combine to create significant economic pressures on businesses (see Fig. 19.1).

- Operational Disruptions: Extreme weather events damage facilities, disrupt supply chains, and delay production. For instance, in manufacturing, extreme weather can cause costly interruptions, from equipment damage, employee absences, to delayed deliveries. In agriculture, changing weather patterns, including rising temperatures and reduced precipitation, threaten crop yields,

Fig. 19.1 Climate related risks, opportunities, and financial impact. *Sources* TCFD (2017)

food production, and farm productivity. Even sectors like tourism are feeling the pressure, as warmer temperatures shorten ski seasons and affect leisure travel.

- Resource and Materials Scarcity: As demand for resources such as water, energy, and raw materials increases, climate change exacerbates the scarcity of these essential inputs. For example, droughts lead to water shortages, while rising temperatures reduce the availability of agricultural products. Additionally, industries relying on critical raw materials, like lithium in technology or cobalt in clean energy, face supply risks.
- Stranded Assets: Stranded assets refer to fossil fuel resources and infrastructure that become unprofitable due to the transition to a low-carbon economy, driven by climate policies or technological changes. This results in significant write-downs on companies' balance sheets. Asset stranding occurs when future profit expectations from these assets collapse, leading to financial losses for investors. Semieniuk et al. (2022) estimate that global stranded fossil fuel assets could reach $1.4 trillion.
- Legal and Liability Risks: As climate regulations evolve, businesses must adapt to new requirements, including emissions reporting and environmental compliance. Failure to comply with these regulations leads to fines, penalties, and even lawsuits. There is also increasing pressure from customers and investors, who demand greater transparency and sustainability in corporate practices. A failure to meet these expectations further leads to reputational damage and financial losses.
- Rising Insurance Costs: Climate change is significantly affecting the insurance industry, leading to higher premiums for flood and storm coverage as extreme

weather events increase. In 2022, natural disasters caused $313 billion in economic losses, with 42% covered by insurance, according to Aon plc, a professional services firm that sells risk mitigation products. As insurers adjust pricing to account for growing risks, businesses face rising insurance costs that could impact their bottom line.
- Changing Consumer Behavior: As extreme weather events and resource scarcity strain household finances, consumers are cutting back on non-essential spending and seeking more sustainable products. This shift in demand requires businesses to adapt or risk losing revenue, as failure to address these changing preferences could lead to reduced sales and profitability.

19.1.3 Turning Risks into Opportunities

To effectively address climate change, businesses must turn risks into opportunities. Efforts to mitigate and adapt to climate change could drive benefits such as improved resource efficiency, cost savings, and access to low-emission energy. Organizations could also capitalize on the development of new products and services, expand into emerging markets, and leverage policies that support clean energy and efficiency. For instance, organizations can reduce costs by improving efficiency in their operations, from production to distribution. Key areas for improvement include energy, water, materials, and waste management. Technological innovations such as LED lighting, energy-efficient heating solutions, electric vehicles, and circular economy models can significantly cut operational expenses while contributing to global emissions reductions. It is also helpful for firms to develop low-emission products and services. Innovating low-emission products and services could improve competitiveness in the market. This includes offering products with a lower carbon footprint and adopting energy-efficient practices throughout the supply chain. By focusing on sustainability in their offerings, organizations can attract environmentally conscious consumers and gain a market edge.

19.2 Business Sustainability

19.2.1 Corporate Social Responsibility

Corporate social responsibility (CSR) refers to the idea that businesses have an obligation to contribute positively to society. It is a form of self-regulation that enables companies to act in ways that benefit the environment, society, and their stakeholders. Businesses engaged in CSR typically incorporate social and environmental issues into their operations and stakeholder relationships, aiming for both social good and profitability. Central to CSR is the "triple bottom line" approach, which urges companies to evaluate their performance across three key dimensions: social responsibility, environmental sustainability, and financial success. The phrase "profit, people, planet" summarizes the essence of CSR, emphasizing a balance between economic goals, social well-being, and environmental health.

CSR is a broad concept encompassing various initiatives that businesses adopt to contribute positively to society and the environment. In general, four key types of CSR are implemented by organizations to create meaningful impact (see Fig. 19.2):

- *Philanthropic Responsibility:* Reflects a company's commitment to giving back to society. This includes donating funds, products, or services to charitable causes or engaging in volunteer activities. Companies may create foundations, sponsor community events, or support nonprofit organizations. The goal is to contribute to social well-being and strengthen relationships with local communities.
- *Ethical and Social Responsibility*: Focuses on fairness, integrity, and respect for human rights. It involves treating employees ethically, ensuring fair wages and safe conditions, and maintaining transparency in business practices. Companies with strong ethical responsibility prevent discrimination, promote inclusivity, and uphold human rights throughout their supply chains.
- *Legal Responsibility:* Requires companies to obey the minimum rules and regulations set by society. This includes complying with laws at national, regional, and

Fig. 19.2 Carroll's framework of CSR

local levels, ensuring fair business practices, and fulfilling obligations as responsible corporate citizens. Legal responsibility forms the foundation for companies to operate ethically and in line with societal expectations.
- *Economic Responsibility:* Focuses on conducting business in a financially responsible and sustainable way. It involves ensuring that financial decisions reflect the well-being of the community, employees, and other stakeholders. Companies may support local businesses, create job opportunities, and contribute to economic development through responsible investments. Economic CSR also demands that businesses strike a balance between generating profits and contributing to wider societal and environmental objectives. Economic CSR also demands that businesses strike a balance between generating profits and contributing to wider societal and environmental objectives.

Many companies embrace CSR not only due to moral convictions but also for strategic benefits. CSR extends beyond merely giving back to society; it is a deliberate approach that offers various advantages. It enhances employee engagement, strengthens corporate reputation, and fosters customer loyalty. CSR significantly boosts a company's reputation by demonstrating its commitment to social and environmental responsibility. When a company aligns its actions with ethical values and supports meaningful causes, it gains the trust and respect of consumers, investors, and the broader community. This positive reputation not only attracts loyal customers but also helps build strong, lasting relationships with stakeholders. Employee satisfaction and retention are improved through CSR. Employees at socially responsible companies tend to feel more engaged and motivated, resulting in higher productivity and lower absenteeism. These employees are also more likely to remain with the company, reducing turnover and fostering innovation. CSR has become a powerful tool for attracting top talent and creating long-term value for both the company and the communities it serves, effectively bridging the gap between doing good and doing well.

Reading

CSR Practices in Leading Companies

In response to the Black Lives Matter movement, Google launched CSR initiatives to address racial inequality in business. In 2020, the company introduced startup accelerators for Black- and women-led businesses and invested $175 million to support economic opportunities for Black entrepreneurs and job seekers. These efforts not only earned Google positive media coverage but also strengthened its reputation as a socially responsible company. By championing racial equity, Google improved its employer brand, attracting a larger pool of qualified applicants who shared its values.

Starbucks has made CSR a cornerstone of its business model, combining a focus on workforce diversity with environmental sustainability. The company's socially responsible hiring practices prioritize veterans, young career starters, and refugees, reflecting its commitment to fostering a diverse and inclusive workplace. Alongside this, Starbucks has invested in its employees' well-being, offering stock grants and enhanced benefits like healthcare, family support, and education. On the environmental front, the company has established ambitious goals to mitigate greenhouse gas emissions, water usage, and waste by 50% by 2030. These CSR initiatives not only strengthen Starbucks' reputation but also create long-term value for employees, communities, and the planet,driving both social impact and business success.

19.2.2 Environmental, Social, and Governance

Environmental, Social, and Governance (ESG) refers to an investment approach that prioritizes environmental sustainability, social responsibility, and effective governance. The growing importance of ESG reflects a shift in business preference, with consumers, employees, and investors. Consumers are favoring products with ESG-related claims, as they hope these products experience notably higher growth rates compared to others. Employees tend to work for companies whose values align with their own, with many considering environmental policies, governance practices, and social impact when evaluating job offers. Similarly, investors recognize that strong ESG practices are indicative of a company's long-term resilience and effective risk management. As a result, ESG has become a critical factor in driving business success, influencing consumer behavior, talent acquisition, and investment decisions.

ESG criteria evaluate a company's impact and operations in three key areas:

- *Environmental:* Focuses on how a company addresses environmental challenges such as climate change, energy use, pollution, and resource management. It includes efforts like reducing carbon emissions, promoting renewable energy, managing water usage, and minimizing waste. Companies that adopt sustainable practices can reduce their environmental impact and enhance long-term business success.
- *Social:* Assesses how a company interacts with its employees, customers, communities, and other stakeholders. It includes ensuring fair labor practices, respecting human rights, promoting diversity and inclusion, and safeguarding customer privacy. By fostering positive relationships and creating a safe, fair environment, companies can improve community well-being and enhance their reputation.
- *Governance:* Refers to the leadership, transparency, and ethical practices within a company. It involves having a diverse and accountable board, fair executive compensation, strict anti-corruption measures, and robust reporting standards. Strong governance practices ensure ethical conduct and protect against risks like bribery or conflicts of interest, contributing to long-term corporate integrity.

ESG as a strategy

ESG investing involves considering environmental, social, and governance factors beyond traditional financial criteria when making investment decisions. This approach, often called sustainable, socially responsible, or impact investing, covers a wide range of issues—from avoiding harmful sectors like tobacco to funding initiatives like clean water projects. ESG investing has evolved from exclusionary practices, focusing on avoiding harmful investments, to actively driving positive change through shareholder engagement and direct intervention.

ESG is important for two main reasons: social responsibility and risk management. Investors, such as pension funds and charities, see their role as more than just seeking returns. They aim to fund initiatives that have a positive societal impact. Additionally, ESG factors help manage investment risks. Companies with poor ESG practices are more likely to face legal, reputational, or operational risks, which can affect their long-term returns. The growing focus on ESG is driven by increased access to information through media and social platforms. As companies become more sensitive to their reputation, institutional investors are increasingly directing their capital toward businesses with strong ESG credentials. With regulatory frameworks tightening and tools for ESG analysis improving, ESG investing is becoming a mainstream approach that balances financial returns with social and environmental responsibility.

19.2.3 ESG Rating Standards

Rating agencies such as the MSCI and S&P Global assess the sustainability and creditworthiness of organizations, including companies and countries, through various ESG criteria. These frameworks offer valuable insights that support informed decision-making for sustainable long-term growth. Below are some examples of rating standards.

MSCI ESG Ratings

It assesses a company's resilience to long-term, financially relevant ESG risks using AI and alternative data for dynamic, investment-focused insights. The ratings are related to a rules-based methodology, ranking companies on a scale from 'AAA to CCC' based on their exposure to ESG risks and how effectively they manage them compared to peers. MSCI also rates countries, mutual funds, and ETFs, providing valuable insights to improve investment decisions and client communication. Table 19.2 shows the framework of the MSCI ESG rating.

The S&P Global ESG Scores package

It uses a double materiality approach, recognizing a sustainability issue as material if it significantly affects society or the environment, as well as the company's value drivers, competitive stance, and long-term shareholder value (see Table 19.3). Key

Table 19.2 MSCI ESG ratings key issue framework

Environment pillar	Climate change	Carbon emission; product carbon footprint; financing environmental impact; climate change vulnerability
	Natural capital	Waster stress; biodiversity and land use; raw material sourcing
	Pollution and waste	Toxic emission and waste; packaging material and waste; electronic waste
	Opportunities	clean tech; green building; renewable energy
Social pillar	Human capital	Labor management; health and safety; Human capital development; supply chain labor standards
	Product liability	Product safety and quality; consumer financial protection; consumer financial protection; responsible investment; chemical safety
	Stakeholder opposition	Controversial sourcing; community
	Social opportunities	Access to finance; access to health care; opportunities in nutrition and health
Governance pillar	Corporate governance	Board; pay; ownership; accounting
	Corporate behavior	Business ethics; tax transparency

Source MSCI (2024)

features include scores, media and stakeholder analysis, disclosure analysis, year-on-year changes, and coverage of approximately 13,000 companies since 2013.

Criticism

The integration of ESG into management and investment has attracted significant attention, yet it faces several challenges that threaten its effectiveness. One of the main issues is the lack of standardization and transparency in ESG reporting, which makes it difficult for investors to assess and compare companies. Berg et al. (2022) highlighted the divergence in ESG ratings from different agencies. Six major ESG rating providers—KLD, Sustainalytics, Moody's ESG, S&P Global, Refinitiv, and MSCI—often produce inconsistent evaluations of the same company. This lack of uniformity stems from differences in scope, measurement, and weight. Notably, measurement divergence accounts for over half (56%) of the inconsistency.

This uncertainty has led to increased cases of greenwashing, where companies make false or exaggerated sustainability claims to attract investment without actual environmental or social improvements. ESG has, in some instances, evolved into a marketing tool aimed at enhancing corporate image rather than fueling real sustainability efforts. This raises criticism among investors who question whether ESG

Table 19.3 The framework of S&P Global ESG scores

E	Biodiversity Climate strategy Decarbonization strategy Energy Environmental policy and management Packaging	Product stewardship Sustainable forestry practices Sustainable raw materials Waste and pollutants Water
S	Community relations Contribution to societal healthcare Customer relations Financial inclusion Health and nutrition Human capital management	Human rights Labor practices Occupational health and safety Privacy protection Transportation safety
G	Business ethics Corporate governance Information security/cybersecurity and system availability Innovation management Market opportunities Materiality Policy influence	Product quality and recall management Risk and crisis management Supply chain management Sustainable finance Tax strategy Transparency and reporting

Source S&P Global (2024)

investments truly yield meaningful results or are merely used to boost profits under the guise of social responsibility.

While ESG investments require time to deliver returns, financial markets often prioritize immediate results, undermining the long-term value of such strategies. Additionally, the link between ESG investments and better financial performance remains inconclusive. The higher fees associated with ESG products further complicate their value proposition. These issues damage the credibility of ESG practices. Without clearer regulations, standardized disclosures, and a real commitment to sustainability, ESG risks become more of a marketing tool than a transition force in business and finance. To unlock its full potential, these challenges must be addressed.

Reading

Is ESG Rating Always Reliable?

RepRisk data reveals a significant rise in greenwashing, as illustrated in Figure 19.5, particularly in the banking and financial services sectors, where climate-related greenwashing incidents increased by 70% from 2022 to 2023. Notably, over 50% of these incidents involved fossil fuels or ties to oil and gas companies. Despite a global downward trend in greenwashing cases in 2024, the report highlights that 30% of companies flagged for greenwashing in 2023 were still linked to it in 2024. This suggests that while public perception and regulation are driving change, more transparent data and regulatory measures are needed to address the growing severity of cases. The EU's Green Claims Directive, mandating proof for environmental claims, contributed to a 20% drop in incidents, while US greenwashing trends reflect a more complex dynamic, potentially influenced by the politicization of ESG issues.

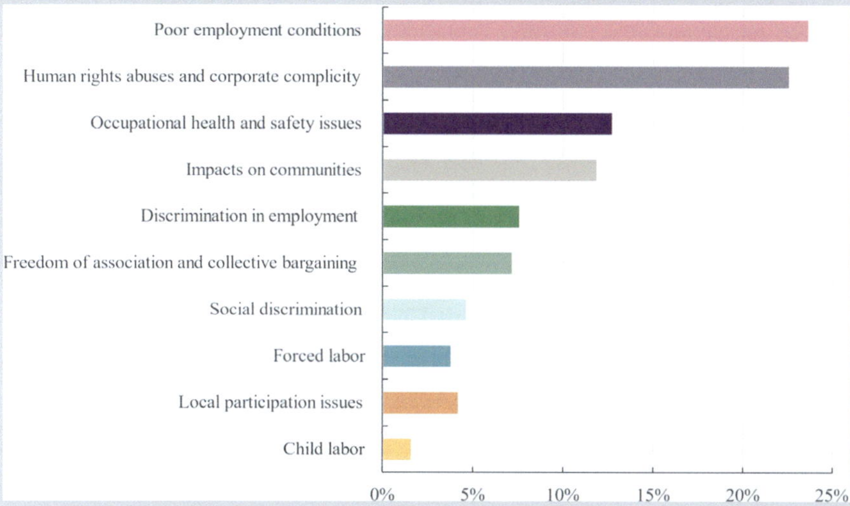

Fig. 19.3 Global ESG risk incidents, Sep 2018–Sep 2023. *Data source* RepRisk (2024)

19.2.4 The Difference Between CSR and ESG

CSR and ESG both aim to promote sustainable practices but differ in focus and approach. CSR involves a company's voluntary, often qualitative efforts to act responsibly, while ESG is a comprehensive, standardized framework centered on measurable, data-driven performance that meets regulatory requirements and aligns with investor and stakeholder expectations.

- *Purpose*: CSR focuses on a company's voluntary initiatives to contribute positively to society and the environment, often through community engagement and ethical practices. It's more about the actions a company takes to demonstrate its social responsibility. ESG, on the other hand, is data-driven, focusing on measuring and disclosing a company's impact on the environment, society, and governance. It aims to quantify these impacts, often for investors and regulators.
- *Voluntary vs. Regulatory*: CSR is a voluntary commitment, driven by a company's values and internal motivations. It's not required by law, though it can enhance

a company's reputation. ESG, however, involves regulatory compliance. Companies are required to report ESG data under international frameworks and standards.

- *Reporting*: CSR reporting is often qualitative, focusing on a company's efforts to build a positive reputation, such as through employee engagement or philanthropy. In contrast, ESG reporting is quantitative, requiring companies to disclose concrete data on sustainability, diversity, and governance practices. ESG data helps investors assess risks and opportunities, and the reporting is standardized for comparability across companies.
- *Implementation*: CSR initiatives are typically more flexible and culture-driven, focusing on initiatives like community outreach and corporate philanthropy. ESG, however, requires a more systematic approach, integrated across the business. It involves departments like finance and legal, setting measurable targets, and continuous monitoring to ensure transparency and accountability.

19.3 An Introduction to Green Finance

Climate challenges continue to pose significant threats to the planet. While innovations, regulations, and global cooperation are essential to address climatic issues, they all rely heavily on financial support. However, traditional finance often prioritizes short-term profits but ignores the broader environmental and social impacts. This profit-driven focus creates a significant funding gap for long-term climate initiatives that prioritize sustainability and resilience. Green finance, on the other hand, offers a solution by directing capital toward projects that advance sustainable development. It involves the targeted allocation of financial resources—across banking, investments, insurance, and microcredit—toward initiatives to address climate change, conserve resources, and promote social well-being. Green financing tools are financial instruments aimed at promoting environmentally sustainable projects. These tools help channel capital towards sustainable development and environmental protection.

19.3.1 Green Credits

Green Credit refers to lending practices that focus on financing environmentally sustainable projects and businesses. It supports initiatives aimed at reducing ecological impact, advancing clean energy, and fostering sustainable development. It primarily funds sectors such as renewable energy, pollution control, energy efficiency, and green infrastructure. Borrowers must typically meet criteria like lowering emissions or improving resource efficiency to qualify. Financial institutions often

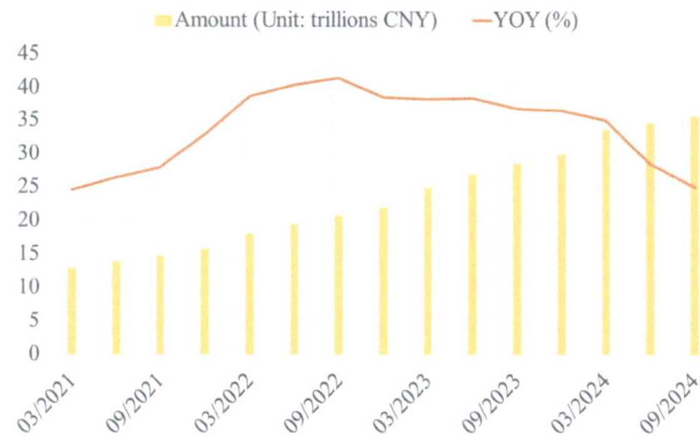

Fig. 19.4 The amount and year-to-year growth of green credits in China. *Data source* Wind database

provide incentives like lower interest rates or flexible repayment terms to encourage participation.

China attaches great importance to the role of green finance, putting in place a multi-tiered green finance market system with a globally leading green credit policy. In January 2012, the China Banking Regulatory Commission (CBRC) introduced the Green Credit Guidelines (GCG), requiring policy banks, commercial banks, rural cooperatives, and other financial institutions to adopt green credit practices. These guidelines aim to redirect financial resources away from high-pollution, high-energy-consuming enterprises and toward sustainable, environmentally friendly projects. Based on national banking laws, the GCG empowered the CBRC to oversee compliance, evaluate green credit performance, and impose penalties for non-compliance. Banks are mandated to formulate and disclose green credit strategies, assess environmental risks, and submit biannual progress reports. The balance of green credit has risen steadily from RMB 4.9 trillion in 2013 to RMB 30.1 trillion in 2023 (see Fig. 19.4), making China the world's largest green credit market. The proportion of green credit to total loan balances also increased from 7.1% in 2013 to 12.7% in 2023.

Reading

India's Green Credit Programme

India's Green Credit Programme is announced by Prime Minister Narendra Modi at COP28. The programme aims to incentivize environmentally beneficial actions by offering monetary rewards, or "Green Credits", to individuals and organizations. Launched in 2023 by the Ministry of Environment, Forest, and Climate Change, the initiative is designed to mobilize citizens toward sustainable practices and positively impact the ecosystem. To fully benefit from the Green Credit Programme, participants must first understand the eligibility criteria and select activities that align with their capabilities while also benefiting the environment. High-impact actions, such as tree planting, sustainable agriculture, waste management, and renewable energy generation, are more likely to yield greater Green Credits. Additionally, engaging in a variety of eco-friendly activities and collaborating across sectors can amplify the overall impact, further enhancing the programme's success in promoting sustainability across the country.

19.3.2 Green Bonds

Green bonds, issued by international organizations, governments, or financial institutions, finance environmentally sustainable projects at preferential rates. Complementing this, the updated 2021 Green Bond Principles (GBP) provide voluntary guidelines to enhance transparency, disclosure, and market integrity. By emphasizing clear processes, accurate reporting, and accountability, the GBP fosters stakeholder confidence and supports the growth of the green bond market. Investors are attracted to green bonds due to their shorter maturities of 3–7 years, which enhance secondary market liquidity, and their eligibility for tax incentives, providing appealing returns. Figure 19.5 demonstrates some important characteristics of the global green bond market.

Since its launch in 2016, China has developed one of the world's largest green bond markets. In 2023, Chinese issuers generated USD 131 billion in labeled green bonds across both onshore and offshore markets. While the total volume of Chinese green bonds decreased by 3.5% in 2023, their quality and credibility improved significantly. Initially dominated by financial institutions, the market now includes contributions from commercial entities, with state-owned enterprises leading, representing about 90% of issuances. Most funds are directed toward renewable energy, low-carbon transport, and buildings. In the same year, Germany secured second place with USD 67.5 billion in green bond issuance, while the UK climbed from seventh to fourth with USD 32.6 billion. Hong Kong SAR experienced the most significant growth, entering the top ten for the first time with a remarkable 173.3% year-on-year increase.

19.3.3 Green Stocks

Green stocks represent companies dedicated to environmentally responsible practices and renewable energy, spanning sectors such as solar, wind, electric vehicles, and waste management. These stocks reflect a growing global commitment to addressing

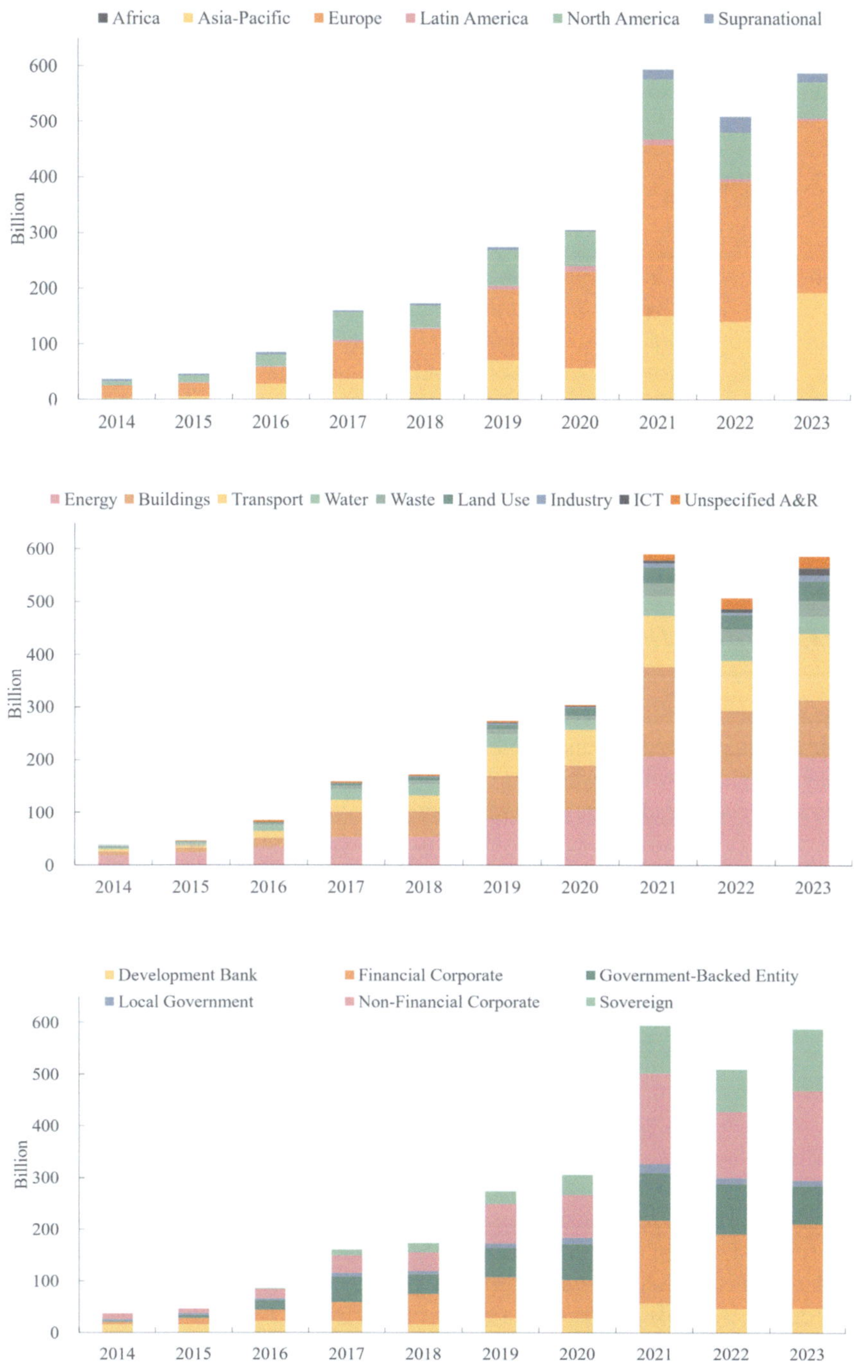

Fig. 19.5 The amount of green bond issuance across regions, use of proceeds, and issuers. *Data source* Climate Bond Initiative (2024)

climate change and fostering sustainability. By investing in green stocks, individuals not only support businesses that prioritize eco-friendly products, services, and technologies but also position themselves to benefit from potential financial returns.

Example

Tesla: The Green Giant of the Stock Market

Tesla (TSLA), valued at $1 trillion, leads the EV industry with cutting-edge innovation under Elon Musk. Following Musk's role in Donald Trump's 2024 campaign, Tesla's stock surged, aligning with the administration's support for market-driven growth, including removing EV purchase tax credits. In Q3 2024, Tesla delivered 462,890 vehicles, up 6.4% year-over-year, while expanding its solar panel and energy storage businesses. A global leader in clean technology, Tesla's advancements in autonomous driving and renewable energy are reshaping the automotive sector.

19.3.4 Green Insurance

Green insurance, also known as eco-friendly insurance, is designed to support environmental protection and combat climate change by incentivizing sustainable behavior. With growing awareness of environmental impacts, more individuals are embracing eco-friendly practices. In response, insurance companies are developing innovative products and services that promote sustainability and encourage the adoption of green technologies.

Example

Allianz's Green Insurance

Allianz provides a range of innovative insurance solutions aimed at managing climate-related risks and promoting sustainability. Its crop insurance, integrated with satellite technology, offers realtime data on crop health and potential weather impacts, serving farmers in regions such as Africa, Australia, and the USA. Weather index insurance safeguards against specific weather events like extreme temperatures or rainfall, with some policies factoring in revenue or price elements, available in Europe and the USA. For forest owners in countries like Brazil and Germany, Allianz offers protection against damages from hazards such as fire, windstorms, and landslides. Additionally, businesses relying on renewable energy can access coverage for revenue loss caused by reduced solar, wind, or water yields, ensuring financial stability across regions including Europe and the USA.

References

Berg, F., Koelbel, J. F., & Rigobon, R. (2022). Aggregate confusion: The divergence of ESG ratings. *Review of Finance, 26*(6), 1315–1344.

Climate Bond Initiative. (2024). *Interactive data platform.* https://www.climatebonds.net/market/data

Morgan Stanley Capital International (MSCI). (2024). *ESG ratings key issue framework.* https://www.msci.com/our-solutions/esg-investing/esg-ratings/esg-ratings-key-issue-framework

RepRisk. (2024). *A turning tide in greenwashing? Exploring the first decline in six years.* https://www.reprisk.com/research-insights/reports/a-turning-tide-in-greenwashing-exploring-the-first-decline-in-six-years?mtm_campaign=pressreleaseq424-greenwashing2024&mtm_kwd=reportsq424&mtm_source=pressrelease-traffic

Semieniuk, G., Holden, P. B., Mercure, J. F., Salas, P., Pollitt, H., Jobson, K., Vercoulen, P., Chewpreecha, U., Edwards, N.R., & Viñuales, J. E. (2022). Stranded fossil-fuel assets translate to major losses for investors in advanced economies. *Nature Climate Change, 12*(6), 532–538.

S&P Global. (2024). *ESG scores and raw data.* https://www.spglobal.com/esg/solutions/esg-scores-data

Task Force on Climate-related Financial Disclosures (TCFD). (2017). *Recommendations of the task force on climate-related financial disclosures.* https://assets.bbhub.io/company/sites/60/2021/10/FINAL-2017-TCFD-Report.pdf

Chapter 20
Non-economic Instruments

The best public policy is made when you are listening to people who are going to be impacted.
Elizabeth Dole

20.1 Non-economic Instruments

Multiple policy tools are needed to achieve the Sustainable Development Goals (Zabala, 2021). In the pursuit of sustainability, economic instruments such as taxes, subsidies, and tradable permits have been widely utilized. However, their effectiveness has proven to be limited in certain scenarios. For example, in some regions, despite the presence of financial incentives like tax breaks for purchasing energy-efficient appliances, the adoption rate has been disappointingly low. This can be attributed to a lack of consumer awareness regarding the long-term benefits and functionality of such products. In other cases, even with significant subsidies for renewable energy projects, bureaucratic hurdles and complex administrative procedures have deterred potential investors.

Moreover, some industries have managed to find loopholes or workarounds in economic regulations. This triggers the inefficiency of economic tools. In carbon emissions and fossil fuel industries, there are carbon taxes or emissions trading schemes designed to reduce the carbon footprint of industries. However, the fossil fuel industry has sometimes found ways to avoid the full impact of these economic regulations. For example, some coal-burning power plants have been known to underreport their actual emissions. By using inaccurate or outdated measurement methods, they can reduce the amount of carbon tax they are liable to pay. In addition, some companies have been found to engage in strategic manipulative practices in emissions trading. They may hoard emissions permits early on to create scarcity, driving up prices, and later sell them at a profit rather than reducing their own emissions.

L. Zhang, *Sustainability Economics*, https://doi.org/10.1007/978-3-031-87829-9_20

The inability of economic measures to comprehensively tackle these issues led to the recognition that alternative, *non-economic approaches* were essential. There was a need to engage with individuals and communities on a more fundamental level, addressing not just the economic rationales but also the psychological, social, and cultural barriers to sustainable behavior. This realization spurred the emergence and implementation of non-economic instruments as a means to fill the gaps and more effectively address the complex and multifaceted challenges of sustainability.

20.2 Command-and-Control Approach

Command-and-control (CAC) policies are regulatory strategies where the government sets specific limits on pollution levels and mandates the use of certain technologies to control emissions. These policies are designed to ensure that firms comply with environmental standards by imposing fines or other penalties for non-compliance. A key feature of CAC policies is their prescriptive nature, which requires firms to adopt specific pollution control measures, such as installing scrubbers or filters on smokestacks.

The CAC policies are widely used in the developed county to control air pollution. One notable example of a CAC is the Clean Air Act in the United States. Enacted in 1963 and significantly amended in 1970, 1977, and 1990, this legislation sets national air quality standards and requires states to develop plans to achieve these standards. The Act mandates the use of specific technologies to reduce emissions from industrial sources and vehicles. For instance, it requires the installation of catalytic converters in cars to reduce harmful emissions. The *Environmental Protection Agency* (EPA) oversees the implementation and enforcement of these standards.

Reading

Clean Air Act

The Clean Air Act (CAA) is a U.S. federal law regulating air emissions from stationary and mobile sources to protect public health and the environment. Administered by the EPA, it establishes frameworks for controlling air pollution through enforceable standards and regulations.

Key provisions of the Clean Air Act include National Ambient Air Quality Standards (NAAQS), which set limits on pollutants like ozone, sulfur dioxide, and particulate matter to ensure clean air (Fig. 20.1). States develop State Implementation Plans (SIPs) approved by the EPA to achieve and maintain these standards. The Act regulates hazardous air pollutants (HAPs), requiring industries to use the best available technologies to reduce emissions. It also addresses mobile sources by setting emission standards for vehicles and engines to curb urban air pollution. Additionally, a permitting system ensures major pollution sources obtain permits detailing emission limits and compliance measures.

Fig. 20.1 Air pollution

The 1990 amendments strengthened provisions to address acid rain, toxic air pollutants, and ozone depletion. The CAA empowers the EPA to enforce compliance, impose penalties, and take legal action.

Since its enactment in 1970, the CAA has significantly improved air quality in the U.S., reducing pollutants responsible for smog, acid rain, and respiratory diseases, while demonstrating the value of strong environmental regulation.

In developing countries like China, CAC policies are also essential for managing air pollution. China's war on air pollution exemplifies CAC policy, focusing on coordinated efforts across multiple regions to address pollution that crosses administrative boundaries. The Air Pollution Prevention and Control Action Plan, introduced in 2013, initially targeted heavily polluted regions such as Beijing-Tianjin-Hebei, the Yangtze River Delta, and the Pearl River Delta. These areas served as pilot regions where strict emission standards and specific pollution control technologies were mandated.

An important aspect of China's CAC approach is its phased implementation. Policies are first tested in these pilot regions to evaluate their effectiveness and feasibility. Based on the outcomes, successful measures are then scaled up and applied nationwide. This method allows for the refinement of strategies, ensuring that the most effective practices are adopted on a larger scale. For instance, the initial success in reducing $PM_{2.5}$ levels in pilot regions led to broader application across other urban areas. This phased approach not only helps in fine-tuning policies but also in managing the regulatory costs associated with continuous monitoring and enforcement, ensuring more efficient and effective pollution control.

Table 20.1 summarizes some CAC policies implemented by major countries or regions. Those policies are designed for particular pollutants and have shown the strength of CAC. However, there are some weaknesses in the process of reducing air pollution. In future, more comprehensive policies are required.

Table 20.1 CAC policy in major countries or regions

Country/region	Scale	Policy	Pollutants	Effectiveness
United States	National	Clean Air Act	$PM_{2.5}$, PM_{10}, NO_x, SO_2	• Reduce air pollution • High policy cost • Require economic incentive
United States	National	Mercury and Air Toxics Standards	Mercury	• Significant reduce mercury emissions • High compliance costs • Low economic incentive
China	National	Air Pollution Prevention and Control Action Plan	$PM_{2.5}$, PM_{10}, NO_x, SO_2	• Reduce air pollution • Promotion of green development • Facing regional disparities
China	National	Volatile Organic Compounds Pollution Control Action Plan	Volatile organic compounds	• Reduce VOC emissions • Enhance regulatory Framework • Requiring long-term sustainability
European Union	Regional	Air Quality Directive	$PM_{2.5}$, PM_{10}, NO_x, SO_2	• Improve air quality • Facing regional disparities
European Union	Regional	Persistent Organic Pollutants Regulation	Persistent organic pollutants	• Reduce POPs emissions • Compliance with international agreements • Require international collaboration
Global	Global	Minamata Convention on Mercury	Mercury	• Promote significant reduction in mercury emissions • Require global cooperation and compliance
North America	Regional	North American Air Quality Cooperation Agreement	$PM_{2.5}$, PM_{10}, NO_x, SO_2	• Reduce transboundary pollution • Improve air quality

Implementing these regulations requires substantial administrative capacity, including technical expertise to set standards, monitoring systems to track compliance, and enforcement capabilities to address violations. This approach provides greater certainty about emission levels compared to market-based methods, ensuring minimum control levels across an industry and delivering predictable environmental outcomes. However, it can lead to higher compliance costs as it does not allow flexibility in achieving environmental goals.

Deciding between market-based instruments and command-and-control regulation involves key trade-offs. Market-based approaches, such as carbon taxes and emissions trading, are cost-effective and foster innovation but require advanced infrastructure. In contrast, command-and-control regulations ensure predictable outcomes and are easier to implement, particularly in regions with limited institutional capacity.

Policymakers must strike a balance between environmental effectiveness and economic efficiency. Effective environmental policies often combine tools tailored to specific needs. For instance, technology standards can establish baseline requirements, while emissions trading or taxes incentivize further reductions. This approach harnesses the reliability of regulations alongside the efficiency of market mechanisms.

20.3 Green Nudge

The concept of "nudging" was popularized by Richard Thaler and Cass Sunstein in their book "*Nudge: Improving Decisions About Health, Wealth, and Happiness*" (Thaler & Sunstein, 2008). A nudge is defined as any aspect of the choice architecture that alters people's behavior in a predictable way without forbidding any options or significantly changing their economic incentives. In other words, it is a gentle push or a small change in the way choices are presented to individuals that can lead them to make more beneficial decisions.

A green nudge, specifically, is a nudge that aims to encourage individuals or organizations to make more environmentally friendly and sustainable choices. It uses the principles of behavioral economics to influence behavior in a way that promotes sustainable development. For example, a green nudge could be as simple as placing recycling bins in more visible and accessible locations to encourage people to recycle more, or using signage in a cafeteria to encourage people to choose more plant-based meals. In the book "*The Little Book of Green Nudges: 40 Nudges to Spark Sustainable Behaviour on Campus*" (United Nations Environment Programme et al., 2020), the UN Environment Programme (UNEP) is now embracing nudging as a strategy to help meet the Sustainable Development Goals and protect the global environment.

20.3.1 Default Options

The utilization of default options represents a highly potent nudging strategy within the realm of sustainability. When it comes to new building construction, the incorporation of energy-efficient systems as the default standard is a prime example. In several countries, new homes are being erected with pre-installed solar panels (Fig. 20.2). While homeowners retain the prerogative to have them removed, the default configuration inherently promotes the utilization of renewable energy sources. This approach capitalizes on human inertia and the tendency to accept the pre-set option, thereby facilitating a seamless transition toward more sustainable energy consumption.

In the domain of waste management, establishing the default behavior for waste collection as the separation of recyclables at the source has proven effective (Fig. 20.3). By streamlining this process and making it the norm, a larger segment of the population is inclined to recycle without the need for deliberate decision-making. This not only simplifies waste management procedures but also significantly augments recycling rates, reducing the burden on landfills and promoting resource conservation.

San Francisco has one of the most successful recycling and composting programs in the country, largely due to setting source separation as the default behavior. The city

Fig. 20.2 Solar panel houses

Fig. 20.3 Recycling trash bins

provides separate bins to households—one for recycling (items like plastic bottles, cans, and paper), one for composting (food scraps and yard waste), and one for landfill waste. This clear separation and the communication that these are the expected ways to dispose of waste have led to high participation rates. In fact, the city now diverts around 80% of its waste from landfills, with a significant portion of that success attributed to residents easily adapting to the default separation system. The city also enforces strict regulations to ensure compliance, but the simplicity of the default setup makes it easier for people to follow the rules.

20.3.2 Framing and Messaging

The manner in which information is presented exerts a profound impact on decision-making processes. In the context of sustainability, positive framing has emerged as a more efficacious approach. Positive framing is a technique that emphasizes the benefits of gains over potential losses or drawbacks. Instead of adopting a negative or punitive tone, such as "Reduce your energy consumption or face higher bills," a more constructive message can be crafted. For instance, "By using energy-efficient appliances, you can save money and help the environment." This framing emphasizes the dual benefits of sustainable behavior, appealing to environmental consciousness.

Yang et al. (2018) employed a contingent valuation method to explore how framing impacts consumers' willingness to purchase green electricity. A sample of 1022 respondents was drawn from an Internet panel and split into two nearly equal-sized sub-samples. One sub-sample was presented with a positively framed version of the questionnaire, while the other received a negatively framed one. The results indicated that consumers exhibited a stronger intention to buy green electricity when it was framed positively rather than negatively. Additionally, the observed framing effect also pointed to the public good aspect of green electricity, as it suggested the existence of a free rider problem.

Example

Green Nudges in Energy Conservation: Smart Meters

The installation of smart meters (Fig. 20.4) constitutes a significant green nudge in the energy conservation landscape. These devices furnish consumers with real-time feedback on their energy consumption patterns. By providing instantaneous data on energy usage, consumers are empowered to make informed decisions and adjust their behavior accordingly. Energy companies implement smart meter-related nudges and play a key role in their effectiveness to reduce energy consumption that their energy consumption surges during a specific time of day, prompting them to take proactive measures such as turning off idle appliances or fine-tuning their thermostat settings. This immediate feedback loop serves as a continuous reminder and incentive for consumers to optimize their energy use and reduce wastage.

Fig. 20.4 Energy smart meter

20.3.3 Integration of Green Nudges into Practice

In recent times, it has become evident that conventional regulatory methods have proven inadequate when it comes to addressing environmental sustainability. As a result, governments have started to incorporate behaviorally informed aspects into their policies and legal frameworks. The existing literature presents a wide array of models that provide guidance on how to formulate policy interventions by leveraging the insights of behavioral sciences. Among these, “The nine principles” model, which was devised in the UK by Darnton (2008), is highly renowned. Sung et al. (2022) developed a modified framework inspired by Darnton (2008) (Table 20.2). This model, grounded in behavioral modeling outcomes, advocates for an iterative, cyclical approach to policy development and implementation.

Policymakers face the complex task of aligning incentives, regulations, and information dissemination strategies to encourage sustainable behavior. One key approach is to conduct a comprehensive review of existing policies to identify areas where green nudges can be seamlessly integrated.

In the realm of energy policy, rather than simply mandating energy efficiency standards, policymakers can introduce a tiered system of incentives. Homeowners who exceed the minimum energy efficiency requirements could be eligible for additional tax credits or rebates. This not only encourages individuals to go above and

Table 20.2 Adapted framework

Stage 1: *understanding consumer behavior*	
Step 1	Identify behavior model for exploration
Step 2	Understand consumer behavior
Stage 2: *Identifying key factors influencing behavior*	
Step 3	Refine the behavior model for operationalization
Step 4	Identify key drivers, facilitators and barriers
Stage 3: *developing interventions*	
Step 5	Design promising interventions
Step 6:	Evaluate and improve the draft interventions
Stage 4: *feedback*	
Step 7	Develop prototype with "actors"
Step 8	Evaluate interventions
Step 9	Process learning

Note The content is based on Sung et al. (2022)

beyond the basic standards but also provides a clear and tangible reward for their efforts.

Another aspect of policy design is the use of information as a nudge. Governments can mandate that energy providers include detailed information about energy consumption patterns and the environmental impact of different energy sources on monthly bills. By presenting this information in an easily understandable format, consumers are nudged to make more informed decisions about their energy usage. For instance, a graph comparing the carbon footprint of different energy consumption levels during peak and off-peak hours can prompt consumers to shift their usage to more environmentally friendly times.

Moreover, policymakers can design policies that target specific sectors or behaviors. In the transportation sector, policies could be designed to nudge consumers towards electric vehicles. This could involve not only financial incentives such as purchase subsidies and tax exemptions but also non-financial nudges. For example, providing preferential parking spaces for electric vehicles in public areas or allowing electric vehicles access to carpool lanes even with a single occupant. These measures not only make electric vehicles more attractive from a convenience and cost perspective but also help to change the social perception of electric vehicles as a viable and desirable mode of transportation.

Case

Reducing Single-use Cutlery with Green Nudge

A study published in *Science* (He et al., 2023) highlights the effectiveness of "green nudges" in significantly reducing single-use cutlery consumption in China's food delivery industry (Table 20.3). Conducted in collaboration with

Alibaba's food-delivery platform Eleme, the research implemented two main interventions to influence consumer behavior:

Table 20.3 Trends in the share of no-cutlery orders

Cutlery choice	
Before green nudges	After green nudges
• "No cutlery" is not listed • Selections only show cutlery number • Cutlery is default option	• "No cutlery" is listed on the top • Other choices of cutlery is put below • *"No Cutlery" is default option*
Proportion of "No Cutlery" in Beijing, Shanghai, and Tianjin	
< 5%	> 20%

- *Default Option Change*: The app's default setting was adjusted to "no cutlery," requiring customers to actively opt in if they wanted disposable utensils;
- *Incentive Mechanism*: Customers who chose to forgo single-use cutlery were rewarded with "green points," which could be redeemed for planting real trees in China.

The study analyzed data from 197,062 randomly selected users across 10 major Chinese cities between 2019 and 2021. Results showed a remarkable *648%* increase in no-cutlery orders. If scaled nationwide, this initiative could save over 21.75 billion sets of disposable utensils annually, equivalent to preventing 3.26 million metric tons of plastic waste and conserving 5.44 million trees used for wooden chopsticks.

These findings suggest that incorporating green nudges is a feasible and effective policy tool for reducing single-use cutlery waste in the rapidly growing online food-delivery industry. The success of this initiative highlights the potential for similar strategies to be implemented in other regions and industries to promote environmentally friendly consumer behaviors.

20.4 Policy Innovation with Digitalization

Non-economic instruments for policy innovation through digitalization focus on leveraging technology to enhance governance, improve transparency, and foster greater public participation. Environmental policy innovation has accelerated significantly since 2015, driven by technological advancement, behavioral insights, and market sophistication. These emerging policy tools mark a transition from traditional approaches toward more dynamic, precise, and behaviorally informed instruments.

20.4.1 Data-Driven Policy Design

The integration of advanced data analytics and artificial intelligence (AI) into environmental policy marks a significant shift toward more precise and responsive governance. Smart environmental monitoring systems now enable real-time pollution tracking and automated response protocols. According to the report of People's Daily Online in 2024, China has recently established the world's largest and most comprehensive scientifically designed monitoring network (MEE). This system currently monitors 33,000 sites, significantly improving monitoring quality and continuously enhancing technical capabilities, providing robust support for ecological and environmental protection efforts.

In Europe, COPERNICUS is the Earth Observation component of the EU's Space Program. It aims at monitoring the planet and benefiting European citizens by providing data, information, and services derived from satellite and in situ (non-space) observations.

Machine learning algorithms and AI are also increasingly employed to optimize policy parameters. These tools can process vast amounts of environmental and economic data to identify optimal intervention points and predict policy outcomes. Gailhofer et al. (2021) illustrate the need for regulatory action to align the design and deployment of AI with the goals of the European Green Deal. In May 2021, the Singapore Exchange (SGX), Singapore's sovereign wealth fund Temasek, DBS Bank, and Standard Chartered jointly launched Climate Impact X (CIX) to expand the voluntary carbon market by enabling the trade of high-quality carbon credits through standardized contracts. Singapore became the first country to use technologies such as satellite monitoring, machine learning, and blockchain to enhance the quality, integrity, and transparency of carbon credits.

20.4.2 Digital Transformation and Environmental Governance

The breakthrough development of digital technology is fundamentally reshaping the paradigm of environmental governance. By leveraging digital technology's capabilities, this transformation may go beyond merely adding technological tools, driving a fundamental innovation in governance philosophy, organizational approaches, and implementation strategies.

Digital infrastructure revolution

The digital transformation of environmental governance has led to significant breakthroughs in infrastructure. The digital infrastructure revolution after 2015 has been marked by significant advancements across multiple technologies, leading to profound changes in how societies, governments, and businesses operate. Key developments in digital infrastructure include the rapid expansion of cloud computing, the rise of 5G networks, the proliferation of IoT (Internet of Things) devices, and breakthroughs in AI, machine learning, and blockchain technologies.

The digital technology revolution has enabled integrated environmental management platforms, breaking down departmental barriers. For example, the EU's COPERNICUS program combines satellite remote sensing, ground monitoring, and socioeconomic data to create a comprehensive environmental information network. This integration improves data acquisition and enables cross-departmental and cross-regional collaboration. China's "Ecological Cloud" platform similarly unifies environmental data from various departments, allowing real-time tracking of pollution, ecosystem changes, and policy effects.

The digital infrastructure revolution enhances regional collaborative governance. The EU's cross-border air quality early warning system provides alerts for transboundary pollution events and coordinates joint actions among member states, improving regional environmental management.

Policy implementation innovation

Digitalization plays a crucial role in driving innovation in policy implementation by enhancing efficiency, transparency, and accessibility. The digital revolution is rapidly transforming the implementation of environmental policies, reshaping enforcement methods and enhancing public participation.

Building on such digital innovations, governments worldwide have been revolutionizing their enforcement systems. In 2015, the U.S. EPA's Next Generation Compliance program marked a significant breakthrough by introducing digital technology into environmental enforcement. In September 2024, the EPA introduced its AI Compliance Plan, detailing strategies to standardize AI requirements across the agency, establish best practices for AI use, and ensure compliance with governance policies. The initiative seeks to integrate AI into environmental enforcement and compliance processes, enhancing the effectiveness and efficiency of the EPA's programs.

Digitalization has enhanced public participation in environmental policy by providing more accessible, transparent, and interactive platforms for engagement. Governments are increasingly using digital platforms to host public consultations and gather feedback on proposed environmental policies. These platforms allow citizens to review documents, submit comments, and participate in discussions, increasing public involvement in decision-making. The European Union uses online platforms like "EU Engage" to involve citizens in discussions on environmental regulations, such as climate change policies and biodiversity protection.

Policy implementation evaluation

Through big data analysis, policymakers can now more effectively assess the cost-effectiveness of various interventions by analyzing large datasets, identifying patterns, and predicting outcomes, leading to more informed decision-making and optimized resource allocation.

Singapore's Smart Nation project combines environmental data with economic and social welfare data to evaluate the comprehensive impact of environmental policies. The Climate TRACE coalition's global greenhouse gas tracking system uses satellite data and AI to precisely locate and quantify major emission sources, enabling targeted interventions and efficient governance.

In carbon pricing and emissions trading systems, big data analysis allows policymakers to evaluate the cost-effectiveness of carbon reduction measures. By analyzing data on emissions, energy consumption, and economic activity, governments can assess which policies achieve the greatest reduction in carbon emissions at the lowest cost. The European Union's Emissions Trading System (EU ETS) utilizes big data to monitor emissions and assess the effectiveness of its carbon market in reducing greenhouse gas emissions.

Reading

China's Blue Map Platform

China's Blue Map platform, launched in 2006 and significantly upgraded in 2019 with 32 new version releases, represents an innovative model for environmental governance through digital public participation. The Blue Map mobile application integrates diverse environmental data, like real-time air and water quality measurements, corporate environmental compliance records, and pollution source information, marking a significant shift from passive information disclosure to active public participation. By 2019, the platform had accumulated over 6 million downloads and documented more than 1.6 million environmental supervision records covering 6.09 million companies. The database expanded to include over 40 different types of environmental data, from corporate benchmarks to safety supervision records.

What makes the Blue Map particularly innovative is its multi-stakeholder ecosystem approach. The platform connects government agencies, corporations, environmental organizations, and the public through an "environmental big data + internet" model. In 2019, 496 companies used the platform to publicly disclose their environmental rectification efforts, while 738 companies published detailed pollution release and transfer (PRTR) data. Over 24,000 companies registered enterprise accounts, actively participating in environmental management through the platform.

The success of Blue Map demonstrates how digital innovation can transform environmental governance from a traditionally government-led approach to a collaborative model engaging all stakeholders in society.

Global governance transformation

Since 2021, the digital revolution has been reshaping global environmental governance, enhancing international cooperation and coordination. The Climate Warehouse program focuses on prototyping, testing, and developing digital infrastructure to enhance transparency, trust, and integrity in carbon markets. Once tested and proven, these digital solutions will be integrated into the market infrastructure. Key initiatives include the newly launched metadata platform for linking and aggregating registry data (the Climate Action Data (CAD) Trust), digital MRV systems, national carbon registries, tokenization tools, and a centralized platform to promote knowledge-sharing and capacity-building.

The UNEP coordinates the United Nations system's response to environmental issues and is a key member of the United Nations Development Group. UNEP's mission is to help the world achieve the 17 Sustainable Development Goals. However, significant barriers remain in utilizing environmental data for decision-making due to ecosystem fragmentation. UNEP is leading the development of the Global Environmental Data Strategy (GEDS) to tackle critical issues such as climate change, pollution, and biodiversity loss. The goal of GEDS is to provide reliable, high-quality, and accessible environmental data to support actions at global, regional, and national levels. This strategy is being created through a collaborative multistakeholder consultation process, ensuring transparency and effectiveness in addressing environmental challenges.

UN Climate Change's Initiative on Artificial Intelligence for Climate Action (#AI4ClimateAction) is a collaborative effort to explore the role of AI in advancing and scaling up transformative climate solutions. The initiative focuses on supporting mitigation and adaptation actions in developing countries, particularly least developed countries (LDCs) and small island developing states (SIDS).

Looking ahead to 2025–2030, these digital innovations will likely continue evolving, fundamentally shifting how nations collaborate on environmental challenges. By promoting transparency, efficiency, and equity, digital technologies are bridging the gap between global environmental goals and effective action.

References

Darnton, A. (2008). *GSR behaviour change knowledge review. Reference report: An overview of behaviour change models and their uses*. HMT Publishing Unit.

Gailhofer, P., Herold, A., Schemmel, J. P., Scherf, C. S., de Stebelski, C. U., Köhler, A. R., & Braungardt, S. (2021). *The role of artificial intelligence in the European green deal*. European Parliament.

Giest, S. (2020). Do nudgers need budging? A comparative analysis of European smart meter implementation. *Government Information Quarterly, 37*(4), 101498.

He, G., Pan, Y., Park, A., Sawada, Y., & Tan, E. S. (2023). Reducing single-use cutlery with green nudges: Evidence from China's food-delivery industry. *Science, 381*(6662), eadd9884.

Sung, K., Cooper, T., & Kettley, S. (2022). Adapting Darnton's nine principles framework for behaviour change: The UK upcycling case study. *Sustainability, 14*(3), 1919.

Thaler, R. H., & Sunstein, C. R. (2008). *Nudge: Improving decisions about health, wealth, and happiness.* Yale University Press.

United Nations Environment Programme, GRID-Arendal, & Behavioural Insights Team. (2020). *The little book of green nudges: 40 nudges to spark sustainable behaviour on campus.* https://wedocs.unep.org/20.500.11822/33578

Yang, Y., Solgaard, H. S., & Ren, J. (2018). Does positive framing matter? An investigation of how framing affects consumers' willingness to buy green electricity in Denmark. *Energy Research & Social Science, 46*, 40–47.

Zabala, A. (2021). Comparing policy instruments for sustainability. *Nature Sustainability, 4*(7), 565–566.

MIX
Papier aus verantwortungsvollen Quellen
Paper from responsible sources
FSC® C105338

If you have any concerns about our products,
you can contact us on
ProductSafety@springernature.com

In case Publisher is established outside the EU,
the EU authorized representative is:
Springer Nature Customer Service Center GmbH
Europaplatz 3, 69115 Heidelberg, Germany

Printed by Libri Plureos GmbH
in Hamburg, Germany